BIOPESTICIDES AND BIOAGENTS

Novel Tools for Pest Management

BIOPESTICIDES AND BIOAGENTS

Novel Tools for Pest Management

Edited by
Md. Arshad Anwer

Apple Academic Press Inc.
3333 Mistwell Crescent
Oakville, ON L6L 0A2 Canada

Apple Academic Press Inc.
9 Spinnaker Way
Waretown, NJ 08758 USA

© 2018 by Apple Academic Press, Inc.

First issued in paperback 2021

No claim to original U.S. Government works

ISBN 13: 978-1-77-463676-3 (pbk)
ISBN 13: 978-1-77-188519-5 (hbk)

Library and Archives Canada Cataloguing in Publication

Biopesticides and bioagents : novel tools for pest management / edited by Md. Arshad Anwer.

Includes bibliographical references and index.
Issued in print and electronic formats.
ISBN 978-1-77188-519-5 (hardcover).--ISBN 978-1-315-36555-8 (PDF)

1. Natural pesticides. 2. Biological pest control agents. 3. Phytopathogenic microorganisms--Control. 4. Weeds--Control. I. Anwer, Md. Arshad, editor

SB951.145.N37B56 2017	632'.95	C2017-903060-4	C2017-903061-2

Library of Congress Cataloging-in-Publication Data

Names: Anwer, Md. Arshad, editor.
Title: Biopesticides and bioagents : novel tools for pest management / editor: Md. Arshad Anwer.
Description: Waretown, NJ : Apple Academic Press, 2017. | Includes bibliographical references and index.
Identifiers: LCCN 2017019826 (print) | LCCN 2017022123 (ebook) | ISBN 9781315365558 (ebook) | ISBN 9781771885195 (hardcover : alk. paper)
Subjects: LCSH: Natural pesticides. | Biological pest control agents. | Phytopathogenic microorganisms--Control. | Weeds--Control.
Classification: LCC SB951.145.N37 (ebook) | LCC SB951.145.N37 B53 2017 (print) | DDC 632/.96--dc23
LC record available at https://lccn.loc.gov/2017019826

Apple Academic Press also publishes its books in a variety of electronic formats. Some content that appears in print may not be available in electronic format. For information about Apple Academic Press products, visit our website at **www.appleacademicpress.com** and the CRC Press website at **www.crcpress.com**

ABOUT THE EDITOR

Md. Arshad Anwer, PhD

Md. Arshad Anwer, PhD, is currently Assistant Professor-cum-Jr. Scientist in the Department of Plant Pathology at Bihar Agricultural University (BAU), Sabour-Bhagalpur, Bihar, India. He is engaged in developing low-cost biopesticides that have increased shelf life. His areas of interest include the development of preliminary information of economically important pathogens associated with maize and the ability to recognize key diseases and their hot spots through survey and surveillance of major crops under agro-ecological condition. He is also working on Panama wilt of banana and host specificity of plant pathogen, evaluation of biocontrol agents against wilts of several crops, and establishment of sick-plots for studies on several soil borne plant pathogens. He is associated with more than eight research groups at BAU, Sabour, including maize research team, viz. organic farming, host parasite research groups, bacteriology, mycology research group, biopesticides and bio-fertilizer research groups.

He teaches undergraduate as well as postgraduate students and has published 24 peer-reviewed research papers as well as one book and seven book chapters, mostly in publications from the USA, UK and Italy.

Dedicated to

my wife Aishatul Bushra, my son Md. Huzaifa Anwer,
and my daughter Aamna Anwer

CONTENTS

LIST OF CONTRIBUTORS

Anil
Department of Entomology, Bihar Agricultural University, Sabour, India

Waheed Anwar
Institute of Agricultural Sciences, University of the Punjab, New Campus, Lahore 54000, Pakistan

Md. Arshad Anwer
Department of Plant Pathology, Bihar Agricultural University, Sabour 813210, India. E-mail: arshad_anwer@yahoo.com

Kahkashan Arzoo
Department of Plant Pathology, GBPUA&T, Pantnagar, Uttara Khand, India

Pranab Barma
Darjeeling Krishi Vigyan Kendra, Uttar Banga Krishi Viswavidyalaya, Kalimpong, Darjeeling 734301, West Bengal 734301, India. E-mail: pranab.barma@gmail.com

Nithya Chandran
Division of Entomology, Indian Agricultural Research Institute, New Delhi 110012, India. E-mail: nithyacr@yahoo.com

Amit Choudhary
Department of Entomology, Punjab Agricultural University, Ludhiana 141004, Punjab, India

Jaipal Singh Choudhary
Research Centre, ICAR-Research Complex for Eastern Region, Research Centre, Plandu, Ranchi 834010, Jharkhand, India

Santhosh J. Eapen
Division of Crop Protection, ICAR-Indian Institute of Spices Research, Kozhikode 673012, Kerala, India

Erayya
Department of Plant Pathology, Bihar Agricultural University, Sabour, Bhagalpur, Bihar, India. E-mail: erayyapath@gmail.com

Tarak Nath Goswami
Department of Entomology, Bihar Agricultural University, Sabour, India. tarakento@gmail.com

Muhammad Saleem Haider
Institute of Agricultural Sciences, University of the Punjab, New Campus, Lahore 54000, Pakistan

Vijay Kumar Jha
Department of Botany, Patna University, Patna, Bihar, India

Lokender Kashyap
Department of Plant Protection, Lovely Professional University, Jalandhar 144410, Punjab, India

Tarak Brambha Maji
Department of Agricultural Entomology, Bidhan Chandra Krishi Viswavidyalaya, Mohanpur 741252, Nadia, West Bengal, India

Kalmesh Managanvi
Department of Entomology, Bihar Agricultural University, Sabour, Bhagalpur 813 210, India. E-mail: kalmesh.managanvi@gmail.com

Pawan Kumar Mehta
Department of Entomology, CSK Himachal Pradesh Krishi Vishvavidyalaya, Palampur, Kangra 176062, Himachal Pradesh, India

Suprakash Pal
Directorate of Research (RRS-TZ), Uttar Banga Krishi Viswavidyalaya, Pundibari, Cooch Behar, West Bengal 736165, India

Vikas Kumar Patel
Department of Entomology, Bihar Agricultural University, Sabour 813210, Bhagalpur, Bihar, India

Chandra Shekhar Prabhakar
Department of Entomology, Veer Kunwar Singh College of Agriculture, Bihar Agricultural University, Dumroan, Buxar 802136, Bihar, India

Nishant Prakash
Department of Plant Pathology, Krishi Vigyan Kendra, Arwal, BAU, Sabour, Bhagalpur, Bihar, India

Bishun Deo Prasad
Department of Molecular Biology and Genetic Engineering, Bihar Agricultural University, Sabour, Bhagalpur, Bihar, India. E-mail: dev.bishnu@gmail.com

Rashid Pervez
Division of Crop Protection, ICAR-Indian Institute of Spices Research, Kozhikode 673 012, Kerala, India. E-mail: rashid_pervez@rediffmail.com

Tushar Ranjan
Department of Basic Science Humanities & Genetics, Bihar Agricultural University, Sabour, Bhagalpur, Bihar, India

Shree Niwas Ray
Department of Entomology, Bihar Agricultural University, Sabour, India

Tamoghna Saha
Department of Entomology, Bihar Agricultural University, Sabour, Bhagalpur 813210, India. E-mail: tamoghnasaha1984@gmail.com

Sangita Sahni
Department of Plant Pathology, Tirhut College of Agriculture, Dholi, RAU, Pusa, Bihar, India

Shyambabu Sha
Department of Entomology, Bihar Agricultural University, Sabour, Bhagalpur 813210, India

Ahmad Ali Shahid
Institute of Agricultural Sciences, University of the Punjab, New Campus, Lahore 54000, Pakistan. E-mail: ahmadali.shahid@gmail.com

Ramesh Kumar Sharma
Department of Horticulture (Veg. & Flori.), Bihar Agricultural University, Sabour 813210, Bhagalpur, Bihar, India

Uma Shankar Singh
International Rice Research Institute, IRRI India office, 1st Floor, NASC Complex, DPS Marg, New Delhi 110012, India

Kundan Singh
Department of Plant Pathology, Bihar Agricultural University, Sabour 813210, India

Raj Narain Singh
Directorate of Extension Education, Bihar Agricultural University, Sabour 813210, India

Pankaj Sood
Krishi Vigyan Kendra, CSK Himachal Pradesh Krishi Vishvavidyalaya, Sundernagar, Mandi, 175019, Himachal Pradesh, India

Ramanuj Vishwakarma
Department of Entomology, Bihar Agricultural University, Sabour, Bhagalpur 813 210, India. E-mail: entoramanuj@gmail.com

Najam Waris Zaidi
International Rice Research Institute, IRRI India office, 1st Floor, NASC Complex, DPS Marg, New Delhi 110012, India. E-mail: n.zaidi@irri.org

LIST OF ABBREVIATIONS

ABC	ATP-binding cassette
ACC	1-aminocyclopropane-1-carboxylic acid
APX	ascorbate peroxidase
CAT	catalase
CWDEs	cell wall degrading enzymes
CWPOX	cell wall bound peroxidises
DAPG	diacetylphloroglucinol
DJ	dauer juvenile
EBPM	ecologically based pest management
EPA	Environmental Protection Agency
EPNs	entomopathogenic nematodes
FYM	farm yard manure
GA	gibberelic acid
GMCs	genetically modified crops
GSP	gossyplure and organic pesticides
GV	granulosis virus
IAA	indole acetic acid
ICM	integrated crop management
ICTV	International Committee on Taxonomy of Viruses
IJs	infective juveniles
IPM	integrated pest management
ISR	induced systemic resistance
KCZ	ketoconazole
LOX	lipoxygenase
MAMPs	microbe-associated molecular patterns
MAPK	mitogen-activated protein kinase
MCZ	miconazole
MIC	minimum inhibitory concentrations
PAL	phenylalanine ammonia lyase
PCNB	pentachloronitrobenzene
PCR	polymerase chain reaction
PDB	potato dextrose broth
PO	peroxidase

PPO	polyphenol oxidase
PR	pathogenesis-related
PSM	phosphate-solubilizing microorganisms
Pvd	pyoverdin
ROS	reactive oxygen species
SAR	systemic acquired resistance
SOD	superoxide dismutase
SPOX	soluble proxidases

PREFACE

Swift changes in the agro ecosystem leave a problem in the establishment of harmony in the discord of a disturbed agro ecosystem. This disturbance is caused by extension of cultivated areas, monoculture, intensive cultivation, faulty cropping pattern, irrational crop husbandry by way of unbalanced fertilizer application, tapping of subsoil water, and indiscriminate application of chemical pesticides, etc. The pesticide consumption has increased 20% per annum although agricultural production increased nominally throughout the globe. The insects, diseases, and weeds cause almost 30% yield loss per annum. This situation has come up because of killing of natural enemies, resistance problem among insects and pathogens to chemicals, residue hazards of used chemicals, disturbing the balance of nature, resurgence of treated insect populations, and emergence of new pathogen races.

When we headed for the green revolution in the past, we had used all kinds of affluent situations like good land, water resources, etc. Even fertilizer and pesticide response was much better than what we can expect today. As a matter of fact, we are now getting some kind of decline in total food productivity. That is the reason why we are saying that there is a need of a different kind of agriculture and different kind of technology. This has to be a major shift and we have to move to a very high level of precision agricultural technology. This technology will enhance the production and the productivity per unit of resource.

Considering the drawbacks of our faulty cropping pattern, it becomes necessary to lay special emphasis on integrated pest management (IPM). To achieve our goal of the next green revolution, we have to phase out the unnecessary consumption of poisonous chemicals and integrated multidisciplinary methodologies in developing agro ecosystem management strategies that are practical, economical, and protective of both human health and environment.

Integrated control or pest management does imply the use of the best combination of controls in organized ways that are designed to avoid harm to anything but pests. The use of biopesticides and bioagents is on the increase to control our crop pests by artificial introduction or increase of their natural enemies such as antagonists, parasites predators, insect pathogens, and plant products.

A number of pathogenic bacteria viz., *Bacillus thuringiensis, B. popillae*, fungi viz., *Trichoderma* spp., *Aspergillus niger, Beauveria bassiana, Metarrhizium anisoplae. Entomophthora* spp. *Verticillium lecanii*, and viruses viz., nuclear polyherosis, cytoplasmic polydedrosis, Ganulosis, the nematode viz., *Steinernema carpocapsae. Rhabditis* spp. *Parasitorhadbitis*, and protozoans viz., *Nosema* spp., *Tetrahymena* spp., are under use to control crop pests.

Besides, a number of important parasites viz., *Trichogramma* spp., *Grniozus* spp., *Tetrastichus* spp. *Telonomus* spp., and predators viz., *Cryptolaemus* spp., *Chrysoperla* spp., *Coccinella* spp., *Mallada boninensis*, and *Chilochorus nigritus*, are being used in farmers' field for pest management.

This book is a standard reference work, offering available basic facts, re-evaluating and reviewing the past research, and providing the new and current discoveries on the subject and up-to-date information on biopesticides and bioagents.

The main features—(1) scope of the biopesticides and bioagents as a tool in integrated pest management, (2) emerging technologies of the subject, (3) sustainability of the technology, and (4) future directions in effective pest management to ensure safety of the environment—have been comprehensively covered in various chapters of the book.

CHAPTER 1

STATUS OF BIOPESTICIDES AND BIOCONTROL AGENTS IN AGRICULTURE: AN OVERVIEW

MD. ARSHAD ANWER

Department of Plant Pathology, Bihar Agricultural University, Sabour 813210, India

E-mail: arshad_anwer@yahoo.com

CONTENTS

ABSTRACT

To achieve sustainable development of agriculture it is a need for advance research and development in the field of biopesticides applications which greatly reduce the environmental pollution caused by the chemical synthetic insecticides residues. Since the advent of biopesticides, a large number of products have been registered and released; some of them have played a leading role in the agriculture market. The development of biopesticide has encouraged replacing the chemical synthetic pesticide in pest management in some extent. Biopesticides and biocontrol agents have proven to be highly effective, species specific and eco-friendly in nature, leading to their adoption in pest management strategies around the world. The microbial biopesticides market constitutes about 92% of total biopesticides and there is sufficient scope for further development in agriculture, although there are challenges as well. This chapter reviews the various microbial especially fungal biopesticides, the different approaches for their production and development, field evaluation, registration and recent technological advances, and the challenges faced by the microbial biopesticide field in the future.

1.1 INTRODUCTION

Biopesticides are the formulations of biocontrol agents in a form which keep the organism at higher count and viable for their introduction or application in the field. Hence, for huge-scale or business use of biocontrol agents, their biopesticides are necessarily produced. Among the total pesticides used in India, more than 60% of the chemical pesticides are used in the agriculture sector. The use of chemical pesticides is highest in Andhra Pradesh with 20%, followed by Punjab with 10%, Tamil Nadu with 9%, and Karnataka and Gujarat with 6%. Cotton with 40–50%, rice with 17–18% and vegetables with 14–16% use maximum quantity of pesticide in the country. In India, among the chemical pesticides, insecticides are used to a large extent of about 60% followed by fungicides and bactericides (20%) herbicides (17%) and other chemicals (3%). While in Western countries, herbicide use is the highest. The world average for herbicide use is about (45%) followed by insecticides (36%), fungicides (17%), and other chemicals (2%) (Wahab, 2003, 2005, 2009).

Till now, 452 biocontrol agents are used in the production of 2000 commercial products worldwide. It includes 149 micro-organisms; 89 natural products; 140 macro-organisms; and 74 semiochemicals. In India, about 16 commercial preparations of *Bacillus thuringiensis*, 38 fungal formulations based on *Trichoderma, Metarhizium, Beauveria* and about 45 baculovirus-based formulations of *Helicoverpa* and *Spodoptera* are available. Microbials are expected to replace at least 20% of the chemical pesticides. Biotic agents are being supplied by about 128 units in the country (80 private companies). Besides, ICAR institutes (8), State Agricultural Universities (10) and Central Integrated Pest Management centers (30), and four parasitoid-producing laboratories are also supplying natural enemies (Wahab, 2003, 2004, 2009).

1.2 DEFINITION AND TYPES OF BIOPESTICIDES

In the European Union, biopesticides have been defined as "a form of pesticide based on micro-organisms or natural products" (Anonymous, 2008). According to the United States Environmental Protection Agency (EPA), they "include naturally occurring substances that control pests (biochemical pesticides), microorganisms that control pests (microbial pesticides), and pesticidal substances produced by plants containing added genetic material (plant-incorporated protectants) or PIPs" (Anonymous, 2012). Biopesticides belong to mainly three categories: (a) living organisms, which include predatory insects, nematodes and micro-organisms, (b) naturally occurring substances which include plant extracts and semiochemicals, for example, insect pheromones, and (c) genetically modified plants that express introduced genes that confer protection against pests or diseases (plant-incorporated products).

1.3 MASS PRODUCTION OF FUNGAL BIOLOGICAL CONTROL AGENTS

Fermentation methods are important for mass production of microorganisms and to harvest a much better yield quantitatively as well as qualitatively. Three methods of fermentation are described by Lewis and Papavizas (1991).

1.3.1 LIQUID FERMENTATION

This technology has been adopted to produce bacterial and fungal biomass. A suitable medium should consist of inexpensive, readily available agricultural byproducts with appropriate nutrient balance. Acceptable materials include molasses, brewer's yeast, corn steep liquor, sulfate waste liquor, cotton seed, and soya flours (Lisansky, 1985). Alegre et al. (2003) proposed liquid fermentation method consisting of molasses, wheat bran, and yeast for large-scale production of *T. harzianum*. A higher produce of *Trichoderma* chlamydospores was harvested through liquid fermentation technology. The preparation based on chlamydospores prevented the disease more effectively than a preparation that contained conidia only (Lewis et al., 1990; Papavizas & Lewis, 1989). Small-scale fermentation in molasses-brewers yeast medium has also resulted in abundant chlamydospore production of *Trichoderma* (Papavizas et al., 1984).

1.3.2 SOLID FERMENTATION

Mass production of antagonists on solid substrates for the production of inoculum of various biocontrol fungi includes straws, wheat bran, and sawdust; bagasse moistened with water or nutrient solutions through fermentation technology is referred to as solid fermentation (Papavizas, 1985). This technology is also effective especially for those organisms, which can multiply on dry substrates.

1.3.3 SEMISOLID FERMENTATION

Semisolid fermentation is done for the fungi which do not sporulate in liquid culture. Diatomaceous earth granules impregnated with molasses (Backman & Rodriguez-kabana, 1975), wheat bran, and vermiculite-wheat bran (Lewis et al., 1989) yield good produce of bioagents. This method, however, requires more area and is labor intensive, and the chances of contamination are high when compared to liquid fermentation.

For commercial production of biological control agents, different technologies have been adopted on industrial scale. Fermented *Trichoderma* consisted mainly of chlamydospores and conidia with some amount of mycelial fragments. Air-dried mats have been grounded and mixed with a commercially available carrier, the formulations thus developed, contained 10^8–10^9 propagules/g (Table 1.1; Chaube et al., 2002; Chaube & Pundhir, 2010).

TABLE 1.1 Base Material/Carriers Used for Mass Production of Fungus Biocontrol Agents.

Antagonists	Base material(s)	Form of formulation
Trichoderma harzianum	Black gram shell, shelled maize cob, coir-pith, peat, gypsum, coffee fruit skin + biogas slurry, coffee husk, coffee-cherry husk, fruit skin and berry mucilage, molasses-yeast, molasses-soy, molasses-NaNO$_3$, mushroom-grown waste, sugarcane straw, wheat bran + biogas manure (1:1), wheat-bran + kaolin.	Powder, pellets
T. viride	Barley grains, black gram shell, shelled maize cob, coir-pith, peat, gypsum, coffee husk, coffee-cherry husk, fruit skin and berry muci-lage, mushroom grown waste, mustard oil cake, neem cake + cow dung, poultry manure, spent tea leaf waste, sugarcane straw, talc, vermiculite + wheat bran + HCL	Pellets
T. virens	Barley grains, coffee husk, coffee-cherry husk, fruit skin and berry mucilage, mushroom-grown waste, neem cake + cow dung, poultry manure, soil, sorghum grains, talc, wheat bran sawdust.	Pellets
T. longibrachiatum	Talc, wheat bran + saw dust	Powder
Aspergillus niger	Citrus pomace (waste from canning industry), talc + cmc	Pellet, powder
A. terreus	Maize-meal + sand	Powder

1.4   IMMOBILIZATION

For field application of a biocontrol agent, an efficient substrate for mass production and an inert immobilizing material are required, which could carry the maximum number of propagules of the organism with minimum volume and necessarily maintain its survival and integrity. An excellent bioinoculant is one that is introduced to an ecosystem, and subsequently survives, proliferates, becomes active, and establishes itself in a new environment (Khan, 2005). For preparing a commercial formulation, these attributes must be considered. In addition, the bioinoculant should be mass cultured on an inexpensive substrate in a short time. Easy application, effectiveness, and consistent results under a variety of environmental conditions are other desirable features required for production of bioinoculant formulations.

Different techniques of cell immobilization have been developed to devise efficient carrier systems to produce commercial formulations of bioinoculants. A number of carriers for immobilization of microorganisms have been used to develop commercial formulations of biocontrol agents viz., peat, seeds, meals, kernels, husks, bran, bagasse, farmyard manure, cow dung cake, compost, oil cakes, wood bark, vermiculite, sand, clay, and liquid carriers. Three types of formulations viz., pellet, granular, and liquid, are widely produced (Mukhopadhyay, 1987; Kousalya & Jeyarajan, 1990; Bhai et al., 1994; Khan & Anwer, 2011). Tiwari et al. (2004) evaluated grains of sorghum, wheat, bajra, wheat bran, rice bran, and sugarcane bagasse for mass multiplication of *T. viride*. Grains of sorghum (cv. swanki) were found the best substrate that provided maximum spore concentration (8×10^9 spores) and spore viability (92.5%) after 15 days of incubation at 27 ± 1 °C. Spores remained viable for 6 months at 5 °C. Wheat bran and sawdust mixture have been used as a carrier media for the mass multiplication of *T. harzianum* (Elad et al., 1980; Mukhopadhyay et al., 1986). Khan et al. (2001) evaluated various agricultural and industrial wastes for mass multiplication of *T. harzianum, T. virens,* and *P. chlamydosporia,* and found highest CFU, 1.2×10^6 on bagasse-soil-molasses for *T. harzianum,* 1.0×10^6 for *T. virens* and 1.1×10^6 on corn meal-sucrose mixture for *P. chlamydosporia*. Vidhyasekaran et al. (1997) developed powder formulations of *P. fluorescens* using talc powder, peat, vermiculite, lignite, and kaolinite. All freshly prepared powdered formulations were effective in controlling pigeon pea wilt, but their efficacy varied depending upon the length of storage. Talc formulation was effective even after 6 months of storage. Bhai et al. (1994) evaluated a number of agricultural wastes which could be used as carrier and multiplication media at the same time. They reported that sterilized tea waste, coffee husk, or a mixture of coffee husk and cattle manure were ideal combinations for the fast growth and multiplication of *T. harzianum* and *T. virens*. Angappan (1992) used molasses yeast medium for growing *T. viride* and mixed it with talc powder to develop commercial formulation. The initial population in the produce was 3×10^8 CFU/g, whereas the product should contain 2×10^6 CFU/g at the time of use. The shelf life of this product was 4 months. Seed treatment of chickpea with this product maintained the rhizosphere population of the bioagent at $11-13 \times 10^3$ CFU/g soil throughout crop. Ranganathan et al. (1995) found that gypsum is a good and cheap substitute for talc. Nakkeeran and Jayrajan (1996) tested two industrial wastes—precipitated silica and calcium silicate as carriers for *Trichoderma* in the place of talc. The material gave a population of 0.99 and 1.04×10^8 CFU/g, respectively compared to 1.4×10^8 CFU/g in talc substrate after 4 months of storage. Both the substrates were much cheaper than talc. Backman and Rodriguez-Kabana (1975) used diatomaceous earth granules impregnated with 10% molasses solution for rearing *T. harzianum*. It was applied to peanut at 140 kg/h on 70 and 100 days after sowing to control *Sclerotium rolfsii*. The disease was reduced by 42% over control and yield increased by 13.5%.

Several researchers have used combination of two or more agricultural materials. Elad et al. (1986) used wheat bran: sawdust: tap water mixture (3:1:4 v/v) for *T. harzianum*. It was applied at the time of sowing and mixed with the soil to a depth of 7–10 cm with a rotatory hoe. It increased yield of beans (15 q/h), tomato (3 q/h), cotton (5 q/h) and potato (4–6 q/h) and controlled *Sclerotium rolfsii* and *Rhizoctonia solani*. Vidhya (1995) applied the formulation of *T. harzianum* based on vermiculate-wheat bran (@ 250 kg/ha) to mungbean and found 41% reduction in root-rot (*Macrophomina phaseolina*) and 91% increase in yield. Papavizas and Lewis (1989) prepared *T. virens* on alginate-bran-fermenter biomass pellets and pyrax–fermenter biomass mixture. Soil application of the product checked the damping off caused by *R. solani*. Several other substrates such as farm yard manure (FYM), biogas plant slurry, press mud, paddy chaff, rice bran, groundnut shell (Kousalya & Jeyarajan, 1988), FYM, FYM + sand, sawdust, wheat bran, pigeon pea leaves, wheat straw, and urdbean straw (Chaudhary & Prajapati, 2004) have been tested to grow *T. viride* and *T. harzianum*. The enumeration of viable CFUs revealed that pigeon pea leaves and urdbean straw were the best substrates showing 3.4 and 3.4×10^5 propagules at 4 months, 1.2 and 1.1×10^5 at 8 months, and 1.5 and 3.0×10^4 at 12 months of storage at room temperature, whereas sorghum seed showed 11.4, 3.8, and 0.6×10^4 propagules at the same intervals, respectively. Next suitable substrates were wheat straw and saw dust. Cabanillas and Bakar (1989) tested some carriers like wheat grains, alginate pellets, and diatomaceous earth granules for soil application of *P. lilacinus.* Kerry et al. (1984) used oat seeds to rear *P. chlamydosporia* for field application. Soil application of the colonized oat kernels @ 0.5 and 1.0% (w/w soil:seed) considerably reduced the population of root-knot and cyst nematodes (Godoy et al., 1983; Rodriguez-kabana et al., 1984). De Leij and Kerry (1991) did encapsulation of liquid suspension of spores and hyphae of *P. chlamydosporia* with sodium alginate containing 10% (w/v) kaolin or wheat bran. On soil application, the fungus proliferated in soil from only those granules which contained wheat bran as energy source. In other study, Kerry (1988) estimated approximately 9×10^4 and 4×10^4 CFUs of the fungus/g soil after 1 and 12 weeks of application of granules, respectively.

1.5 SHELF-LIFE TEST

The limiting factor in commercialization of a biocontrol agent preparation is its loss of viability of the biocontrol agents over time. Considerable efforts have been made in India itself to determine the viability of biocontrol agents in their preparations when stored at room temperature and in refrigerator. Most of the results are variable and therefore it appears that shelf life is also dependent upon species/isolate/strain.

1.6 EVALUATION FOR EFFECTIVENESS

In general, the antagonist multiplied in an organic food base has greater shelf life than that on an inert or inorganic food base (Jeyarajan & Nakkeeran, 2000). A talc-based preparation of *T. virens* conidia retained 82% viability at 5 °C in refrigerator after 6 months, while at room temperature (25–35 °C), same level of viability was observed only up to 3 months. Shelf life was same, when *T. virens* treated chickpea or soybean seeds were stored at room temperature (Tiwari et al., 2004). Seed coating with biocontrol agents has emerged as a feasible was of delivering the antagonists, that is, supplying the coated seeds to the farmers directly by the seed companies/agencies. The time gap between coating seeds and sowing such seeds by farmers is critical. Mukherjee (1991) quantitatively assessed the viability of *T. virens* on coated chickpea seeds when the seeds were stored at low temperature (5 °C) and at room temperature (25–35 °C), 88% of the propagules remained viable for up to 4 months. Conidia of *Trichoderma* in pyrophyllite survived better than fermenter biomass propagules alone at −5 to 30 °C. The most suitable temperature to prolong shelf life of conidia and fermenter biomass propagules in pyrophyllite were −5 to 5 °C (Mukherjee, 1991). Storage at 5 °C increased the shelf life of *T. virens* and *T. hamatum* in granular formulations of pre gelatinizing starch flour up to 6 months. Chlamydospore-based formulations of *T. virens* and *T. harzianum* exhibited longer shelf life (80% viability for 9 months) than conidia-based formulations (80% viability after 4 months) at room temperature and a preparation of *T. virens* (mainly in the form of chlamydospores was from peat moss Czapek broth culture) stored at 25 °C for 6 months without loss of viability (Mishra, 2002). Lewis et al. (1995) reported that among the different carriers tested; the shelf life of *B. subtilis* in soybean flour was increased up to three months. Ranganathan et al. (1995) also reported 4-month shelf life of *T. viride* in gypsum-based formulations. Talc-based formulations of *T. harzianum* can retain more than 10^6 viable propagules per gram up to 90 days (Prasad & Rangeshwaran, 2000). Jeyarajan et al. (1994) developed talc-, peat-, lignite-, and kaolin-based formulations of *T. viride,* which had a shelf life of 4 months. Studies on storage temperature revealed that 20–30°C was optimum to store vermiculite fermenter biomass of *Trichoderma* up to 75 days without losing the viability (Nakkeeran et al., 1997). Commercial formulations of *Aspergillus niger* AN27 showed an extraordinary long shelf life of more than two years at room temperature (25–35 °C) when packed in polyethylene bags and stored under less than 80% relative humidity (Sen, 2000).

1.7   BIOSAFETY ANALYSIS

Quality control is the most essential aspect of biopesticide production. A good quality of the preparation is necessarily required to retain the confidence of farmers on the efficacy of biocontrol formulation. Being living agents their population in a product may be influenced by storage. The other contaminating microorganisms in the product should also be within permissible limits.

1.8 REGISTRATION

Current legislation demands that new products are subjected to detailed study of their environment impact and toxicological effects and they are registered. As current legislation stands, there are certain categories of biocontrol agents that have an easier and quicker passage for registration. Indigenous microorganisms that are specific to a defined group of targets have a comparatively straightforward progress. Under Section 9(3) of Pesticide Act of India (1968), information required for registration of any biopesticides includes: systematic name and common name; natural occurrence and morphological descriptions; details of manufacturing process (active and inert ingredients of formulation); test methods (dual culture of pathogenicity); quantitative analysis (CFU on selective medium, absence of Gram negative bacteria contaminants); moisture content; shelf life; mammalian toxicity; bioefficacy; environmental toxicity and residue analysis. Because of less awareness of growers toward biocontrol programs, the Indian Biopesticide Industry involves more than 15–20% expenditure on marketing compared to only 1–2% marketing expenses in the case of conventional pesticides (Singhal & Sharma, 2003). Therefore, there is a need for simplification of registration requirements and government subsidies should be granted to farmers to promote biopesticide use.

The registration policies may vary with the country. In the United States, registration of microbial pesticide requires toxicological tests for oral, dermal, eye, and other health hazards using test animals or fish. If these tests show no adverse effects and the biocontrol agent is not a pathogen, it is registered and can be sold. The cost required for research and development for biopesticide is only US$ 0.8–1.6 million as against US$ 20 million for chemical pesticides. The toxicological tests for a biocontrol agent cost US$ 0.5 million as against US$ 10 million for chemical pesticides. The number of candidates to be tested to develop one biocontrol product will be in 100s as against 20,000 for a chemical pesticide. It was estimated that the market size required for profit for a biocontrol agent is US$ 1.6 million per year as against US$ 4 million per year for a chemical pesticide (Cook, 1993).

1.9 CONCLUSION AND FUTURE PROSPECTS

It would be impractical to expect that biocontrol products will replace chemical pesticides in the near future. There will always be a need for chemical pesticides, given that the world population continues to increase at a swift rate and worldwide food production must keep balance with it. However, there is likely to be some decline in the use of synthetic pesticides, especially in developed countries and this will provide an opportunity for microbial products to satisfy markets.

Developing a safe, easy to use, cost-effective formulation that will keep the microorganism alive is one of the most important steps in developing a biological product. Formulation is the blending of active ingredients such as fungal spores with inert carriers such as diluents and surfactants in order to improve the physical characteristics. A final formulation must have a long shelf life at room temperature, be easy to handle, be insensitive to abuse, and must be stable over a range of −5 to 35 °C. The most needed technique is drying techniques, which allows retention of maximum number of viable propagules in dried product. Most important steps of production of biopesticides are mass production of biocontrol agents, their immobilization, shelf-life test, evaluation for effectiveness, biosafety analysis, registration, etc.

KEYWORDS

- biopesticides
- biocontrol agents
- chemical pesticides
- bacterial and fungal biomass
- liquid fermentation

REFERENCES

Alegre, R. M.; Rigo, M.; Joekes, I. Ethanol Fermentation of a Diluted Molasses Medium by *Saccharomyces cerevisiae* Immobilized on Chrysotile. *Braz. Arch. Biol. Technol.* **2003,** *46* (4), 751–757.

Angappan, K. Biological Control of Chickpea Root Rot Caused by *Macrophomina phaseolina* (Tassi) Goid. M. Sc.(Ag.)Thesis, TNAU, Coimbatore, 1992, pp. 114.

Anonymous. Encouraging Innovation in Biopesticide Development. European Commission, 2008. (accessed April 20, 2012).

Anonymous. Regulating Biopesticides. Environmental Protection Agency of the USA, 2012. (accessed April 20, 2012).

Backman, P. A.; Rodriguez-Kabana, R. A System for the Growth and Delivery of Biological Control Agent to the Soil. *Phytopathology.* **1975,** *65,* 819–821.

Bhai, S. R.; Thomas, J.; Naidu, R. Evaluation of Carrier Media for Field Application of *Trichoderma* Spp. In Cardamon Growing Soils. *J. Plant. Crops.* **1994,** *22* (1), 50–52.

Cabanillas, E.; Barkar, K. R. Impact of *Paecilomyces lilacinus* Inoculum Level and Application Time on Control of *Meloidogyne incognita* on Tomato. *J. Nematol.* **1989,** *21,* 115–120.

Chaube, H. S.; Mishra, D.S.; Varshney, S.; Singh, U. S. Biological Control of Plant Pathogens by Fungal Antagonists: Historical Background, Present Status and Future Prospects. In *Annual Review of Plant Pathology;* Indian Society of Mycology and Plant Pathology: Rajasthan, India, 2002; vol. II.

Chaube, H. S.; Pundhir, V. S. Crop Diseases and their Management; PHI: New Delhi, India, 2010; p 724.

Chaudhary, R. G.; Prajapati, R. K. Comparative Efficacy of Fungal Bioagents Against *Fusarium udum. Ann. Plant Protect. Sci.* **2004,** *12,* 75–79.

Cook, R. J. Making Greater Use of Introduced Microorganisms for Biological Control of Plant Pathogens. *Ann. Rev. Phytopathol.* **1993,** *31,* 53–80.

De Leij, F. A. A. M.; Kerry, B. R. The Nematophagous Fungus *Verticillium chlamydosporium* as a Potential Biological Control Agent for *Meloidogyne arenaria. Rev. Nematol.* **1991,** *14,* 157–164.

Elad, Y.; Chet, I.; Katan, T. *Trichoderma harzianum* a Biological Agent Effective against *Sclerotium rolfsii* and *Rhizoctonia solani. Phytopathology.* **1980,** *70,* 119–121.

Elad, Y.; Zuiel, Y.; Chet, I. Biological Control of *Macrophomina phaseolina* Tassi Gold by *Trichoderma harzianum. Crop Prot.* **1986,** *5,* 288–292.

Godoy, G.; Rodriguez-Kabana, R.; Morgan-Jones, G. Fungal Parasites of *Meloidogyne arenaria* Eggs in an Alabama Soil. A Mycological Survey and Green House Studies. *Nematropica.* **1983,** *13,* 201–213.

Jeyarajan, R.; Nakkeeran, S. Exploitation of Biocontrol Potential of *Trichoderma* for Field Use. In *Current Trend in Life Sciences;* Manibhushan, R. K., Mahadevan, A., Eds.; Today and Tomorrow Printers and Publishers: New Delhi, India, 2000; Vol. XXI, pp 61–66.

Jeyarajan, R.; Ramakrishnan, G.; Dinakaran, D.; Sridar, R. Development of Products of *Trichoderma viride* and *Bacillus subtilis* for Biocontrol of Root Rot Diseases. In *Biotechnology in India;* Dwivedi, D. K., Ed.; Bioved Research Society: Allahabad, India, 1994; p 25.

Kerry, B. R. In *Two Micro-Organisms for the Biological Control of Plant Parasitic Nematodes,* Proceedings of the Vrighton Crop Protection Conference 2, England, 1988; pp 603–607.

Kerry, B. R.; Simon, A.; Rovira, A. D. Observations on the Introduction of *Verticillium chlamydosporium* and Other Parasitic Fungi into Soil for Control of the Cereal Cyst Nematode *Heterodera avenae. Ann. Appl. Biol.* **1984,** *105,* 509–516.

Khan, M. R. Biological Control of Fusarial Wilt and Root-Knot of Legumes; Government of India Publication, Department of Biotechnology, Ministry of Science and Technology: New Delhi, India, 2005; p 50.

Khan, M. R.; Anwer, M. A. Fungal Based Bioinoculants for Plant Disease Management. In *Microbes and Microbial Technology: Agricultural and Environmental Applications;* Pichtel, J., Ahmad, I., Eds.; Springer: USA, 2011.

Khan, M. R.; Khan, N.; Khan, S. M. Evaluation of Agricultural Materials as Substrate for Mass Culture of Fungal Biocontrol Agents of Fusarial Wilt and Root-Knot Nematode Diseases. *Ann. Appl. Biol.* (TAC-21 Suppl.) **2001,** *21,* 50–51.

Kousalya, G.; Jeyarajan, R. In *Techniques for Mass Multiplication of Trichoderma viride and T. Harzianum rifai,* National Seminar on Management of Crop Diseases with Plant Products/Biological Agents, Agricultural college and Research Institute Madurai: Tamil Nadu, India, 1988; pp 32–33.

Kousalya, G.; Jeyarajan, R. Mass Multiplication of *Trichoderma* spp. *J. Biol. Control.* **1990,** *4* (1), 1–10.

Lewis, J. S.; Barksdale, T. H.; Papavizas, G. C. Green House and Field Studies on the Biological Control of Tomato Rot Caused by *Rhizoctonia solani. Crop Prot.* **1990,** *9,* 8–14.

Lewis, J. A.; Fravel, R. D.; Lumsden, R. D.; Shasha, B. S. Application of Biocontrol Fungi in Granular Formulations of Pregelatinized Starch Flour to Control Damping off Diseases Caused by *Rhizoctonia solani. Biol. Control.* **1995,** *5,* 397–404.

Lewis, J. A.; Larkin, R. P.; Rogers, D. L. A Formulation of *Trichoderma* and *Gliocladium* to Reduce Dampings-Off Caused by *Rhizoctonia solani* and Saprophytic Growth of the Pathogen in Soil Mixture. *Plant Dis.* **1989,** *82,* 501–506.

Lewis, J. A.; Papavizas, G. C. Biocontrol of Plant Diseases: The Approach of Tomorrow. *Crop Prot.* **1991,** *10,* 95–105.

Lisansky, S. G. Production and Commercialization of Pathogens. In *Biological Pest Control.* Proceedings of International Phytopathology Congress, Indonesia, Martosupon, N. W. M., Prayudi, B., Eds.; Root Disease Control on Tea by Fungicides, V. Malang, Indonesia: 1985.

Mishra, D. S. Development of Mixed Formulation of Fungal (*Trichoderma*) and Bacterial (*Pseudomonas*) Biocontrol Agents for Management of Plant Disease. Ph.D. Thesis, GB Pant University of Agriculture and Technology, Pantnagar, 2002; pp 185.

Mukherjee, P. K. Biological Control of Chickpea Wilt Complex. Ph.D. Thesis, GB Pant University of Agriculture and Technology, Pantnagar, 1991; p 188.

Mukhopadhyay, A. N. Biological Control of Soil Borne Plant Pathogens by *Trichoderma* spp. *Indian J. Mycol. Plant Pathol.* **1987,** *17,* 1–10.

Mukhopadhyay, A. N.; Brahamabhatt, A.; Patel, G. J. *Trichoderma harzianum* a Potential Biocontrol Agent of Tobacco Damping-off. *Tob. Res.* **1986,** *12,* 26–35.

Nakkeeran, S.; Gangadharan, K.; Renukadevi, P. Seed Borne Microflora of Pigeon Pea and their Management. *Plant Dis. Res.* **1997,** *12,* 103–107.

Nakkeeran, S.; Jeyarajan, R. In *Exploitation of Antagonistic Potential of Trichoderma for Field Use,* Paper presented in National Symposium on 'Disease of Plantation Crops and their Management' at Institute of Agriculture, Sriniketan, West Bengal, India, 1996.

Papavizas, G. C. *Trichoderma* and *Gliocladium:* Biology, Ecology and Potential for Biocontrol. *Ann. Rev. Phytopathol.* **1985,** *23,* 23–54.

Papavizas, G. C.; Dunn, M. T; Lewis, J. A.; Beagle-Ristaino, J. Liquid Fermentation Technology for Experimental Production of Biocontrol Fungi. *Phytopathol.* **1984,** *74,* 1171–1175.

Papavizas, G. C.; Lewis, J. A. Effect of *Gliocladium* and *Trichoderma* on Damping-off of Soybean caused by *Sclerotium rolfsii* in the Greenhouse. *Plant Pathol.* **1989,** *38,* 277–286.

Prasad, R. D.; Rangeshwaran, R. Shelf Life and Bioefficacy of *Trichoderma harzianum* Formulated in Various Carriers Materials. *Plant Dis. Res.* **2000,** *15,* 38–42.

Ranganathan, K.; Sridar, R.; Jeyarajan, R. Evaluation of Gypsum as a Carrier in the Formulation of *Trichoderma viride. J. Biol. Control.* **1995,** *9,* 61–62.

Rodriguez-Kabana, R.; Morgan-Jones, G.; Godoy, G.; Gintis, B. O. Effectiveness of Species of *Gliocladium, Paecilomyces* and *Verticillium* for Control of *Meloidogyne arenaria* in Field Soil. *Nematropica.* **1984,** *14,* 155–170.

Sen, B. Biocontrol: A Success Story. *Indian Phytopathol.* **2000,** *53* (3), 243–249.

Singhal, V.; Sharma, M. C. Promoting IPM at a Faster Pace. In *Technology Transfer in Developing Countries*, 6th International PGPR Workshop, Oct 5–10, 2003; Calicut, India, 2003; pp 68–73.

Tiwari, A. K.; Kumar, K.; Razdan, V. K.; Rather, T. R. Mass Production of *Trichoderma viride* on Indigenous Substrates. *Ann. Plant Protect. Sci.* **2004,** *12* (1), 71–74.

Vidhya, R. Studies on Biological Control of Mungbean Root Rot *Macrophomina phaseolina* Tessi Gold. By *Trichoderma viride.* M.Sc(Ag.) Thesis, TNAU, Coimbatore, India, 1995; p 143.

Vidhyasekaran, P.; Sethuraman, K.; Rajappan, K.; Vasumathi, K. Powder Formulations of *Pseudomonas fluorescens* to Control Pigeonpea Wilt. *Biol. Control.* **1997,** *8* (3),166–171.

Wahab, S. Biotechnological Approaches in the Management of Plant Pests, Diseases and Weeds for Sustainable Agriculture. *J. Biopestic.* **2009,** *2* (2), 115–134.

Wahab, S. Biotechnological Approaches in Plant Protection. In *Biopesticides and Pest Management;* Koul, O., Dhaliwal, G. S., Marwaha, S. S., Arora, J. K., Eds.; Campus Books International: New Delhi, India, 2003; Vol.1, pp 113–127.

Wahab, S. Biotechnological Approaches in the Management of Plant Pests, Diseases and Weeds for Sustainable Agriculture. In *Deep Roots, Open Skies: New Biology in India*; Basu, S. K., Batra, J. K., Salunke, D. M., Eds.; Narosa Publishing House: New Delhi, India, 2004; Vol. 200, pp 113–129.

Wahab, S. Development and Uses of Biopesticides in India. In *Biopesticides: Emerging Trends;* Koul, O., Dhaliwal, G. S., Shanker, A., Raj, D., Kaul, V. K., Eds.; Campus Books International: New Delhi, India, 2005.

PART I
Plant Health Promoting Biocontrol Agents

CHAPTER 2

TRICHODERMA—AN IMPECCABLE PLANT HEALTH BOOSTER

NAJAM WARIS ZAIDI* and UMA SHANKAR SINGH

Department of Plant Breeding, International Rice Research Institute, IRRI India Office, 1st Floor, NASC Complex, DPS Marg, New Delhi 110012, India

Corresponding author. E-mail: n.zaidi@irri.org

CONTENTS

ABSTRACT

Species of *Trichoderma* belong to one of the most useful group of microbes to have had an impact on human welfare in recent times. *Trichoderma* (teleomorph *Hypocrea; Hypocreaceae, Hypocreales,* and *Ascomycota*) spp. have had a major influence on agriculture too. *Trichoderma* spp. are one of the most plentiful fungal genera found in many ecosystems. They are most widespread and popular research tools as microbial inoculants that are largely used against several plant pathogenic fungi causing soil borne, foliar, and post-harvest diseases of plants. This is well reflected by the completion of whole genome sequencing of as many as seven species of *Trichoderma* by the Joint Genomic Institute (JGI) (Mukharjee et al., 2013). Having highly antagonistic and mycoparasitic potential in laboratory conditions, they have shown to reduce the severity of plant diseases by inhibiting plant pathogens in the soil. *Trichoderma* can have direct effects on plants, increasing their biomass and nutrient uptake, fertilizer use efficiency, seed germination, and stimulation of plant defenses by enhanced photosynthetic ability, against biotic and abiotic stresses and hence is a famous plant growth promoter. Some species can kill plant parasitic nematodes and have the potential to be used as bio-nematicides.

Most *Trichoderma* are endophytic plant symbionts. These changes alter plant physiology and may result in the improvement of nitrogen uptake, resistance to pathogen, photosynthetic efficiency, as well as abiotic stress tolerance. According to an estimate, about 60% of all the registered biofungicides worldwide are *Trichoderma* based with more than 250 *Trichoderma*-based formulations being commercially sold in India alone (Verma et al., 2007). *Trichoderma* strains interact with the plant by colonizing roots, establishing chemical communication, and systemically altering the expression of numerous plant genes.

So far major bottleneck with the use of *Trichoderma* is their uneven field performance. This is due to the fact that they are biotic agents. Their field performance depends upon a number of factors like, strain/host selectivity, environmental conditions, inoculum density, method of delivery, and so forth. By far, application of *Trichoderma* both through seed and colonized compost is the most effective delivery system for the management of seedling and root diseases in different crops. Also they can help alleviate abiotic stresses in plants, including water deficit (drought), salinity, and temperature.

2.1 INTRODUCTION

The taxonomy of *Trichoderma* is relatively young compared with that of Hypocrea. *Trichoderma* spp. are free-living fungi that are common in soil, root, and foliar ecosystems throughout the world and are highly interactive. They are present in nearly all types of soils and other natural habitats especially those containing high organic matter. Being secondary colonizers, they are often isolated from well-decomposed organic matter like decaying barks and from sclerotia and fruiting bodies of other fungi. *Trichoderma* generally exhibit a preference for wet soils but individual species exhibit different preferences for soil temperature, moisture, and pH. While species like *Trichoderma viride* and *Trichoderma polysporum* are generally found in areas with low temperature, *Trichoderma harzianum* most commonly occurs in warm climatic regions. *Trichoderma hamatum* and *Trichoderma koningii* can occur in diverse climatic conditions (Zaidi & Singh, 2013). *Trichoderma* spp. have the ability to utilize a wide range of compounds as sole carbon and nitrogen source and can utilize monosaccharides, disaccharides, and polysaccharides for carbon with ammonia being the most preferred source of nitrogen.

The presence of carbon dioxide has been reported to favor growth of *Trichoderma*. Apart from these factors, the iron content of the soil, HCO_3^- salt and organic matter content, and presence or absence of other microbes are also important determinants of microsite preference by *Trichoderma* spp. (Papavizas, 1985).

Trichoderma spp. are adapted to various ecological niches as fungal antagonists, opportunistic, avirulent plant symbionts, and endophytes and saprophytes on bark and dead wood (*Harman* et al., *2004*). Beneficial actions of *Trichoderma*/plant interactions include direct antagonism against plant pathogens mainly fungi and nematodes (Singh et al., 2003), increased plant growth, especially of roots, and particularly under stress (Harman, 2000; Shoresh *et al.*, 2010), systemic resistance to diseases (Lorito *et al.*, 2010; Bae *et al.*, 2011), enhanced tolerance to abiotic plant stresses, including water deficit (drought), salinity, and temperature, enhancement of the vigor of poor-quality seeds (Mastouri *et al.*, 2010; Shoresh *et al.*, 2010), decomposition of organic matter resulting in increased concentration of humic acid (Singh et al., 2003), solubilization of phosphorus and increased availability of micronutrients (Altomare et al., 1999; Singh et al., 2003; Harman, 2011), and improved nitrogen use efficiency (NUE) by plants (Harman, 2006, 2011; Shoresh *et al.*, 2010).

Trichoderma spp. also produce many extra-cellular enzymes like cellulases, chitinases, glucanases, proteases, and so forth. They are used in foods and textiles and also in poultry feed. *Trichoderma* chitinases are used in generating disease resistant transgenic plants, in plant disease control and improvement of plant growth.

Species of *Trichoderma* have been reported to be excellent colonizers of cow dung compost/farm yard manure (FYM). Reports also suggest that it helps in faster decomposition of cattle dung to convert it into compost (Zaidi & Singh, 2004a). The antifungal action which is driven by mechanisms that comprise both enzymatic and antibiotic activities, allows them to switch between biotrophic and saprophytic lifestyles. Current knowledge suggests that all species are fungicolous (Jacklitsch, 2009; Druzhinina & Kubieck, 2013).

Trichoderma spp. have the ability to reduce the severity of plant diseases by inhibiting plant pathogens, mainly in the soil or on plant roots, through their high antagonistic and mycoparasitic potential (Viterbo & Horwitz, 2010). Studies by Advanced Biological Marketing, Inc. (ABM), Van Wert, Ohio and Cornell University showed that *Trichoderma* strains induce changes in the microbial composition on roots, enhance nutrient uptake, stabilize soil nutrients, promote root development, and increase root hair formation (Harman, 2006). Since *Trichoderma* is a living entity and can be affected by the surroundings, it will not be justifiable enough to compare its efficacy with the chemical pesticides and fertilizers. Success of *Trichoderma* spp. as a biopesticide under field condition depends not on its antagonistic activity alone but a blend of several other characteristics and soil and climatic conditions and plant host. The consistency in performance is still considered as a constraint to the widespread adaptation with high variability across sites and seasons, most likely as a result of the differential impact of fluctuating biotic and abiotic factors. *T. harzianum* and *T. viride* are the widely used species and have been exploited on about 87 different crops and about 70 soil borne and 18 foliar pathogens, respectively (Sharma et al., 2014).

2.2 *TRICHODERMA* GENERAL PHYSIOGNOMIES

2.2.1 *TRICHODERMA AS A MYCOPARASITE*

Trichoderma has been known as a successful biofungicide on a commercial scale. This success is mainly attributed to its ability to parasitize and kill other fungi (Mukherjee et al., 2012a). Several *Trichoderma* spp. have been long known for their potent necrotrophic mycoparasitic abilities (Elad 1995; Druzhinina et al., 2011; Benítez et al., 2004) and are therefore used as model systems to study the mechanisms of mycoparasitism. Mycoparasitism in itself is a complicated process that involves sequential events, including recognition, attack and subsequent penetration, and killing of the host.

Trichoderma can detect its host from a distance and branches atypically toward the host fungus (Chet et al., 1997). The coiling of *Trichoderma* is not merely a contact stimulus (Dennis & Webster, 1971). Inbar and Chet (1992, 1994) provided direct evidence for the role of lectins in mycoparasitism. Once *Trichoderma* is attacked, it coils around the pathogen and forms the appresoria-like structures. The following step consists of the production of cell wall degrading enzymes (CWDEs) and secondary metabolites (Howell, 2003), which facilitate both the entry of *Trichoderma* hypha into the lumen of the parasitized fungus and the assimilation of the cell wall content. The significance of lytic enzymes has been demonstrated by overexpression and deletion of the respective genes (Viterbo et al., 2002; Mukherjee et al., 2012a).

Mitogen-activated protein kinase (MAPK) cascades and G-protein α-subunits transduce a large variety of signals, including those associated with pathogenesis. Signal transduction pathways triggering the genes involved in biocontrol and mycoparasitism have been studied in considerable depth and include heterotrimeric G-protein signaling, MAPK cascades, and the cAMP pathway (Zeilinger & Omann, 2007). Especially the MAP-kinase TVK1/TmkA, characterized in *Trichoderma virens* (Mendoza-Mendoza et al., 2003; Mukherjee et al., 2003; Mendoza-Mendoza et al., 2007) as well as its orthologs in *Trichoderma asperellum* (Viterbo et al., 2005) and *Trichoderma atroviride* (Tmk1; Reithner et al., 2007), is important in regulation of signaling mechanisms targeting output pathways relevant for efficient biocontrol. The seven-transmembrane G-protein coupled receptor, Gpr1 is involved in sensing the fungal prey; silencing of the *Gpr1* gene in *T. atroviride* rendered the mycoparasite unable to respond to the presence of the host fungus (Omann et al., 2012). Binding of a ligand to such receptors leads to downstream signaling events via activation of G-protein cascades. Indeed, deletion of the Tga3 G alpha protein-encoding gene affected the mycoparasitic abilities of *T. atroviride* in a similar way to loss of Gpr1 (Zeilinger et al., 2005). Deletion of the adenylate cyclase gene *tac1* severely impaired growth and mycoparasitic abilities of *T. virens* (Mukherjee et al., 2007). Like most other filamentous fungi, *Trichoderma* spp. have three MAPK cascades comprising MAPKKK, MAPKK, and MAPK (Schmoll, 2008) and MAPK pathways may act in mycoparasitism and biocontrol (Reithner et al., 2007; Kumar et al., 2010).

Cell wall degradation during mycoparasitism is mediated by a set of hydrolytic enzymes secreted by *Trichoderma* including chitinases, β-(1,4)-, β-(1,3)- and β-(1,6)-glucanases, and proteases (Benítez et al., 2004). Genome analysis enabled assessment of the total numbers of CWDEs encoded in the genomes of *Trichoderma* spp. and unraveled even more complex enzymatic degradation machinery for fungal cell walls than previously anticipated (Kubicek et al., 2011).

Diverse DNA array experiments have determined that an expansin-like protein, aspartyl proteases, and hydrophobins, among others, are involved in the biocontrol and mycoparasitism of *Trichoderma* spp. (Brotman et al., 2008; Samolski et al., 2009).

Trichoderma genes can be expressed functionally in plants to control of plant diseases. High expression levels of the *T. harzianum* endochitinase gene, *ech42* were obtained in different plant tissues, with no visible effect on plant growth and development (Lorito et al., 1998). *Trichoderma* chitinase genes are used to generate transgenic plants resistant to fungal diseases. Transgenic cotton plants expressing the *T. virens* endochitinase gene *Tv-ech42* showed significant resistance to *Alternaria alternata* and *Rhizoctonia solani* (Emani et al., 2003; Kumar et al., 2009). The expression of *ech42* in lemon enhanced resistance to *Phoma tracheiphila* and *Botrytis cinerea* (Gentile et al., 2007; Distefano et al., 2008). The homologous *ech42* gene from *T. virens* can enhance resistance against *R. solani* when it was expressed in rice, tobacco, and tomato (Shah et al., 2009, 2010). *Ech42* expression enhanced resistance to *Venturia inaequalis*, but reduced plant growth in apple (Bolar et al., 2000) and significantly increased the resistance of broccoli to *Alternaria brassicicola* (Mora &

Earle, 2001). Liu et al. (2004) reported the multiple expression of rice transgenes encoding two chitinases (*ech42* and *nag70*) and one β-1,3-glucanase (*gluc78*) of *T. atroviride* resulted in resistance to *R. solani* and *Magnaporthe grisea* in rice. The expression of the endochitinase *chit36* gene of *T. harzianum* in carrot significantly enhanced tolerance to *Alternaria radicina* and *B. cinerea* (Baranski et al., 2008).

Although no quantitative studies have been performed previously, *Trichoderma* spp. is a suitable genus and has been recognized for their extreme facility in producing a large variety of extra-cellular enzymes and the degradation of lignocellulose (Kirk & Farrell, 1987). Furthermore, *T. viride* and *Trichoderma reesei* are the most extensively studied fungi in the field of cellulosic material degradation (Cullen & Kersten, 1992).

2.2.2 TRICHODERMA AS A PRODUCER OF SECONDARY METABOLITES

Trichoderma is known to produce volatile and non-volatile toxic metabolites that impede colonization by competing microorganisms. Some of them are harzianic acid, tricholin, peptaibols, 6-penthyl-alpha-pyrone, massoilactone, viridin, gliovirin, glisoprenins, and heptelidic acid (Lumsden et al., 1992; Mukherjee et al., 2012b). One of the first characterized peptide antibiotics of *Trichoderma* spp. was Paracelsin (Bruckner et al., 1984). Mycotoxins and more than 100 metabolites derived from amino acids, with antibiotic activity including antibiotics like gliotoxin and glioviridin from *T. virens,* viridin, pyrones, alkyl pyrones, isonitriles, terpenes, polyketides, peptaibols, diketopiperazines, sesquiterpenes polypeptides (Sivasithamparam & Ghisalberti, 1998), and some steroids from other *Trichoderma* species have been suggested to be used for chemotaxonomy of these species.

2.2.3 COMPETITION AND RHIZOSPHERE COMPETENCY OF TRICHODERMA

Trichoderma is known to be a good competitor that involves competition between antagonist and plant pathogen for space and nutrients (Chet, 1987). Usually competition is thought to occur when there is absence of any evidences that mycoparasitism or antibiosis has occurred during a particular *Trichoderma*–host fungus interaction.

Howell (2003) used ultraviolet light irradiation to produce mutants of *T. virens*, deficient for both mycoparasitism and antibiotic production. However, the mutants still retained biocontrol efficacy equal to that of the parent strain against both *Pythium ultimum* and *R. solani* causing cotton seedling diseases. This indicated that neither mycoparasitism nor antibiosis is the principal mechanism involved in the biocontrol of seedling diseases in cotton.

Most fungi excrete siderophores that are low-molecular-weight ferric iron-specific chelators to mobilize environmental iron (Eisendle et al., 2004). Some *Trichoderma* spp. produce highly efficient siderophores that chelate iron and stop the growth of other fungi (Chet & Inbar, 1994). Siderophores also help in improving antagonistic activities, rhizospheric competence, and plant growth. *T. virens* produces three types of hydroxy-mate siderophores: a monohydroxamate (*cis*- and *trans*-fusarinines), a dipeptide of *trans*-fusarinine (dimerum acid), and a trimer disdepsipeptide (coprogen) (Mukherjee et al., 2012b).

Trichoderma spp. are also highly efficient in mobilizing and taking up soil nutrients. The efficient use of available nutrients is based on the ability of *Trichoderma* to obtain ATP from the metabolism of different sugars, such as those derived from polymers widespread in environments: cellulose, glucan, and chitin (Chet et al., 1997).

2.2.4 *TRICHODERMA AS A HYPER PARASITE OF NEMATODES*

Trichoderma has been proved to parasitize nematodes and inactive pathogen enzymes and therefore can be exploited as a bio nematicide to control diseases caused by plant parasitic nematodes (Sharon, 2011).

T. harzianum colonized egg masses, eggs, and second stage juveniles of *Meloidogyne incognita*. It formed loops and trapped second stage juveniles of *M. incognita*. *Trichoderma* penetrated nematode body by forming haustoria like structures and colonized internally replacing all internal organs with fungal mycelia resulting in death of the nematode. Culture filtrate of *T. harzianum* and *T. virens* suppressed hatching and release of second stage juveniles of *Meloidogyne*.

Reduced root galling was reported by Sharon et al. (2001) under green-house experiments on the potential of *T. harzianum* to control root knot nematode, *Meloidogyne javanica* associated with protease production by *T. harzianum*. Prasad and Anes (2008) reported that ethyl acetate and methanol extracts of *T. viride* and *T. harzianum* significantly reduced the total number of galls and *M. incognita* population in Okra. Windham et al. (1989) reported the reduction of egg production by the root knot nematodes *M. incognita* following soil treatment with a *Trichoderma* conidial suspension. Saifullah and Thomas (1996) have also reported direct interactions between *T. harzianum* and potato cyst nematode *Globodera rostochiensis*.

However, a lot of experimentations under field conditions need to be done to fully harness the potential of *Trichoderma* for the control of plant parasitic nematodes.

2.3 *TRICHODERMA* AS AN ENDOPHYTE AND DEFENSE INDUCER

Trichoderma isolates form intimate relationships with plants including the colonization of internal plant tissues. It is the internal colonization of plant tissues by *Trichoderma* without causing harm to the plant that supports the consideration of *Trichoderma* as an endophyte. In recent years it has become apparent that *Trichoderma* isolates not only penetrate and survive inside root tissues but can also internally colonize above ground plant tissues, including woody tissues. The colonization of below and above ground plant tissues have similarities, being influenced by factors such as nutrient availability, water availability, host tissue structural characteristics, host plant genetics, and the *Trichoderma* isolates involved (Bailey & Melnick, 2013).

In addition to direct mycoparasitism of soil-borne fungal pathogens, *Trichoderma* spp. interact with roots. This interaction induces systemic resistance (ISR), which reduces disease in above ground parts of the plant. In the molecular dialog between fungus and plant leading to ISR, proteins secreted by *T. virens* provide signals (Lamdan et al., 2015). *Trichoderma* spp. and other beneficial microorganisms do not cause disease, but result in a stronger immune response upon subsequent challenge by a pathogen (Van Vees et al., 1999; Prime et al., 2006; Conrath, 2011).

Recent discoveries indicate *Trichoderma* spp. can also induce systemic and localized resistance to a variety of plant pathogens and act as opportunistic, avirulent plant symbiont. *Trichoderma* spp. produce or release a variety of compounds that induce localized or systemic resistance responses. Few strains can establish robust and long-lasting colonization of root surfaces, penetrate into the epidermis and a few cortical cells below (Harman et al., 2004). This restricts further advance of the *Trichoderma* and make the plants resistant to other diseases. Induced resistance as a result of activated plant defense in response to *Trichoderma* colonization has been reported by several workers (Alfano et al., 2007; Djonovic et al., 2006; Harman et al., 2004; Korolev et al., 2007; Meyer et al., 1998; Segarra et al., 2009; Shoresh & Harman, 2008; Shoresh et al., 2005). Root colonization by *Trichoderma* spp. frequently enhances root growth and development, crop productivity, resistance to abiotic stresses, and the uptake and use of nutrients (Harman et al., 2005). *Trichoderma* strains become endophytic in roots, but the greatest changes in gene expression occur in shoots. These changes alter plant physiology and may result in the improvement of abiotic stress resistance, nitrogen fertilizer uptake, resistance to pathogens, and photosynthetic efficiency. Typically, the net result of these effects is an increase in plant growth and productivity (Hermosa et al., 2012). The induction of defense response in plants by *Trichoderma* spp. is mostly associated with accumulation of various antimicrobial compounds like phytoalexins, pathogenesis-related (PR) proteins along with the strengthening of cell walls and other barriers in the plant cells.

Various groups of compounds secreted by *Trichoderma* spp. may act as elicitors for the induction of defense responses in plants. Xylanase from *Trichoderma* spp. induce systemic resistance in cotton, tobacco, grapevine, and so forth (Anderson et al., 1993; Bailey & Lumsden 1998; Calderpn et al., 1993; Elad 2000; Howell 2003). Cellulase produced by *T. harzianum* acts as elicitor for systemic acquired resistance (SAR) by triggering peroxidase and chitinase activity (Elad, 2003; Elad & Kapat, 1999). *Fusarium oxysporum* f. sp. *melonis* and powdery mildew infections on greenhouse melon plants were reduced in plants treated with cellulases.

Recently chitin oligosaccharides were found to act as elicitors of defense response in plants, and the scavenging of such oligomers is fundamental to the lifestyle of fungal pathogens upon colonization of their hosts (de Jonge et al., 2010). Effective *Trichoderma* strains produce a variety of microbe-associated molecular patterns (MAMPs). The proteome and transcriptome of plant leaves are observed to be systemically affected as a result of *Trichoderma* root colonization and MAMP interactions (Shoresh et al., 2010).

Xylanase and peptaibols (peptaibiotics with high content of α-aminoisobutyric acid) like alamethicin and trichovirin II which are produced by *Trichoderma* spp. can elicit an immune response in plants (Leitgeb et al., 2007; Druzhinina et al., 2011). Recently, a PKS/NRPS hybrid enzyme involved in defense responses in maize was identified (Mukharjee et al., 2012a). The best characterized elicitor produced by *Trichoderma* spp. is Sm1/Epl1, an abundantly secreted, small cysteine-rich hydrophobin-like protein of the cerato-platanin (CP) family (Djonovic et al., 2006; Seidl et al., 2006). Deletion of this *Trichoderma* gene impairs elicitation of ISR in maize (Djonovic et al., 2006).

Trichoderma can colonize root intercellular spaces by recognizing and adhering to the plant roots. Once inside the plant, it can tolerate the toxic metabolites produced by the plant as a defense in response to fungal invasion. Plant cell-wall-degrading enzymes are involved in active root colonization. Hermosa et al. (2012) reported that adherence of *Trichoderma* to the root surface is facilitated by hydrophobins, which are small hydrophobic proteins of the outermost cell wall layer that coat the fungal cell surface, and expansin-like proteins related to cell wall development. *T. asperellum* produces the class I hydrophobin TasHyd1, which support the colonization of plant roots, possibly by enhancing its attachment to the root surface and protecting the hyphal tips from plant defense compounds (Viterbo & Chet, 2006), and the swollenin TasSwo, an expansin-like protein with a cellulose-binding domain was able to recognize cellulose and modify the plant cell wall architecture, to help in root colonization (Brotman et al., 2008). In *Trichoderma*, this resistance has been associated with the presence of ATP-binding cassette (ABC) transport systems, which are key factors in the multiple interactions established by *Trichoderma* biocontrol strains with other microbes in a potentially toxic or antagonistic environment with rapid degradation of the phenolic compounds exuded from plants (Chen et al., 2011), and with the suppression of phytoalexin production, as detected in *Lotus japonica* during colonization with *T. koningii* (Masunaka et al., 2011).

2.4   *TRICHODERMA* AS A PLANT GROWTH STIMULANT

Trichoderma has recently emerged out as a successful plant growth promotor and is reported to enhance plant growth (Hermosa et al., 2012; Harman et al., 2004; Mc Lean et al., 2005; Bae et al., 2009; Joshi et al.,

2010; Rojan et al., 2010; Singh, 2010; Masunaka et al., 2011; Sharma et al., 2012). Seed treatment with *Trichoderma* spp. usually improves seedling vigor. The growth promotion effect is dependent not only on isolate of *Trichoderma* but also on plant species and/or cultivar involved (Singh et al., 2011).

Trichoderma spp. have a positive effect on plant growth and yield in some vegetable crops by promoting efficiency of basal organic fertilizer application. The increased growth response is mainly due to hydrolysis of cellulose in soil, mineral solubilization and uptake of minor and other minerals as well as improvement in the root morphology enabling the roots to explore a large volume of soil (Maral et al., 2012).

The colonization of cucumber roots by *T. asperellum* (*asperelloides*) colonization in cucumber roots enhances the availability of P and Fe to plants resulting in significant increases in dry weight, shoot length, and leaf area (Yedidia et al., 2001). The uses of *T. harzianum* improves the growth and development of maize plants (Akladious & Abbas, 2012) causing an increase in all measured parameters which include growth, chlorophyll content, starch content, nucleic acids content, total protein content, and phytohormones content of maize plants when applied to the soil or the seeds but the magnitude of these increases was much more pronounced in case of plants developed from seeds treated with various concentrations of metabolic solution of *T. harzianum*. Even if planted in poor nitrogen soil, corn seeds pre-treated with *T. harzianum* (T-22) were greener and larger initially and gave increased yields at maturity (Harman, 2000).

Trichoderma is also reported to increase the production of humic acid. According to an unpublished report, almost six-fold increase in water-soluble humic matter content was reported in colonized FYM as compared to non-colonized FYM that led to a significant increase in growth of tomato and okra plants as compared to non-colonized FYM and/or control. This report suggests an additive effect of microorganism with humic acid in better plant growth.

2.5 *TRICHODERMA* AS AN ABIOTIC STRESS RELIEVER OF CROP PLANTS

Apart from biocontrol of plant diseases, one of the major attribute of the fungus is its ability to enhance abiotic stress tolerance of plants, including to water and/or salt stress (Singh et al., 2011). Recently there have been several reports where *Trichoderma* is shown to enhance root growth and helps in water absorption and nutrient uptake under osmotic stress (Howell, 2003; Harman et al., 2004).

The root colonization by *Trichoderma* increases the growth of roots and of the entire plant, thereby increasing plant productivity and tolerance to abiotic stresses like drought (Singh et al., 2011). When plants are under stress, the content of reactive oxygen species (ROS) may increase to toxic levels. A common factor that adversely affects plants under these stress conditions is the accumulation of toxic ROS resulting in increased concentration of lipid peroxides. Treatment of seeds reduced accumulation of lipid peroxides in seedlings due to stress by scavenging ROS.

Several studies have shown that root colonization by *T. harzianum* results in increased level of plant enzymes, including various peroxidases, chitinases, β-1,3-glucanases, lipoxygenase-pathway hydroperoxide lyase, and such changes in plant metabolism can lead to accumulation of compounds like phytoalexins and phenols (Harman, 2006; Hoitink et al., 2006; Gachomo & Kotchtoni, 2008). These compounds act as scavengers of ROS. Almost similar mechanism is involved in better germination of *T. harzianum* strain T-22 treated seeds of tomato exposed to biotic (seed and seedling disease caused by *P. ultimum*), abiotic (osmotic, salinity, chilling, or heat stress), or physiological (poor seed quality induced by seed aging) stresses (Mastouri et al., 2010).

Enhanced resistance of colonized plants to water deficit is at least partly due to higher capacity to scavenge ROS and recycle oxidized ascorbate and glutathione, a mechanism that is expected to enhance tolerance to abiotic and biotic stresses (Mastouri et al., 2012). Several pathways in plants convert and *Trichoderma* strains enhance the activity of many pathways that convert oxidized glutathione and ascorbate to the reduced form (Mittler, 2002). Enhancement of these pathways in chloroplasts would increase photosynthetic efficiency by reducing damage by the superoxide anion and other reactive species involved in photosynthesis. At least part of the stress resistance, and probably the increased photosynthetic efficiency, is because the fungi improve the redox status of the plant.

Recent studies indicate that abiotic stress tolerance in plants is accompanied by growth inhibition after over-expression of heat-shock genes of plant origin (Cazalé et al., 2009). Transgenic tobacco plants expressing a *T.*

virens glutathione transferase gene imparted tolerance to different concentrations of cadmium demonstrated to be through amelioration of oxidative stress (Dixit et al., 2011). *T. harzianum hsp*70 gene helps in increasing the fungal resistance to heat and other stresses such as salt tolerance, osmotic, and oxidative stresses (Montero Barrientos et al., 2008). Recent examples of biotechnological solutions from *Trichoderma* are the *T. harzianum Thkel1* gene, encoding a kelch-repeat protein involved in the modulation of glucosidase activity that enhanced seed germination and plant tolerance to salt and osmotic stresses when it was expressed in *Arabidopsis* (Hermosa et al., 2012). The expression of a *T. harzianum hsp70* gene in *Arabidopsis* enhanced tolerance to high temperatures, high salinity, and drought without loss of vigor and growth or developmental alterations (Montero-Barrientos et al., 2010).

2.6 DELIVERY OF *TRICHODERMA*

Success in biocontrol depends on understanding and use of delivery system. The research on delivery system is well below that of chemical pesticides. The awareness on the application technology can improve biopesticides performance. The success of biological control of plant pathogens also depends on the ability of the introduced microorganisms to competitively colonize the rhizosphere of the host plant, which is mostly influenced by the availability of nutrients from the substrate or a carrier medium through which, the biocontrol agent is applied. Therefore, the delivery systems greatly affect the field performance of a biocontrol agent. Where and when to deliver the biocontrol agent, depends on the microbe, the pathosystem, and the cropping system. Since these biocontrol agents are present in low populations in native soils, further augmentations of their population density to reach a higher level through artificial application is necessary. Additionally, the product must be in a formulation that is compatible with the grower and his practices.

The most common methods of application of *Trichoderma* are seed treatment, seed biopriming, seedling dip, soil application, and wound dressing. *Multiple delivery systems:* Reports suggest that protection of plants through all sites from where plant pathogens establish host parasite relationships such as spermosphere, rhizosphere, and phyllosphere can offer a better means for disease management through biocontrol agents. Seed and foliar application of *T. viride* reduced alternaria blight in linseed under field conditions (Singh et al., 2013). Likewise seed treatment and soil application (2.5 kg/ha) of *T. viride* followed by foliar spray of azadirachtin @ 3 mL/L on 30 and 45 days after sowing (DAS) can be significantly effective in reducing the disease incidence and increasing the seed yield in sesame (Jeyalakshmi et al., 2013).

2.7 *TRICHODERMA* AS AN EFFICIENT COLONIZER OF COMPOST

Trichoderma can colonize and decompose dead organic matter. Recent studies have revealed that it enhances the decomposition of organic composts like cow dung, poultry manure, and press mud.

Trichoderma multiplies very well on cow dung/FYM not only under laboratory condition but also on farm in the compost pits. The population of *T. harzianum* on colonized cow dung may reach as high as 2.46×10^{12} cfu.g^{-1} air-dried samples at 30% moisture level, 32 °C temperature, and two weeks of incubation (Zaidi & Singh, 2004b). Population of *T. harzianum* in decomposed FYM at farmer's field could go up to 10^{10} cfu. g^{-1} air-dried colonized FYM (Zaidi & Singh, 2005). This population is better than most of the commercial formulations available in the market. This colonized compost is good for use not only as soil amendment but also as seed treatment or foliar application. When added through colonized FYM, these bioagents have the advantage of substrate colonization and are able to thrive in soil much better as compared to their application through any other means.

Trichoderma colonized FYM is superior to non-colonized FYM in providing protection and improving the growth and overall health of vegetable seedlings (tomato and okra) as compared to non-colonized FYM under greenhouse condition (Zaidi & Singh, 2004b). Chemical analysis showed that both total and water soluble content of a number of macro- and micro-nutrients like P, K, S, Zn, Cu, and Fe were significantly higher in *T. harzianum* colonized FYM as compared to non-colonized FYM. These nutrients probably get released during the process of accelerated decomposition. *Trichoderma* helps in degrading complex organic molecules into simpler molecules that may help to improve fertility. Humic acid is an organic fertilizer which enhances the plant growth. There was almost six-fold increase in water-soluble humic matter content in colonized FYM as compared to non-colonized FYM . Since humic matter is reported to have growth promoting effect, in addition to better availability of macro- and micro-nutrients, higher soluble humic matter content might also be responsible for the better plant growth in *T. harzianum* colonized FYM (Zaidi et al., unpublished).

Composting with *Trichoderma* as an activator is mainly utilized in rice as organic fertilizer in the Philippines. The use of this technology can reduce fertilizer use by 30–50% and an increase in rice and corn yield by 20% (Cuevas, 1997). In Bangladesh commercial availability of *Trichoderma* colonized compost has started by the name as "Tricho-compost."

2.8 LIMITATIONS

A thorough understanding of the bioagent-pathogen- host and environment interaction and their impact on disease development is necessary for any biocontrol program to become a commercial success. These aspects determine the method of application, quality of bioagent, and the critical stage of application. The other important side that is not paid much attention is the quality of bioagent formulation and the efficiency and viability of the antagonistic strain used. A strict regulation for quality control must be in place to check the poor quality products available in the market. Being living entities, proper care should be taken to check their shelf life, viability of propagules, and purity of formulations. Proper awareness of both farmers and dealers needs to be done to inform them about the suitable storage conditions, shelf life, and mode of action of biocontrol agents. Importantly, the thrust must be on the fact that their effects may not be as dramatic as chemical pesticides, but can be sustainable to yield long-term benefits if properly used, with least possible environmental hazards.

2.9 CONCLUSION

As mentioned above, *Trichoderma* has multiple beneficial effects on plants. The widespread occurrence of *Trichoderma* indeed suggests their significant ecological role with considerable potential to improve agricultural productivity and sustainability, with particular significance in stressed environments. However, being living entities (unlike chemical pesticides), it will not be reasonable to compare their potential with chemical pesticides. *Trichoderma* is relatively slow acting and more dependent on environmental conditions. This might result in inconsistent field performance. Because of these reasons it may be challenging to target them as a stand-alone treatment for the management of any disease, may be except seedling diseases. This can be partly improved by promoting it as a component of integrated pest management (IPM) or integrated crop management (ICM). Use of *Trichoderma* in combination with varieties resistant/tolerant to biotic and/or abiotic stresses may be an effective strategy under field condition.

Combination of seed treatment and use of *Trichoderma* colonized compost can be a very effective strategy for the management of seedling and root diseases in soil. Even in absence of disease this combination treatment can significantly improve the overall plant health. These two treatments may be promoted as general agricultural practices irrespective of the crop and disease.

KEYWORDS

- *Trichoderma*
- fungal antagonist
- mycoparasitism
- biofungicides
- colonization

REFERENCES

Akladious, S. A.; Abbas, S. M, Application of *Trichoderma harziunum* T22 as a Biofertilizer Supporting Maize Growth. *Afr. J. Biotechnol.* **2012**, *11*, 8672–8683.

Alfano, G.; Lewis Ivey, M. L.; Cakir, C.; Bos, J. I. B.; Miller, S. A.; Madden, L. V.; Kamoun, S. Hoitink, H. A. J. Systemic Modulation of Gene Expression in Tomato by *Trichoderma hamatum* 382. *Phytopathology.* **2007**, *97*, 429–437.

Altomare, C.; Norvell, W. A.; Bjorkman, T.; Harman, G. E. Solubilization of Phosphates and Micronutrients by the Plant Growth Promoting Biocontrol Fungus *Trichoderma harzianum* Rifai 1295-22. *Appl. Environ. Microbiol.* **1999**, *65*, 2926–2933.

Anderson, R. D.; Bailey, B. A.; Taylor, R.; Sharon, A.; Avni, A.; Matto, A. K.; Fuchs, Y. Fungal Xylanase Elicits Ethylene Biosynthesis and Other Defense Responses in Tobacco. In *Cellular and Molecular Aspects of the Plant Hormone Ethylene*; Pech, J.

Latche, C. A., Balague, C., Eds.; Kluwer Academic Publishers: Netherlands, 1993; pp 197–204.

Bae, H.; Roberts, D. P.; Lim, H. S.; Strem, M. D.; Park, S. C.; Ryu, C. M.; Melnick, R. L.; Bailey, B. A. Endophytic *Trichoderma* Isolates from Tropical Environments Delay Dsease Oset and Induce Resistance Against *Phytophthora capsici* in Hot Pepper Using Multiple Mechanisms. *MPMI.* **2011,** *24,* 336–351.

Bae, H.; Sicher, R. C.; Kim, M. S.; Kim, S. H.; Strem, M. D.; Melnick, R. L.; Bailey, B. A. The Beneficial Endophyte *Trichoderma hamatum* Isolate DIS 219b Promotes Growth and Delays the Onset of the Drought Response in *Theobroma cacao. J. Exp. Bot.* **2009,** *60* (11), 3279–3295.

Bailey, B. A., Melnick, R. L The Endophytic *Trichoderma.* In *Trichoderma, Biology and Applications*; Mukharjee, P. K., Horwitz, B. A., Singh, U. S., Mukharjee, M., Schmoll, M., Eds.; CABI: UK, 2013; pp 152–172.

Bailey, B. A.; Lumsden, R. D.; Direct Effect of *Trichoderma* and *Gliocladium* on Plant Growth and Resistance to Pathogens. In *Trichoderma and Gliocladium;* Harman, G. E., Kubicek, C. P., Eds.; Taylor & Francis: London, 1998; Vol. 2, pp 185–204.

Baranski, R.; Klocke, E.; Nothnagel, T. Chitinase CHIT36 from *Trichoderma harzianum* Enhances Resistance of Transgenic Carrot to Fungal Pathogens. *J. Phytopathol.* **2008,** *56,* 513–521.

Benítez, T.; Rincón, A. M.; Limón, M. C.; Codón, A. C. Biocontrol Mechanisms of *Trichoderma* Strains. *Int. Microbiol.* **2004,** *7,* 249–260.

Bolar, J. P.; Norelli, J. L.; Wong, K. W.; Hayes, C. K.; Harman, G. E.; Aldwinckle, H. S. Expression of Endochitinase from *Trichoderma harzianum* in Transgenic Apple Increases Resistance to Apple Scab and Reduces Vigor. *Phytopathology.* **2000,** *90,* 72–77.

Brotman, Y.; Briff, E.; Viterbo, A.; Chet, I. Role of Swollenin, an Expansin-Like Protein from *Trichoderma*, in Plant Root Colonization. *Plant Physiol.* **2008,** *147,* 779–789.

Bruckner, H.; Graf, H.; Bokel, M. Paracelsin; Characterization by NMR Spectroscopy and Circular Dichroism, and Hemolytic Properties of a Peptaibol Antibiotic from the Cellulolytically Active Mold *Trichoderma reesei.* Part B *Experientia.* **1984,** *40,* 1189–1197.

Jeyalakshmi, C.; Rettinassababady, C.; Sushma, N. Integrated Management of Sesame Diseases. *J. Biopestic.* **2013,** *6*(1), 68–70.

Calderon, A. A.; Japata, J. M.; Munoj, R.; Pedreno, M. A.; Barcelo, A. R. Resveratrol and Their Production as a Part of the Hypersensitive Like Response of Grapevine Cells to an Elicitor from *Trichoderma viride. New Phytol.* **1993,** *124,* 424–463.

Cazalé, A. C.; Clément, M.; Chiarenza, S.; Roncato, M. A.; Pochon, N.; Creff, A.; Marin, E.; Leonhardt, N.; Noë l, L. D. Altered Expression of Cytosolic/Nuclear HSC70-1 Molecular Chaperone Affects Development and Abiotic Stress Tolerance in Arabidopsis Thaliana. *J. Exp. Bot.* **2009,** *60,* 2653–2664.

Chen, L. L.; Yang, X.; Raza, W.; Li, J.; Liu, Y.; Qiu, M.; Zhang, F.; Shen, Q. *Trichoderma harzianum* SQR-T037 Rapidly Degrades Allelochemicals in Rhizospheres of Continuously Cropped Cucumbers. *Appl. Microbiol. Biotechnol.* **2011,** *89,* 1653–663.

Chet, I.; Inbar, J. Biological Control of Fungal Pathogens. *Appl. Biochem. Biotechnol.* **1994,** *48,* 37–43.

Chet, I.; Inbar, J.; Hadar, I. Fungal Antagonists and Mycoparasites. In *Environmental and Microbial Relationships;* Wicklow, D. T., Söderström, B., Eds.; The Mycota IV, Springer-Verlag: Berlin, 1997; pp 165–184.

Chet, I. *Trichoderma-* Application, Mode of Action and Potential as a Biocontrol Agent of Soil Borne Plant Pathogenic Fungi. In *Innovative Approaches to Plant Disease Control*; Chet, I., Ed.; John Wiley & Sons: New York, NY, 1987; pp 137–160.

Conrath, U. Molecular Aspects of Defence Priming. *Trends Plant Sci.* 2011, *16,* 524–553

Cuevas, V. C. Rapid Composting Technology in the Philippines: Its Role in Producing Good Quality Fertilizers. *Food Fertilizer Technical Center Extension. Bulletin.* 1997, *444,* 1–13.

Cullen, D.; Kersten, P. In *Applied Molecular Genetics of Filamentous Fungi;* Kinghorn, J. R., Turner, G., Eds.; Chapman & Hall: New York, NY, 1992; p 100.

de Jonge, R.; van Esse, H. P.; Kombrink, A.; Shinya, T.; Desaki, Y.; Bours, R.; van der Krol, S.; Shibuya, N.; Joosten, M. H. A. J.; Thomma, B. P. H. J. Conserved Fungal LysM Effector Ecp6 Prevents Chitin-Triggered Immunity in Plants. *Science.* 2010, *329,* 953–955.

Dennis, C.; Webster, J. Antagonistic Properties of Species- Groups of *Trichoderma* I. Production of Non-Volatile Antibiotics. *T. Brit. Mycol. Soc.* 1971, 57, 25–31.

Distefano, G.; La Malfa, S.; Vitale, A.; Lorito, M.; Deng, Z.; Gentile, A. Defence-Related Gene Expression in Transgenic Lemon Plants Producing an Antimicrobial *Trichoderma harzianum* Endochitinase During Fungal Infection. *Transgenic Res.* 2008, *17,* 873–879.

Dixit, P.; Mukherjee, P. K; Ramachandran, V.; Eapen, S. Glutathione Transferase from *Trichoderma virens* Enhances Cadmium Tolerance without Enhancing its Accumulation in Transgenic *Nicotiana tabacum*. *PLoS ONE,* 2011, *6*(1), 1–15.

Djonovic, S.; Pozo, M. J.; Dangott, L. J.; Howell, C. R.; Kenerley, C. M. Sm1, a Protein-aceous Elicitor Secreted by the Biocontrol Fungus *Trichoderma virens* Induces Plant Defense Responses and Systemic Resistance. *Mol. Plant-Microbe Interact.* 2006, *19,* 838–853.

Druzhinina, I. S.; Seidl-Seiboth, V.; Herrera-Estrella, A.; Horwitz, B. A.; Kenerley, C. M.; Monte, E.; Mukherjee, P. K.; Zeilinger, S.; Grigoriev, I. V.; Kubicek, C. P *Trichoderma—* the Genomics of Opportunistic Success. *Nature Rev. Microbiol.* 2011, *9,* 749–759.

Druzhinina, I. S.; Kubieck, C. P Ecological Genomics of *Trichoderma*. In *Ecological Genomics of Fungi*; Martinn, F., Ed.; Wiley-Blackwell: Oxford, UK. 2013.

Eisendle, M.; Oberegger, H.; Buttinger, R.; Illmer, P.; Haas, H. Biosynthesis and Uptake of Siderophores is Controlled by the Paccmediated Ambient-Ph Regulatory System in *Aspergillus nidulans*. *Eukaryotic Cell.* 2004, *3,* 561–563.

Elad, Y.; Kapat, A.; Role of *Trichoderma harzianum* Protease in the Biocontrol of *Botrytis Cineria*. *Eur. J. Plant Pathol.* 1999, *105,* 177–189.

Elad, Y.; Mycoparasitis: Pathogenesis and Host Specificity in Plant Diseases. In *Histopathological, Biochemical, Genetic and Molecular Basis;* Kohmoto, K., Singh, U. S., Singh, R. P., Eds.; Eukaruotes, Pergamon, Elsevier Science: Oxford, 1995; Vol. II, pp 289–307.

Elad, Y. Biological Control of Foliar Pathogens by Means of *Trichoderma harzianum* and Potential Modes of Action. *Crop Prot.* 2000, *19,* 709–714.

Emani, C.; García, J. M.; Lopata-Finch, E.; Pozo, M. J.; Uribe, P.; Kim, D. J.; Sunil-kumar, G.; Cook, D. R.; Kenerley, C. M.; Rathore, K. S. Enhanced Fungal Resistance in

Transgenic Cotton Expressing an Endochitinase Gene from *Trichoderma virens*. *Plant Biotech. J.* **2003,** *1,* 321–336.

Gachomo, E. W.; Kotchoni, S. O. The Use of *T. harzianum* and *T. viride* as Potential Biocontrol Agents against Peanut Microflora and Their Effectiveness in Reducing Aflatoxin Contamination of Infected Kernels. *Biotechnology.* **2008,** *7,* 439–447.

Gentile, A.; Deng, Z.; La Malfa, S.; Distefano, G.; Domina, F.; Vitale, A.; Polizzi, G.; Lorito, M.; Tribulato, E. Enhanced Resistance to *Phoma tracheiphila* and *Botrytis cinerea* in Transgenic Lemon Plants Expressing a *Trichoderma harzianum* Chitinase Gene. *Plant Breeding.* **2007,** *126,* 146–151.

Harman, G. E. *Trichoderma*—Not Just for Biocontrol Anymore. *Phytoparasitica.* **2011,** *39,* 103–108.

Harman, G. E.; Howell, C. R.; Viterbo, A.; Chet, I.; Lorito, M. *Trichoderma* Species – Opportunistic Avirulent Plant Symbionts. *Nat. Rev.* **2011,** *2,* 43–56.

Harman, G. E. Myths and Dogmas of Biocontrol: Changes in Perceptions Derived from Research on *Trichoderma harzianum* T-22. *Plant Dis.* **2000,** *84,* 377–393.

Harman, G. E.; Howell, C. R.; Vitrevo. A.; Chet, I.; Lorito, M. *Trichoderma* Species – Opportunistic Avirulent Plant Symbionts. *Nat. Rev.* **2005,** *2,* 43–56.

Harman, G. E. Overview of Mechanisms and Uses of *Trichoderma* Spp. *Phytopathology.* **2006,** *96,* 190–194.

Hermosa, R.; Viterbo, A.; Chet, I.; Monte, E. Plant-Beneficial Effects of *Trichoderma* and of its Genes. *Microbiology.* **2012,** *158,* 17–25.

Hoitink, H. A. J.; Madden, L. V.; Dorrance, A. E. Systemic Resistance Induced by *Trichoderma* Spp.: Interactions between the Host, the Pathogen, the Biocontrol Agent, and Soil Organic Matter Quality. *Phytopathology.* **2006,** *96,* 186–189.

Howell, C. R. Mechanisms Employed by *Trichoderma* Species in the Biological Control of Plant Diseases: The History and Evolution of Current Concepts. *Plant Dis.* **2003,** *87,* 4–10.

Inbar, J.; Chet, I. Biomimics of Fungal Cell–Cell Recognition by Lectin-Coated Nylon Fibres. *J. Bacteriol.* **1992,** *174,* 1055–1059.

Inbar, J.; Chet, I. A Newly Isolated Lectin from the Plant Pathogenic Fungus *Sclerotium roltsii*: Purification, Characterization and Role in Mycoparasitism. *Microbiology.* **1994,** *140,* 651–657.

Jacklitsch, W. M. Europian Species of Hypocrea Part I. the Green Spored Species. *Stud. Mycol.* **2009,** *63,* 1–91.

Joshi, B. B.; Bhatt, R. P.; Bahukhandi, D. Antagonistic and Plant Growth Activity of *Trichoderma* Isolates of Western Himalayas. *J. Envir. Biol.* **2010,** *31,* 921–928.

Kirk, T. K.; Farrell, R. L. Enzymatic Combustion: The Microbial Degradation of Lignin. *Annu. Rev. Microbiol.* **1987,** *41,* 465–505.

Korolev, N.; David, R.; Elad, Y. The Role of Phytohormones in Basal Resistance and *Trichoderma*-Induced Systemic Resistance to *Botrytis cinerea* in *Arabidopsis thaliana*. *Biocontrol.* **2007,** *53,* 667–668.

Kubicek, C. P.; Herrera-Estrella, A.; Seidl-Seiboth, V.; Martinez, D. A.; Druzhinina, I. S.; Thon, M.; Zeilinger, S.; Casas-Flores, S.; Horwitz, B. A.; et al. Comparative Genome Sequence Analysis Underscores Mycoparasitism as the Ancestral Life Style of *Trichoderma*. *Genome Biol.* **2011,** *12,* R40.

Kumar, A.; Scher, K.; Mukherjee, M.; Pardovitz-Kedmi, E.; Sible, G. V.; Singh, U. S.; Kale, S. P.; Mukherjee, P. K.; Horwitz, B. A. Overlapping and Distinct Functions of Two *Trichoderma virens* MAP Kinases in Cell-Wall Integrity, Antagonistic Properties and Repression of Conidiation. *Biochem. Biophys. Res. Commun.* **2010**, *398,* 765–770.

Kumar, V.; Parkhi, V.; Kenerley, C. M.; Rathore, K. S. Defence Related Gene Expression and Enzyme Activities in Transgenic Cotton Plants Expressing an Endochitinase Gene from *Trichoderma virens* in Response to Interaction with *Rhizoctonia solani*. *Planta.* **2009**, *230,* 277–291.

Lamdan, N. L.; Shalaby, S.; Ziv, T.; Kenerley, C. M.; Horwitz, B. A. Secretome of the Biocontrol Fungus *Trichoderma virens* Co-Cultured with Maize Roots: Role in Induced Systemic Resistance. *MCP.* **2015**, *14,* 1054–1063. doi:10.1074/mcp.M114.046607

Leitgeb, B.; Szekeres, A.; Manczinger, L.; Vagvolgyl, C.; Kredics, L. The History of Alamethicin: A Review of Most Extensively Studied Peptaibol. *Chem. Biodivers.* **2007**, *4,* 1027–1051.

Liu, M.; Sun, Z. X.; Zhu, J.; Xu, T.; Harman, G. E.; Lorito, M. Enhancing Rice Resistance to Fungal Pathogens by Transformation with Cell Wall Degrading Enzyme Genes from *Trichoderma atroviride*. *J. Zhejiang Univ. Sci.* **2004**, *5,* 133–136.

Lorito, M.; Woo, S. L.; Harman, G. E.; Monte, E. Translational Research on *Trichoderma*: From 'Omics to the Field. *Annu. Rev. Phytopathol.* **2010**, *48,* 395–417.

Lorito, M.; Woo, S. L.; Fernandez, I. G.; Collucci, G.; Harman, G. E.; Pintor-Toros, J. A.; Filippone, E.; Muccifora, S.; Lawrence, C. B.; Zoina, A.; Tuzun, S.; Scala, F. Genes from *Mycoparasitic* Fungi as a Source for Improving Plant Resistance to Fungal Pathogens. *Proc. Natl. Acad. Sci. USA.* **1998**, *95,* 7860–7865.

Lumsden, R. D.; Locke, J. C.; Adkins, S. T.; Walter, J. F.; Rido, C. J. Isolation and Localization of the Antibiotic Gliotoxin Produced by *Gliocladium virens* from Alginate Prill in Soil and Soilless Media. *Phytopathology.* **1992**, *82,* 230–235.

Maral, S.; Sozer, Y.; Gezgin, C.; Kara, S.; Sargın, R.; Eltem, F.; Vardar, S. Evaluation of the Properties of *Trichoderma* Sp. Isolates as a Biocontrol Agent and Biofertilizer. *EEMJ.* **2012**, Vol. 11, No. 3, Supplement, S154 http://omicron.ch.tuiasi.ro/EEMJ/

Mastouri, F.; Björkman, T.; Harman, G. E. *Trichoderma harzianum* Enhances Antioxidant Defense of Tomato Seedlings and Resistance to Water Deficit. *Mol. Plant. Microbe. Interact.* **2012**, *25* (9), 1264–1271. http://dx.doi.org/10.1094/MPMI-09-11-0240

Masunaka, A.; Hyakumachi, M.; Takenaka, S. Plant Growth-Promoting Fungus, *Trichoderma. Koningii* Suppresses Isoflavonoid Phytoalexin Vestitol Production for Colonization on/in the Roots of Lotus Japonicus. *Microbes Environ.* **2011**, *26,* 128–134.

Mc Lean, K. L.; Swaminathan, J.; Frampton, C. M.; Hunt, J. S.; Ridgway, H. J.; Stewart, A. Effect of Formulation on the Rhizosphere Competence and Biocontrol Ability of *Trichoderma atroviride* C52. *Plant Pathol.* **2005**, *54,* 212–218.

Mendoza-Mendoza, A.; Pozo, M. J.; Grzegorskid, D.; Martinez, P.; Garsia. J. M.; Olmedo-Monfil, V.; Cortes, C.; Kenerley. C.; Herrera-Estrella, A. Enhanced Biocontrol Activity of *Trichoderma* Through Inactivation Of Mitogen-Activated Protein Kinases. *Proc. Natl. Acad. Sci. USA.* **2003**, *100,* 15965–15970. doi:10.1073/pnas.2136716100.

Mendoza-Mendoza, A.; Rosales-Saavedra, T.; Cortes, C.; Castellanos-Juarez, V.; Martinez, P.; Herrera-Estrella, A. The MAP Kinase TVK1 Regulates Conidiation, Hydrophobicity and the Expression of Genes Encoding Cell Wall Proteins in the Fungus *Trichoderma virens*. *Microbiology.* **2007**, *153,* 2137–2147.

Meyer, G. D.; Bigirimana, J.; Elad, Y.; Hofte, M. Induced Systemic Resistance in *Trichoderma harzianum* T39 Biocontrol of *Botrytis cinerea*. *Eur. J. Plant Pathol.* **1998,** *104,* 279–286.

Mittler, R. Oxidative Stress, Antioxidants and Stress Tolerance. *Trends Plant Sci.* **2002,** *7,* 405–410.

Montero-Barrientos, M.; Hermosa, R.; Nicolas, C.; Cardoza, R. E.; Gutierrez, S.; Monte, E. Overexpression of a *Trichoderma* HSP70 Gene Increases Fungal Resistance to Heat and Other Abiotic Stresses. *Fungal Genet. Biol.* **2008,** *45,* 1506–1513.

Montero-Barrientos, M.; Hermosa, R.; Cardoza, R. E.; Gutie´ rrez, S.; Nicola´ s, C.; Monte, E. Transgenic Expression of the *Trichoderma harzianum* Hsp70 Gene Increases Arabidopsis Resistance to Heat and Other Abiotic Stresses. *J. Plant Physiol.* **2010,** *167,* 659–665.

Mora, A.; Earle, E. D. Resistance to Alternaria Brassicicola in Transgenic Broccoli Expressing a *Trichoderma harzianum* Endochitinase Gene. *Mol. Breed.* **2001,** *8,* 1–9.

Mukherjee, M.; Mukherjee, P. K.; Kale, S. P cAMP Signalling is Involved in Growth, Germination, Mycoparasitism and Secondary Metabolism in *Trichoderma virens*. *Microbiology*. **2007,** *153,* 1734–1742.

Mukherjee, P. K.; Buensanteai, N.; Moran-Diez, M. E.; Druzhinina, I. S.; Kenerley, C. M. Functional Analysis of Non-Ribosomal Peptide Synthetases (Nrpss) in *Trichoderma virens* Reveals a Polyketide Synthase (PKS)/NRPS Hybrid Enzyme Involved in Induced Systemic Resistance Response in Maize. *Microbiology*. **2012,** *158,* 155–165.

Mukherjee, P. K.; Latha, J.; Hadar, R.; Horwitz, B. A. TmkA, A Mitogen-Activated Protein Kinase of *Trichoderma virens*, is Involved in Biocontrol Properties and Repression of Conidiation in the Dark. *Eukaryot. Cell.* **2003,** *2,* 446–455.

Mukherjee, M.; Mukherjee, P. K.; Horwitz, B. A.; Zachow, C.; Berg, G.; Zeilinger, S. *Trichoderma–Plant–Pathogen Interactions: Advances in Genetics of Biological Control. Indian J. Microbiol.* **2012a,** 1–8. DOI 10.1007/s12088-012-0308-5.

Mukherjee, P. K.; Horwitz, B. A.; Singh, U. S.; Mukharjee, M.; Shmoll, M. *Trichoderma* in Agriculture, Industry and Medicine: An Over view. In *Trichoderma Biology and Applications*; Mukherjee, P. K., Horwitz, B. A., Singh, U. S., Mukharjee, M., Shmoll, M., Eds.; CABI: UK, 2013; p 5.

Omann, M. R.; Lehner, S.; Escobar Rodriguez, C.; Brunner, K.; Zeilinger, S. The Seven-Transmembrane Receptor Gpr1 Governs Processes Relevant for the Antagonistic Inter-action of *Trichoderma atroviride* with its Host. *Microbiology*. **2012,** *158,* 107–118.

Papavizas, G. C. *Trichoderma* and *Gliocladium*: Biology and Potential for Biological Control. *Annu. Rev. Phytopathol.* **1985,** *23,* 23–54.

Prasad, D.; Anes, K. M. Effect of Metabolites of *Trichoderma harzianum* and *T. Viride* on Plant Growth and *Meloidogyne incognita* on Okra. *Ann. Plant Prot. Sci.* **2008,** *16,* 461–465.

Prime A. P. G.; Conrath, U.; Beckers, G. J.; Flors, V.; Garcia-Agustin, P.; Jakab, G.; Mauch, F.; Newman, M. A.; Pieterse, C. M.; Poinssot, B.; Pozo, M. J.; Pugin, A.; Schaffrath, U.; Ton, J.; Wendehenne, D.; Zimmerli, L.; Mauch-Mani, B. Priming: Getting Ready for Battle. *MPMI,* **2006,** *19,* 1062–1071.

Reithner, B.; Schuhmacher, R.; Stoppacher, N.; Pucher, M.; Brunner, K.; Zeilinger, S. Signaling Via The*trichoderma atroviride* Mitogen-Activated Protein Kinase Tmk 1 Differentially Affects Mycoparasitism and Plant Protection. *Fungal Genet. Biol.* **2007,** *44,* 1123–1133.

Rojan, P. J.; Tyagi, R. D.; Prévost, D.; Brar, S. K.; Pouleur, S.; Surampalli, R. Y. Mycoparasitic *Trichoderma viride* as a Biocontrol Agent Against *Fusarium oxysporum* f. Sp. Adzuki and *Pythium arrhenomanes* and as a Growth Promoter of Soybean. *Crop Prot.* **2010,** *29,* 1452–1459.

Saifullah, S. M.; Thomas, B. J. Studies on the Parasitism of Globodera Rostochiensis by *Trichoderma harzianum* Using Low Temperature Scanning Electron Microscopy. *Afro Asian J. Nematol.* **1996,** *6,* 117–112.

Samolski, I. A.; de Luis, J.; Vizcaino, E.; Monte, E.; Suarez, M. B. Gene Expression Analysis of the Biocontrol Fungus *Trichoderma harzianum* in the Presence of Tomato Plants, Chitin, or Glucose Using a High-Density Oligonucleotide Microarray. *BMC Microbiol.* **2009,** *9,* 217.

Schmoll, M. The Information Highways of a Biotechnological Workhorse Signal Transduction in Hypocrea Jecorina. *BMC Genomics.* **2008,** *9,* 430.

Segarra, G.; Van der Ent, S.; Trillas, I.; Pieterse, C. M. J. MYB72, a Node of Convergence in Induced Systemic Resistance Triggered by a Fungal and a Bacterial Beneficial Microbe. *Plant Biol.* **2009,** *11,* 90–96.

Seidl, V.; Marchetti, M.; Schandl, R.; Allmaier, G.; Kubicek, C. P. EPL1, the Major Secreted Protein of *Hypocrea atroviridis* on Glucose, is a Member of a Strongly Conserved Protein Family Comprising Plant Defense Response Elicitors. *FEBS J.* **2006,** *273,* 4346–4359.

Shah, J. M.; Raghupathy, V.; Veluthambi, K. Enhanced Sheath Blight Resistance in Transgenic Rice Expressing an Endochitinase Gene from *Trichoderma virens*. *Biotechnol. Lett.* **2009,** *31,* 239–244.

Shah, M.; Mukherjee, P. K.; Eapen, S. Expression of a Fungal Endochitinase Gene in Transgenic Tomato and Tobacco Results in Enhanced Tolerance to Fungal Pathogens. *Physiol. Mol. Biol. Plants.* **2010,** *16,* 39–51.

Sharma, P.; Sharma, M.; Raja, M.; Shanmugam, V. Status of *Trichoderma* Research in India: A Review. *Indian Phytopathol.* **2014,** *67*(1), 1–19.

Sharma, P.; Patel, A. N.; Saini, M. K.; Deep, S. Field Demonstration of *Trichoderma harzianum* as a Plant Growth Promoter in Wheat (*Triticum aestivum* L). *J. Agri. Sci.* **2012,** *4*(8), 65–73.

Sharon, E.; Chet, I.; Spiegel, Y. *Trichoderma* as Biological Control Agent. In *Biological Control of Plant Parasitic Nematodes: Building Coherence Between Microbial Ecology and Molecular Mechanisms;* Davies, K., Spiegel, Y., Eds.; Springer: Berlin, 2011; pp 183–202.

Sharon, E.; Bar-Eyal, M.; Chet, I.; Herrera-Estrella, A.; Kleifeld, O.; Spiegel, Y. Biological Control of the Root Knot Nematode *Meloidogyne javanica* by *Trichoderma harzianum*. *Phytopathology.* **2001,** *91,* 687–693.

Shoresh, M.; Harman, G. E. The Molecular Basis of Shoot Responses of Maize Seedlings to *Trichoderma harzianum* T22 Inoculation of the Root: A Proteomic Approach. *Plant Physiol.* **2008,** *147,* 2147–2163.

Shoresh, M.; Harman, G. E.; Mastouri, F. Induced Systemic Resistance and Plant Responses to Fungal Biocontrol Agents. *Annu. Rev. Phytopathol.* **2010,** *48,* 21–43.

Shoresh, M.; Yedidia, I.; Chet, I. Involvement of Jasmonic Acid/Ethylene Signaling Pathway in the Systemic Resistance Induced in Cucumber by *Trichoderma asperellum* T203. *Phytopathology.* **2010,** *95,* 7.

Singh, R. K. *Trichoderma*: A Bio-Control Agent for Management of Soil Borne Diseases. **2010**, http://agropedia.iitk.ac.in

Singh, U. S.; Mishra, D. S.; Singh, A.; Rohilla, R.; Vishwanath. Induced Resistance: Present Status and Future Prospects as Disease Management Strategy. In *Biopesticides and Pest Management;* Koul, O., Dhaliwal, G. S., Marwaha, S. S., Arora, J. K., Eds.; Campus Book International: New Delhi, India, 2003; Vol. I, pp 262–302.

Singh, R. B.; Singh, H. K.; Parmar, A. Integrated Management of *Alternaria blight* in Linseed. *Biol. Sci.* **2013,** *83* (3), 465–469.

Singh, U. S.; Zaidi, N. W.; Singh, H. B. Use of Microbes and Host Tolerance for Abiotic Stress Management in Plants. In *Plant Pathology in India: Vision 2030;* Indian Pyhto-pathological Society, Division of Plant Pathology, IARI: New Delhi, India, 2011; pp 265–275.

Sivasithamparam, K.; Ghisalberti, E. L. Secondary metabolism in *Trichoderma* and Glio-cladium. In *Trichoderma and Gliocladium;* Harman, G. E., Kubicek, C. P., Eds.; Taylor & Francis: London, 1998; pp 139–192.

Van Wees, S. C.; Luijendijk, M.; Smoorenburg, I.; van Loon, L. C.; Pieterse, C. M. Rhizo-bacteria-Mediated Induced Systemic Resistance (Isr) in Arabidopsis is Not Associated with a Direct Effect on Expression of Known Defense-Related Genes but 23 Stimulates the Expression of the Jasmonate-Inducible Gene Atvsp upon Challenge. *Plant Mol. Biol.* **1999,** *41,* 537–549.

Verma, M.; Brar, S. K.; Tyagi, R. D.; Surampalli, R. Y.; Val'ero, J. R. Antagonistic Fungi, *Trichoderma* Spp.: Panoply of Biological Control. *Biochem. Eng. J.* **2007,** *37,* 1–20.

Viterbo, A.; Ramot, O.; Chemin, L.; Chet, I. Significance of Lytic Enzymes from *Tricho-derma* Spp. in the Biocontrol of Fungal Plant Pathogens. *Anton. Leeuw.* **2002,** *81,* 549–556.

Viterbo, A.; Chet, I. TasHyd1, a New Hydrophobin Gene from the Biocontrol Agent *Trich-oderma asperellum*, is Involved in Plant Root Colonization. *Mol. Plant Pathol.* **2006,** *7,* 249–258.

Viterbo, A.; Horwitz, B. A. Mycoparasitism: In *Cellular and Molecular Biology of Fila-mentous Fungi;* Borkovich, K. A., Ebbole, D. J., Eds.; American Society for Microbi-ology: Washington, DC, 2010; vol. 42, pp 676–693.

Viterbo, A.; Harel, M.; Horwitz, D. A.; Chet. I. Mukherjee, P. K. *Trichoderma* MAP Kinase Signaling is Involved in Induction of Plant Systemic Resistance. *Appl. Environ. Micro-biol.* **2005,** *71,* 6241–6246.

Windham, G. L.; Windham M. T.; Williams, P. W. Effects of *Trichoderma* Spp. on Maize Growth and *Meloidogyne arenaria*re Production. *J. Plant Dis.* **1989,** *73,* 493–495.

Yedidia, I.; Srivastava, A. K.; Kapulnik, Y.; Chet, I. Effect of *Trichoderma harzianum* on Micro Elements Concentration and Increased Growth of Cucumber Plants. *Plant Soil.* **2001,** *235,* 235–242.

Zaidi, N. W.; Singh, U. S. *Trichoderma* in Plant Health Management. In *Trichoderma Biology and Applications;* Mukherjee, P. K., Horwitz, B. A., Singh, U. S., Mukharjee, M., Shmoll, M., Eds.; CABI: UK, 2013; pp 230–246.

Zaidi, N. W.; Singh, U. S. Mass Multiplication of *Trichoderma harzianum* on Cow Dung. *Indian Phytopathol.* **2004a,** *57,* 189–192.

Zaidi, N. W.; Singh, U. S. Use of Farmyard Manure for Mass Multiplication and Delivery of Biocontrol Agents, *Trichoderma harzianum* and *Pseudomonas fluorescens. Asian Agri. History.* **2004b,** *8,* 297–304.

Zaidi, N. W.; Singh, U. S. Mass Multiplication and Delivery of *Trichoderma* and *Pseudo-monas. J. Mycol. Plant Pathol.* **2005,** *34,* 732–741.

Zeilinger, S.; Omann, M. *Trichoderma* Biocontrol: Signal Transduction Pathways Involved in Host Sensing and Mycoparasitism. *Gene Regul. Syst. Bio.* **2007,** *1,* 227–234.

Zeilinger, S.; Reithner, B.; Scala, V.; Peiss, I.; Lorito, M.; Mach, R. L. Signal Transduction by Tga3, a Novel G Protein Alpha Subunit of *Trichoderma atroviride. Appl. Environ. Microbiol.* **2005,** *71,* 1591–1597.

CHAPTER 3

ASPERGILLUS NIGER: A PHOSPHATE SOLUBILIZING FUNGUS AS BIOCONTROL AGENT

MD. ARSHAD ANWER[1*], KUNDAN SINGH[1], and
RAJ NARAIN SINGH[2]

[1]*Department of Plant Pathology, Bihar Agricultural University, Sabour 813210, India*

[2]*Directorate of Extension Education, Bihar Agricultural University, Sabour 813210, India*

*Corresponding author. E-mail: arshad_anwer@yahoo.com

CONTENTS

ABSTRACT

One of the biggest ecological challenges facing plant pathologists is to develop environmental friendly alternatives to the extensive use of chemical synthetic pesticides for combatting crop diseases. In this regard scientists are more concerned to develop such practices that alone or in integration with other practices could bring about a reasonably good degree of reduction of disease potential coupled with sustainability of production, cost effectiveness, and eco-friendliness. A large number of microorganisms from fungi and bacteria have shown some ability to antagonize plant diseases. Some plant growth promoting organisms also suppress plant diseases. These microorganisms may suppress plant pathogens through antibiosis, production of hormones, solubilization of minerals, and induction of host resistance. Among phosphate-solubilizing microorganisms, *Aspergillus niger* is the most important plant growth promoting fungus which efficiently suppress plant pathogens.

3.1 INTRODUCTION

At present, awareness is growing to develop novel management practices that alone or in integration with other practices could bring about a reasonably good degree of reduction of disease potential coupled with sustainability of production, cost effectiveness, and eco-friendliness. A large number of microorganisms from fungi and bacteria have shown some ability to antagonize plant diseases. As most of the soil-borne plant pathogens are fungi, biocontrol through mycoparasitism has been attempted extensively (Henis et al., 1979; Baker, 1987; Suarez et al., 2004; Sant et al., 2010). The microbial fungus can also work through other mechanisms (Khan & Anwer, 2011). Plant growth promoting organisms may also suppress plant diseases but they have not been adequately evaluated for disease management (Papavizas, 1985; Nair & Burke, 1988). These microorganisms may suppress plant pathogens through antibiosis (Vey et al., 2001), production of hormones (Osiewacz, 2002), solubilization of minerals (Harman et al., 2004; Benitez et al., 2004), and induction of host resistance (Harman et al., 2004). Among phosphate-solubilizing microorganisms (PSM), *Aspergillus niger* (Sen, 2000; Medina et al., 2007; Pandya & Saraf 2010; Khan et al., 2009, 2011), *Aspergillus awamori* (Khan & Khan, 2002; Mittal et al., 2008), *Penicillium digitatum* (Asea et al., 1998;

Mittal et al., 2008), and so forth are most important microorganisms and may prove to be efficient biocontrol agents of plant pathogens if exploited properly (Khan et al., 2009).

A critical analysis of the relevant information on the biological control of wilt and root-knot diseases has revealed that majority of the research efforts have dealt with *Trichoderma* spp., *Paecilomyces lilacinus* and *Pochonia* (=*Verticillium*) *chlamydosporia* (Benitez et al., 2004; Federico et al., 2007; Khan & Anwer, 2011), whereas other antagonists have not been duly tested. Researchers have also tested antagonism of *A. niger* against plant pathogens and have found the fungus effective than *Trichoderma* spp. (Vassilev et al., 1996; Chet et al., 1997; Vassilev et al., 2006). However, these studies are of primitive type and lack some important biochemical (ochratoxin A (OTA) negative/positive, production of HCN, NH_3, H_2S, etc.) and molecular approaches. Moreover, due attention has not been paid toward genetic variability in *A. niger* aggregate in order to identify efficient strains with regard to antagonism and phosphate solubilization, and least efforts have been made to select environmentally safe strains of *A. niger* (OTA negative).

The antagonism by *A. niger* involves multiple mechanisms. *A. niger* competes with the pathogens for nutrients (Vassilev et al., 1996, 2006) and space (Mondal et al., 2000), modifies the microenvironment (Domich, et al., 1980), produces plant growth promoting substances (Mondal et al., 2000; Vaddar & Patil, 2007; Khan & Anwer, 2008) and elicit plant defense mechanisms (Bai et al., 2004; Gomathi & Gnanamanickam, 2004) and produce antibiotic (Sen, 2000; Benitez et al., 2004; El-Hasan et al., 2007), or directly parasitize the pathogenic fungi and hyphal lysis (Mondal et al., 2000). These indirect and direct mechanisms may operate together during the disease suppression (Howell, 2003). Effectiveness of *A. niger* against a pathogen depends on several factors such as the target fungus, plant types and environmental conditions, including soil nutrition, pH, temperature, and iron concentration. Activation of each mechanism implies the production of specific compounds and metabolites such as plant growth factors (Harman et al., 2004), hydrolytic enzymes (Howell, 1998; Monte, 2001), siderophores (Eisendle et al., 2004), antibiotics (Chet et al., 1997) and carbon and nitrogen permeases (Mach & Zeilinger, 2003; Eisendle et al., 2004). There are some other most important attributes which are found in *A. niger*. The fungus does not produce mycotoxins and ribotoxin (Campbell, 1994). Interestingly the fungus is reported to decrease aflatoxin contamination (Wicklow et al., 1980; Horn & Wicklow, 1983). It is xerotolerant (Cooke & Whips, 1993) and grow at a wide range of temperatures (10–50 °C), pH (2.0–11.0), and osmolarity (from nearly pure water up to 34% salt) (Kis-Papo et al., 2003) can also improve its thermo-stability (Zhang et al., 2007). The fungus is extremely resistant to herbi-cides, fungicides, and pesticides at very high concentrations (Braud et al., 2006). *A. niger* isolates not only survive high concentration of many toxic heavy metals but also adsorb these metals (Ahmad et al., 2006).

The application of *A. niger* @ 8 g/kg seed, and soil with *A. niger* @ 30 g/pit in a field where muskmelon and watermelon crops were suffering from *Fusarium* wilt, *Rhizoctonia solani,* and *Pythium* spp. resulted in 81% control of the disease. The vines were more vigorous, and even with 15% incidence of disease, yield was ~5% greater as compared to that in disease-free areas (Chattopadhyay & Sen, 1996). In an experiment, seed treatment with *A. niger* @ 8 g/kg effectively controlled the blast (*Pyricu-laria oryzae*), sheath blight (*R. solani*), brown spot (*Helminthosporium oryzae*), and false smut (*Ustilaginoides virens*) of rice in the field (Sehgal et al., 2001). Problems of pre and post-emergence damping-off incited by *Pythium aphanidermatum* and *R. solani* in fruit and vegetable farms were successfully overcome by a combined treatment of seed and soil appli-cation of *A. niger* (Majumdar & Sen, 1998). Similarly, 93% control of charcoal rot of potato in a *Macrophomina phaseolina*-infested field was obtained with *A. niger* (Mondal, 1998). In another study, 87% control of black scurf of potato (*R. solani*) and a 10% increase in the yield by application of *A. niger* @ 8 g/kg on infected seed tubers grown in worst affected fields was reported (Sen et al., 1998). Lodhi (2004) has reported the control of soil borne diseases of potato by the application of *A. niger* alone or in combination with VAM fungi.

Winter sorghum can be severely damaged by *Macrophomina* infection; however, *A. niger* seed treatment brought down incidence of the disease from 30 to 7% (Das, 1998). *Fusarium* wilt of *Hibiscus* was satisfactorily controlled by the application of *A. niger* in the soil after uprooting the dead plants and soil population of the pathogen was found non-detectable level after a month of *A. niger* application (Sen, 2000). Malformation of mango caused by *Fusarium moniliforme* was controlled by spraying conidial suspension of *A. niger* over dead necrotic malformed panicles (Chand et al., 2007). Guava wilt caused by *Fusarium oxysporum* f. sp. *psidii* and *F. solani* were controlled by the soil application of *A. niger* formulation prepared on the field wastes (Misra, 2007). Root-dip applications of *A. niger* resulted in significant decline in the rhizosphere population of *F. oxysporum* f. sp. *lycopersici* and increase in the tomato yield (Khan & Khan, 2001). In another study, direct soil inoculation with *A. niger* decreased the rhizosphere population of *F. oxysporum* f. sp. *lycopersici* by 23–49% while the tomato yield increased by 28–53% in a field experiments (Khan & Khan, 2002). Application of *A. niger* has also managed root-knot of tomato, significantly increased the plant growth and reduced the soil population of *Meloidogyne incognita* (Goswami et al., 2008).

3.2 BIOLOGICAL CONTROL OF PLANT DISEASES

Continuous increase in global human population has put two-fold pressure on agriculture. Precious agricultural lands are being diverted from crop production to urbanization and industrialization. As a result, the net area under crop production is shrinking whereas demand for food products continues to increase at an alarming pace. According to one estimate, the present global land area under crop production would produce much greater quantities of food than present requirements if pest and disease-free crops were grown (Khan & Jairajpuri, 2010). Hence, the primary requirement to meet food requirements of both present and future populations is to integrate plant protection techniques with crop production systems. Numerous methods of pest and disease management are available including chemical, cultural, physical, and biological, which are used depending on crop, pathogen, availability of material, and demand of the situation.

Consensus is developing that chemical-based farming is not ecologically sustainable and economically viable. As a result, ecological approaches are being researched more intensively. The most obvious environment-friendly alternative to pesticide application for managing agriculturally important diseases is the use of biological approaches. Biological control is based on the phenomenon that every living entity has an adversary in nature to keep its population in check (Khan, 2005). Baker and Cook (1974) defined biological control as the "reduction of inoculum density or disease producing activities of a pathogen or parasite in its active or dormant state, by one or more organisms, accomplished naturally or through manipulation of the environment, host, or antagonists, or by mass introduction of one or more antagonists." In 1983 Baker and

Cook revised the definition to "the reduction of the amount of inoculum or disease producing activity of a pathogen accomplished by one or more organisms other than man."

Biological control can be achieved either by introducing bioinoculants (biocontrol agents) directly into a field or by adopting cultural practices which stimulate survival, establishment, and multiplication of the bioinoculants. Hence, more scientifically, biological control of pests and diseases can be defined as: reduction in disease severity, crop damage, population or virulence of the pest or pathogen in its active or dormant state by the activity of microorganisms that occur naturally through altering cultural practices which favors survival and multiplication of the microorganisms or by introducing bioinoculants.

3.3 PLANT GROWTH PROMOTING MICROORGANISMS AS ANTAGONISTS OF PATHOGEN

A large number of biocontrol agents have been investigated to harness their beneficial effects on crop productivity. Biological control agents are primarily fungal and bacterial in origin. Fungal biological agents basically work through parasitism (Papavizas, 1985; Stirling, 1993) against plant pathogenic fungi and nematodes (Khan, 2005). The important genera of biocontrol fungi which have been tested against plant pathogenic fungi and nematodes include *Trichoderma, Aspergillus, Chaetomium, Penicillium, Neurospora, Fusarium* (saprophytic), *Rhizoctonia, Dactylella, Arthrobotrys, Catenaria, Paecilomyces, Pochonia,* and *Glomus.* All these biocontrol agents of plant pathogens have been divided into two kinds of microorganisms. Classical parasites (e.g., *Trichoderma* spp., *P. lilacinus, Pasteuria penetrans,* etc.) which have been used in the disease control since old times (Khan & Khan, 1995). In recent years, more interest has been developed in using plant growth promoting microorganisms. Of these PSM, *A. niger, Penicillium* spp., *Bacillus subtilis, Bacillus polymyxa, Pseudomonas fluorescens, Pseudomonas stutzeri, Pseudomonas Striata,* and so forth are efficient solubilizers and may prove efficient biocontrol agents of plant pathogens (Gaur, 1990; Rao, 1990; Khan et al., 2009). The PSM may suppress rhizospheric population of pathogens by promoting host growth, inducing systemic resistance, and/or producing toxic metabolites (Kirkpatric et al., 1964). Pathogen management employing phosphate-solubilizing fungi or bacteria has advantage over classical biocontrol agents as the former provide essential nutrients to plants in addition to their antagonistic ability.

3.4 *ASPERGILLUS NIGER,* EFFICIENT PHOSPHATE SOLUBILIZER AND ANTAGONIST OF PLANT PATHOGEN

Among various efficient phosphate solubilizers, *A. niger* has most important attributes possesses. The fungus is not included in the toxigenic species of *Aspergillus* responsible for mycotoxins (OTA) (Khan & Anwer, 2007). Production of ribotoxin by *A. niger* was negative (Campbell, 1994) and the fungus is reported to decrease aflatoxin contamination (Boller & Schroeder, 1974; Wicklow et al., 1980; Horn & Wicklow, 1983), in addition of being highly antagonistic to a variety of pests and pathogens.

3.4.1 *BIOCONTROL THROUGH ASPERGILLUS NIGER VAN TIEGHEM, 1867*

3.4.1.1 *SYSTEMATIC POSITION*

Domain	:	Eukaryota
Kingdom	:	Fungi
Phylum	:	Ascomycota
Sub phylum	:	Pezizomycotina
Class	:	Eurotiomycetes
Order	:	Eurotiales
Family	:	Trichocomaceae
Genus	:	*Aspergillus*
Species	:	*niger*

3.4.1.2 *MACROSCOPIC MORPHOLOGY*

Colonies on potato dextrose agar at 25 °C are initially white, quickly becoming black with conidial production (Fig. 3.1). Reverse is pale yellow and growth may produce radial fissures in the agar (Raper & Fennell, 1965; Gilman, 2001; Alexopoulos et al., 2002).

3.4.1.3 MICROSCOPIC MORPHOLOGY

Hyphae are septate and hyaline. Conidial heads are radiate initially, splitting into columns at maturity. The species is biseriate (vesicles produces sterile cells known as metulae that support the conidiogenous phialides). Conidiophores are long (400–3000 μm), smooth, and hyaline, which become darker at the apex and terminating in a globose vesicle (30–75 μm in diameter). Metulae and phialides cover the entire vesicle. Conidia are brown to black, very rough, globose, and measure 4–5 μm in diameter (Fig. 3.1) (Raper & Fennell, 1965; Sutton et al., 1998; de Hoog et al., 2000).

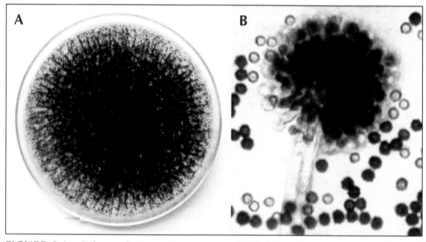

FIGURE 3.1   Colony of *Aspergillus niger* on PDA (A), microscopic morphology of conidial head with conidia (B).

3.4.1.4 MOLECULAR CHARACTERIZATION

The introduction of molecular biology techniques has been a major force in the area of systematics, genetic diversity, and population biology of microorganisms. One of the first applications of polymerase chain reaction (PCR) in mycology was described by White et al. (1990) to establish the taxonomic and phylogenetic relationships of fungi. Internal transcribed spacers (ITS) sequences are generally constant, or show little variation within species but vary between species in a genus, and so these sequences have been widely used to develop rapid procedures for the identification of fungal species by PCR-RFLP analysis (Vigalys & Hester, 1990; Goodwin & Annis, 1991; Nazar et al., 1991; Anderson & Stasovski, 1992; Buscot et al., 1996). Individual strains within a particular population, some examples being toxic producing strains of *Aspergillus flavus* (Bayman & Cotty, 1993), and in strain authentication in species of *Trichoderma* can be precisely differentiated by the RAPD-PCR method. Although the black *Aspergillus* species, *Aspergillus carbonarius,* and the uniseriate species (*Aspergillus aculeatus*, *Aspergillus japonicus*) can be microscopically distinguished by vesicle and conidial size, and ornamentation. However, the black Aspergilli (section Nigri) form a subgroup of the genus *Aspergillus.* Traditionally, fungal trains are assigned to a particular species based on their morphological characteristics such as, color, shape, size, ornamentation of the conidia, and the length of the conidiophore. Using these techniques, the black Aspergilli were divided into 12 species (Raper & Fennell, 1965) and later into seven species (Al-Musallam, 1980). The phenotype of a fungal strain can vary significantly depending on the growth conditions (Varga et al., 2000), indicating that errors may occur when assigning strains by morphological methods. The development of molecular techniques (RFLP, RAPD PCR, DNA fingerprinting, and nucleotide sequencing) for the identification of fungal strains has resulted in a reclassification of black Aspergilli (Kusters-van Someren et al., 1991; Varga et al., 1994; Accensi et al., 1999; Parenicová et al., 1997, 2001; Abarca et al., 2004).

3.4.1.5 MECHANISM OF DISEASE SUPPRESSION

A. niger suppresses plant pathogens by direct parasitism, lysis, competition for food, direct or indirect antibiosis through production of volatile substances. Activity of biocontrol agents mainly depends on the physicochemical environmental conditions to which they are subjected. These mechanisms are complex, and what has been defined as biocontrol is the final result of varied mechanisms acting antagonistically to achieve disease control. *A. niger* can even exert positive effects on plants with an increase in plant growth (mineralization) and the stimulation of plant defense mechanisms. Mechanism of disease suppression may be due to competition, antibiosis, or mycoparasitism.

3.4.1.5.1   Fungistatic

An effective antagonist is usually able to survive in the presence of metabolites produced by other microorganisms and plants, and multiply under extreme competitive conditions. *Aspergillus* spp. were found to be most resistant to herbicides, fungicides, pesticides and many toxic heavy metals at minimum inhibitory concentrations (MIC) of 125–850 µg/mL (Baytak et al., 2005; Yuh-Shan, 2005; Ahmad et al., 2006; Braud et al., 2006). Dose-response relationships of fungicide resistance in agar growth tests were examined with *A. niger* and *Aspergillus nidulans* to pentachloronitrobenzene (PCNB), 3-phenylindole, benomyl or thiabendazole, and resistance was measured at high concentrations of these chemicals (van Tuly, 1977).

In an experiment adaptation of *A. niger* to short-term stress induced by three antifungal agents [amphotericin B (AMPH), miconazole (MCZ), and ketoconazole (KCZ)] was observed and evaluated quantitatively using individual hyphae and found that exposure to AMPH (0.075 µg/mL) stopped the growth of the hypha. After washing with potato dextrose broth (PDB), the same concentration of AMPH was applied again. The growth of the test hypha was not inhibited. This phenomenon was defined as adaptation to the short-term stress of AMPH. Similarly, adaptation was observed with MCZ (0.01 µg/mL) and KCZ (0.5 µg/mL). The time required for the *A. niger* hypha to restart growth after washing with PDB depended upon the concentration of MCZ or KCZ, but not upon the concentration of AMPH (Park et al., 1994).

3.4.1.5.2   Competition for Nutrients

Starvation or shortage of nutrients is one of the most common causes of death of microorganisms (Chet et al., 1997). Competition resulting in limiting the nutrient supply to fungal pathogens results in their biological control (Chet et al., 1997). For instance, in most filamentous fungi iron (Fe) uptake is essential for viability (Eisendle et al., 2004) and under Fe-deficient condition most fungi excrete low molecular-weight ferric iron-specific chelators termed siderophores to mobilize environmental Fe (Eisendle et al., 2004). Siderophores play a considerable role in biocontrol of soil-borne plant pathogens (Leeman et al., 1996) and as a supplier of Fe nutrition to crop plants (Jadhav et al., 1994). Since plant pathogens may not have the cognate ferri–siderophore receptor for uptake of the Fe–siderophore complex they are prevented from proliferating in the immediate vicinity because of Fe deficiency (O'Sullivan & O'Gara, 1992). Hence, siderophore-producing bioinoculants can confer a competitive advantage in interactions in the rhizosphere (Raijmakers et al., 1995). One of the most sensitive stages for nutrient competition in the life cycle of *Fusarium* is chlamydospore germination (Scher & Baker, 1982). In soil the chlamydospores of *F. oxysporum* require adequate nutrition to maintain a germination rate of 20–30%. Germination may decrease due to sharing of nutrients with other microorganisms. Root exudates are a major source of nutrients in soil. Thus, colonization in the rhizosphere by an antagonist might reduce infection by *Fusarium*-like pathotypes (Cook & Baker, 1983). *A. niger* AN27 was found to produce both hydroxamate and catecholate groups of siderophores (Sen, 1997; Mondal et al., 2000). In another *in vitro* study Vassilev et al. (1996, 2006) also demonstrated that *A. niger* produced siderophore on modified CAS medium.

3.4.1.5.3 *Antibiosis*

Antibiosis is the phenomenon of suppression of one organism by another due to release of toxic substances/metabolites into the environment. Antibiosis is important in determining the competitive saprophytic and necrotrophic ability of antagonists. The *A. niger* may suppress plant parasitic nematodes and pathogenic fungi through antibiosis and by stimulating host defense. Low molecular-weight compounds and antibiotics (both volatile and non-volatile) produced by *Aspergillus* spp. impede colonization of harmful microorganisms such as bacteria, fungus including nematodes in the root zone (Buchi et al., 1983; Fujimoto et al., 1993; Eapen & Venugopal, 1995). Harzianic acid, alamethicins, tricholin, peptaibols, 6-penthyl-α-pyrone, y-lactone, massoilactone, viridin, gliovirin, glisoprenins, heptelidic acid, oxalic acid, and enzymes are some of the chemicals possessing antibiotic properties produced by *Aspergillus* species (Mankau, 1969a, 1969b; Benitez et al., 2004; El-Hasan et al., 2007).

   A. niger, that parasitize nematode eggs, prefer eggs which are deposited in cyst or a gelatinous matrix. The oviposition nature of *Heterodera* spp. and *Meloidogyne* spp. makes them more vulnerable to attack by the fungi. As soon as the fungi identify a cyst or an egg mass they rapidly grow and colonize those eggs where larval formation is not complete.

However, when larva is formed the egg becomes less vulnerable. It has been suggested that this differential vulnerability of egg and larval stage is due to chitinolytic activity of the fungi. Chitin is a major constituent of the eggshell, which is lacking in the larval cuticle.

3.4.1.5.4 Mycoparasitism

Mycoparasitism involves direct parasitism of one fungus by another and involves recognition, attack and subsequent penetration and killing of the host fungus (Harman et al., 2004). In a necrotrophic association there is direct contact between two fungi, and a nutrient exchange channel is established between them. Typical examples are the association of *Arthrobotrys oligospora* with *R. solani* (Persson et al., 1985); *Trichoderma hamatum* with species of *Phythium, Rhizoctonia* and *Sclerotium* (Bruckner & Pryzybylski, 1984). *A. niger* also shows mycoparasitism (Sen et al., 1995).

Observations using scanning electron microscopy revealed that *A. niger* coiled around the pathogen hyphae and penetrated within. Presence of *A. niger* hyphae inside pathogen hyphae using fluorescent microscopy has been confirmed repeatedly in *F. oxysporum* f. sp. *melonis, ciceri* and other pathogens (Sen et al., 1997; Sharma & Sen, 1991a, 1991b). Further studies have revealed that *A. niger* could kill *M. phaseolina,* several species of *Pythium, R. solani, Sclerotinia sclerotiorum* (Sen et al., 1995), and *Sclerotium rolfsii* (Palakshappa et al., 1989). The dead hyphae of the pathogens were eventually invaded. These observations confirm that *A. niger* is a contact and invasive necrotroph (Mondal et al., 2000).

3.4.1.5.5 Plant Growth Promotion

Two growth promoting compounds, 2-carboxymethyl 3-n-hexyl maleic acid (compound 1) and 2-methylene-3-hexylbutanedioic acid (compound 2) isolated from *A. niger* are responsible for increasing root and shoot length and biomass of crop plants (Mondal et al., 2000). Both the compounds increased germination and improved crop vigor. Compound 1 was more effective for increase in germination and shoot length, whereas compound 2 had relatively greater role in increasing the root length and biomass of cauliflower seedlings (Mondal et al., 2000). *A. niger* frequently enhances root growth and development, crop productivity, resistance to abiotic stresses and the uptake and use of nutrients (Sen, 2000). Plant growth production of tomato can increase up to 467% after the addition of *A. niger* (Vassilev et al., 1996, 2006) greater than increased by application of *T. hamatum* or *Trichoderma koningii* (300%) of field crops (Chet et al., 1997). In an *in vitro* test, *A. niger* PSBG-12 strain produced plant growth promoting substances such as indole acetic acid (IAA), gibberelic acid (GA) and under pot culture conditions showed better effect on plant growth parameters and nutrient uptake compared to uninoculated control (Vaddar & Patil, 2007). In a field where muskmelon and watermelon crops were suffering from *Fusarium* wilt (sometimes *R. solani* and *Pythium* spp. were associated with the disease), treatment of seeds with *A. niger* (Kalisena SD) @ 8 g/kg and soil with *A. niger* (Kalisena SL) @ 30 g/pit resulted in 81% control of the disease. The vines were more vigorous, and even with 15% incidence of disease, yield was ~5% greater as compared to that in disease-free areas (Chattopadhyay & Sen, 1996). Recently, Anwer and Khan (2013) proved that *A. niger* enhanced the quality of tomato fruit (*Lycopersicum esculentum* Mill.) and also promoted plant health.

3.4.1.5.6 Phytohormone Production

Several strains of *B. subtilis* and *P. fluorescens* are well known to synthesize phytohormones such as indole acetic acid, gibberellins, cytokinins, and zeatin which promote plant growth at various stages (Gracia de Salamone et al., 2001). Evidences exists which indicate that some *A. niger* isolate also produce IAA and other phytohormones (Mostafa & Youssef, 1962). Youssef and Mankarios (1975) reported that soil isolates of *A. niger* produced auxins, gibberllins, and gibberellins like substances in their culture filtrates. Khan and Anwer (2008) have recorded *in vitro* production of IAA by *A. niger* which resulted to significant increase in eggplant yield.

3.4.1.5.7 Phosphorus Solubilization

The production of organic acid in the microenvironment around the root or in the culture media is considered the most important parameter to measure phosphate solubilization by microorganisms (Sperber, 1958; Hayman, 1975; Gaur, 1985a, 1985b). *A. niger* have been found to synthesize citric, gluconic, glycolic, oxalic, and succinic acids (Sperber, 1958;

Blumenthal, 2004; Ramachandran et al., 2008). The role of organic acids in solubilizing mineral phosphates and phosphorylated minerals is attributed to the lowering pH which helps in the formation of stable complex and later forms are more soluble and available form for plants (Jennings, 1989; Li et al., 1991). It has been reported that *A. niger* reduce medium pH to 4.0, sufficiently low enough to solubilize phosphorus (Domich, et al., 1980; Medina et al., 2007).

Different mechanisms have been suggested by some scientists for the solubilization of inorganic phosphates by *A. niger*.

Organic acids effect: Solubilization of tricalcium phosphate in liquid medium by organic acids such as citric, gluconic, glycolic, oxalic, and succinic acids depends on pH and formation of soluble Ca-complexes (Azcon et al., 1986). These organic acids lowers down soil pH and with change in soil pH, calcium phosphate $[Ca_3 (PO_4)_2]$ changed to calcium biphosphate $[CaHPO_4]$ or calcium triphosphate $[Ca(H_2PO_4)_2]$ which are more soluble and become available to plants (Jennings, 1989). The nature of the acids released is more important to the amount of acid produced because the amount of phosphate solubilized by reaction with organic acids depends primarily on the strength of the acid (Li et al., 1991).

Chelating effect: Chelations of cations (bound to P) of the organic acids has been shown to be an important mechanism for P-solubilization (Bajpai & Sundra Rao, 1971a, 1971b). The extent to which an organic acid is capable of chelating metal ions is governed by its molecular structure and vary from acid to acid. The ability of organic acids to chelate Fe-organic complex is in the order of citrate > oxalate > malonate > tartrate > aceate (Lal, 2002).

Carbon dioxide effect: Carbon dioxide produced by *A. niger* (Ramachandran et al., 2008) in the rhizosphere increases the availability of phosphate and its uptake by crop plants (Morean, 1959; Knight et al., 1989).

Siderophores effect: In well-aerated soils, ferric ion (Fe^{3+}) is dominant form of iron which reacts with soluble phosphorus forming insoluble ferric phosphate. Nonporphyrin metabolite secreted by *A. niger* forms highly stable coordination compounds (chelate) with iron (Sen, 1997), a high affinity iron-binding compound called siderophore. These siderophores bind iron tightly to prohibit its reaction with soluble phosphates, and rather help release phosphate fixed as ferric phosphate. Specific tests confirmed that *A. niger* produced both hydroxamate and catecholate group of siderophores (Mondal et al., 2000). By this mechanism *A.*

niger may able to increase the solubility and availability of phosphorus to plants.

3.4.2 MYCOTOXINS FROM ASPERGILLUS NIGER

Aflatoxins are toxic fungal metabolites that can inhibit human development, cause cancer, and even induce death such toxins are mainly produced by *A. flavus* (Hell et al., 2008; Cotty et al., 2009). The mycotoxins, OTA and diketopiperazine asperazine are currently the most problematic compounds, especially in foods and feeds such as coffee, nuts, dried fruits, and grape-based products. For chemical differentiation/identification of the less toxic species the diketopiperazine asperazine can be used as a positive marker since it is consistently produced by *A. tubingensis* (177 of 177 strains tested) and *A. acidus* (47 of 47 strains tested) but never by *A. niger* (140 strains tested) (Nielsen et al., 2009).

3.4.2.1 OCHRATOXIN A

Ochratoxin A (OTA) is a toxic metabolite produced primarily by *Aspergillus*, but also by *Penicillium* and other molds. It is a white crystalline powder. Recrystallized from xylene, it forms crystals that emit green (acid solution) and blue (alkaline solution) fluorescence in ultraviolet light, the melting point of these crystals is 169 °C. The free acid of OTA is soluble in organic solvents (IARC, 1983, 1993). The sodium salt is soluble in water. OTA is unstable to light and air, degrading and fading even after brief exposure to light, especially under humid conditions. Ethanol solutions are stable for longer than one year if kept refrigerated and in the dark. OTA is fairly stable to heat; in cereal products, up to 35% of the toxin survives autoclaving for up to three hours (IARC 1976).

OTA is reasonably anticipated to be a human carcinogen classified 2B (for human) by IARC based on sufficient evidence of carcinogenicity in experimental animals and possible teratogenic effects on fertility (fetotoxicity and post-implantation mortality) and birth defects. It may cause cancer (hazardous in case of skin contact, permeator, or irritant), adverse reproductive effects (paternal effects and genetic material), conjunctivitis, respiratory tract irritation, and may be fatal if swallowed. Otherwise may cause digestive tract irritation with possible ulceration or bleeding from stomach. OTA may affect the endocrine system, liver (necrosis, fatty liver degeneration), blood behavior (somnolence, altered sleep time, and ataxia), respiration (acute pulmonary edema), metabolism, cardiovascular system and may damage kidney (chronic potential health effects) (NTP, 1989; IARC, 1993).

Acute oral toxicity (LD_{50}): 20 mg/kg bw (rat), 46 mg/kg bw (mouse), 0.2 mg/kg bw (dog), 1.0 mg/kg bw (pig), 3.3 mg/kg bw (chicken) (Harwing et. al., 1983). Several assays have been reported for genotoxicity with OTA (Creppy et al,. 1985; Dorrenhaus et al., 2000; Zepnik et al., 2001).

Until recently, only 6–10% of members of the *A. niger* group isolated from corn, peanuts, raisins, onions, mango, apples, and dried meat products (Perrone et al., 2007) are known to produce OTA in least amounts (Abarca et al., 1994; Varga et al., 2000; Esteban et al., 2006). Production of ribotoxin by *A. niger* was negative in Eastern and Southern blot tests (Campbell, 1994). Interestingly the fungus is reported to decrease aflatoxin contamination (Wicklow et al., 1980; Horn & Wicklow, 1983). This effect of *A. niger* has been attributed to lowering of substrate pH and an inhibitory substance produced by the fungus for degradation of aflatoxin (Boller & Schroeder, 1974).

3.4.3 EFFECTIVENESS OF ASPERGILLUS NIGER AGAINST PLANT PATHOGENIC FUNGI

3.4.3.1 IN VITRO STUDIES

In an *in vitro* study, Muhammad and Amusa (2003) reported that *A. niger* inhibited 40–60% growth of some seedling blight inducing pathogens viz., *F. oxysporum, S. rolfsii, P. aphanidermatum, Helminthosporium maydis, M. phaseolina,* and *R. solani*. In another study culture filtrate of *A. niger* inhibited the mycelium growth of *Fusarium udum* by 66.6% in dual culture test (Singh et al., 2002). *A. niger* also restricted the development of colonies of the *F. moniliforme* by its over growth and finally by parasitizing over the *Fusarium* mycelia (Chand et al., 2007). In dual culture test, *A. niger* significantly suppressed all *Fusarium* spp. of head blight of wheat, viz. *F. avenaceum, F. culmorum, F. graminearum,* and *F. poae* isolated from wheat kernels (Mullenborn et al., 2008). In another dual inoculation assay, *A. niger* significantly inhibited the growth of anthracnose (*Colletotrichum gloeosporioides*) of Indian bean (*Lablab purpureus* L.) mycelia by 57% (Deshmukh et al., 2010).

3.4.3.2 POT CULTURE STUDIES

Muskmelon seeds were soaked overnight in *A. niger* AN 27 (Kalisena SD) spore suspension and grown in sand for six days. The roots of seedlings (with fully opened cotyledonary leaves) were washed thoroughly in water to remove *A. niger* spores. The seeds were suspended in *F. oxysporum melonis* (aqueous) spore suspension. Muskmelon seedlings raised from *A. niger*-treated seeds showed 56% resistance to *F. oxysporum melonis* without physical presence of *A. niger* in the root zone (Radhakrishna & Sen, 1986). These seedlings were 58, 26, and 2% higher in peroxidase, polyphenol oxidase, and phenylalanine ammonia lyase activity, respectively, over controls (Angappan et al., 1996). The lignin content was also higher in the tissues of treated plants and resulted in the induced resistance (Kumar & Sen, 1998). When *A. niger* was amended in unsterilized and sterilized pot soils at the rate of 1, 2, or 3% (w/w), controlled the Fusarial wilt of pigeon-pea (*F. udum*) by 48.3, 52.9, and 68.2% in unsterilized soil and 74.0, 83.0, and 88.4% in sterilized soil, respectively (Singh et al., 2002). In a greenhouse condition the damping off of chilli (*Capsicum frutescens*) caused by *P. aphanidermatum* was controlled by using seed coating with *A. niger* resulted increased germination percentage (82%), and reduced the pre (12%) and post emergence (34.2%) damping off as compared to sick soil treated as control (Chakraborty et al., 2007).

3.4.3.3 FIELD CONDITIONS

A good deal of work has been conducted in field trials with *A. niger* against soil-borne fungal pathogens. Dhillion (1994) suggested that biocontrol of root diseases by the soil amendment of *A. niger* applied with arbuscular mycorrhizal inoculum, enhanced nutrient uptake through increase solubilization of phosphate and increased stress tolerance, and yield of maize, wheat, millet, sorghum, barley, and oat plants (Dhillion, 1994). In a field where muskmelon and watermelon crops were suffering from *Fusarium* wilt (sometimes *R. solani* and *Pythium* spp. were associated with the disease), treatment of seeds with *A. niger* (Kalisena SD) @ 8 g/kg and soil with *A. niger* (Kalisena SL) @ 30 g/pit resulted in 81% control of the disease. The vines were more vigorous, and even with 15% incidence of disease, yield was ~5% greater as compared to that in disease-free areas (Chattopadhyay & Sen, 1996). Seed treatment with Kalisena SD also provided 30% less sheath blight disease rice over control plants (Kumar & Sen, 1998). Problems of pre- and post-emergence damping-off incited by *P. aphanidermatum* and *R. solani* in fruit and vegetable farms was successfully overcome by a combined treatment of seed and soil application of Kalisena SD and Kalisena SL (Majumdar & Sen, 1998). Similarly, 93% control of charcoal rot of potato (*Solanum tuberosum* L.) in *M. phaseolina*-infested field was obtained with *A. niger* (Kalisena SD and Kalisena SL) (Mondal, 1998). In another study 87% control of black scurf of potato (*R. solani*) and a 10% increase in yield by the application of *A. niger* @ 8g/kg on infected seed tubers and sown in worst affected fields was reported (Sen et al., 1998). In a study, Lodhi (2004) has reported the control of soil borne diseases of potato by the application of *A. niger* alone and in combination with VAM fungi. Winter sorghum can be strongly damaged by *Macrophomina* infection; however, *A. niger* (Kalisena SD) seed treatment brought down incidence of the disease from 30 to 7% (Das, 1998). Fusarial wilt of *Hibiscus* was controlled by the application of *A. niger* in the soil after uprooting the dead plants and soil population of the pathogen was found non-detectable level after a month of *A. niger* application (Sen, 2000).

In an experiment seed treatment with *A. niger* @ 8 g/kg seed effectively controlled the blast (*P. oryzae*), sheath blight (*R. solani*), brown spot (*H. oryzae*), and false smut (*U. virens*) of rice in the field (Sehgal et al., 2001). In another study, a plant shows a disease of international importance, malformation disease of mango caused by *F. moniliforme* was controlled by spraying conidial suspension of *A. niger* over dead necrotic malformed panicles (Chand et al., 2007). Stem canker of Aonla *(Emblica officinalis)* caused by *R. solani* was controlled by the application of *A. niger* (Kalisena SL) in soil around the tree trunk revived the partially diseased trees and brought them to normal bearing in an orchard (Sen, 2000). Guava (*Psidium guajava* L.) wilt caused by *F. oxysporum* f. sp. *psidii* and *F. solani* was controlled by the soil application of *A. niger* multiply on the field wastes (Misra, 2007). Wheat loose smut (*Ustilago tritici*) and spot blotch (*Chaetomium globosum*) was effectively controlled be the application of *A. niger* (Birthal & Sharma, 2004). In an experiment

A. niger was found potent antagonist against *F. udum,* both *in vivo* and *in vitro* (Upadhyay & Rai, 1981). Under field conditions, 38% control of Fusarial wilt of pigeon-pea was observed with *A. niger* application amended at the rate of 42 g/m^2 (150 g/1.8 × 2 m^2) of *A. niger* cultured on wheat (Singh et al., 2002).

In field trials, Khan and Khan (2001, 2002) tested the effectiveness of *A. niger* against *Fusarium* wilt of tomato (*F. oxysporum* f. sp. *lycopersici*). Root-dip treatment with *A. awamori* or *A. niger* resulted in a significant decline in the rhizosphere population of *F. oxysporum* f. sp. *lycopersici.* Tomato yield was greatly enhanced by *A. awamori* and *A. niger.* Direct soil-plant inoculation with *A. niger* and *A. awamori* decreased the rhizosphere population of the pathogen by 23–49% while the tomato yield increased by 28–53% in field experiments. Vassilev et al. (1996, 2006) reported an efficient biotechnological scheme for preparing a material with biocontrol and plant-growth-promoting functions. Sugar beet wastes were mineralized by an acid-producing strain of *A. niger* with a simultaneous solubilization of rock phosphate under conditions of solid-state fermentation. The product of this process, used as soil amendment, resulted in 347 and 467% higher (vs unamended control) plant biomass in plant–soil experiments contaminated or not with *F. oxysporum* f. sp. *lycopersici,* respectively. Disease severity on tomato and number of *F. oxysporum* f. sp. *lycopersici* colony-forming units reached the lowest levels, particularly when plants were mycorrhized with *Glomus deserticola.*

3.4.4   *ASPERGILLUS NIGER* AGAINST PLANT NEMATODES

P. lilacinus, Dactylaria candida and *P. penetrans* have been widely used in nematode control studies during last few decades (De Bach, 1964), but now a days PGPR including *A. niger* are receiving attention of scientists. Effects of *A. niger* has been evaluated against different nematodes under *in vitro,* pot and field conditions.

3.4.4.1   *IN VITRO STUDIES*

A. niger is reputed to produce nematoxic metabolites (Dahiya & Singh, 1985). Culture filtrates of *A. niger* caused mortality to *Radopholus similis* and *M. incognita* after 48 h of immersion (Molina & Davide, 1986). In an *in vitro* trial, *A. niger* (isolated from egg-masses of *M. incognita*) was identified to be toxic, egg parasitic, or opportunistic against *M. incognita* (Goswami et al., 2008). In another study culture filtrates of *A. niger* soil isolates AnC2 and AnR3 efficiently suppressed hatching of eggs and mortality of juveniles of *M. incognita* (Khan & Anwer, 2008). Culture filtrates of *Aspergillus terreus, A. nidulans, A. niger,* and *Acremonium strictum* were very effective against the nematode in regards to egg parasitism (up to 53%), egg hatching inhibition (up to 86%), and mortality (up to 68%) compared to controls (Singh & Mathur, 2010).

3.4.4.2   POT CULTURE STUDIES

The majority of studies exploring the potential of biocontrol against plant nematodes have been carried out under pot conditions (Khan, 2007). In a pot experiment, chilli (*Capsicum annum*) seedlings were inoculated with *Meloidogyne javanica, A. niger,* and *R. solani* alone or in various combinations. All growth parameters were significantly greater with *A. niger* and lower with *M. javanica* or *R. solani* (Shah et al., 1994). Singh et al. (1991) showed that application of *A. niger* decreased the damage caused by *M. incognita* and *R. solani* singly or together on tomato cv. perfection. Sharma et al. (2005) demonstrated in pot conditions that inoculation of *A. niger* in preinoculated juveniles of *M. incognita* in okra significantly decreased the galling, eggmass production and soil population and increased the growth parameters of the plant. Similarly, inoculation with *A. niger, Epicoccum purphurascum, Penicillium vermiculatum,* and *Rhizopus utgricans* effectively diluted the adverse effect of *R. solani* and *M. incognita* resulting in an increase in germination of tomato cv. Pusa Ruby (Rekha & Saxena, 1999). Pant and Pandey (2001) reported maximum reduction in populations of *M. incognita* with *T. harzianum, Purpureocillium lilacinum* and *A. niger* applied in sterilized soil in pots @ 5000 spores/pot. In an experiment second-stage larvae of root-knot nematode were incubated in microbial filtrates of many fungal and bacterial (isolated from agricultural soil) metabolites for 48 h and then introduced to healthy tomato seedlings in green house or controlled chamber, after two months it has been found that several bacterial and fungal species reduced the ability of second-stage larvae of root-knot nematode from penetrating, infecting, and/or developing on the tomato roots with highest *A. niger* isolate (Abdel-Rahman et al., 2004). In a pot experiment application of *A. niger* isolates (AnC$_2$ and AnR$_3$) significantly suppressed galling, egg mass production, and soil populations of *M. incognita*. The isolates AnC$_2$ and AnR$_3$ produced the greatest quantities of siderophores, HCN, and NH$_3$, and solubilized the greatest quantity of soil phosphorus (Khan & Anwer, 2008).

3.4.4.3   FIELD CONDITIONS

Relatively few field trials have been conducted to evaluate the effectiveness of *A. niger* against nematode infestations. The studies so far conducted have demonstrated nematode control to a level that can be exploited commercially (Khan, 2005). In an experiment, soil treatment by *A. niger* in castor beans abated the population of *Rotylenchulus reniformis* up to 71% (Das, 1998). A field trials was conducted by Goswami et al. (2008) to manage root-knot nematode disease infecting tomato by dual treatment of *A. niger* and *T. harzianum* both alone and together showed significant increase in health improvement of tomato plant with a remarkable reduction in *M. incognita* population (Goswami et al., 2008). In another field experiment application of *A. niger* in combination with cattle manure resulted in a significant increase (62%) in the growth of nematode-inoculated plants and also resulted in a high reduction in galling and nematode multiplication (Siddiqui & Kazuyoshi, 2009).

3.5   *ASPERGILLUS NIGER* IN IPM AND CHEMICAL COMPATIBILITY

The term integrated control describes the combined use of more than one approach to suppress losses due to pests, with greater overall effect than would result from the use of any one of the controls applied singly (Rahe & Utkhede, 1985). The integrated control of plant pathogens has been discussed by several workers, including Lewis and Papavizas (1991). A combination of broad-spectrum fungicides with biocontrol fungi is more commonly used than any other method of integrated control. The rationale for using fungi in combination with decreased amounts of pesticides stems from the need to decrease pesticide use in disease control (Lewis & Papavizas, 1991). Integrated mixture of chemicals and biological agents are used when inconsistency and low efficiency of biocontrol occur in the field. Integration of biological control and chemicals should be based on a dose of pesticide sufficiently low enough not to cause environmental contamination and other adverse effects.

Aspergillus spp. are naturally resistant to heavy metals and have been found to be quite compatible with broad-spectrum herbicides, nematicides, and fungicides (Bhatnagar, 1995; Baytak et al., 2005; Yuh-Shan, 2005; Ahmad et al., 2006; Braud et al., 2006; Anwer & Bushra, 2015). The effect of five fungicides, benomyl (1 mg/L), dodine (50 mg/L), manzate (100 mg/L), cupric sulfate (200 mg/Ll), and thiabendazole (4 mg/L) was tested under *in vitro* conditions on the development of *Aspergillus* spp. and found highly resistance to tested concentrations of the fungicides. In another experiment when *A. niger* included with two fungicides, Foltaf 80W (Captafol 80%) and Blue Copper-50, for the treatment of pigeon-pea wilt, the disease was more effectively controlled than the fungicides were used alone to control the disease without adversely affecting the growth of pigeon-pea (Bhatnagar, 1995). Seed treatment of chickpea with *A. niger* (10^6 spores/mL/10 g seed) in combination with carboxin (2 g/kg seed) enhanced seed germination by 10.0–13.0% and grain yields by 33–63% and reduced wilt incidence (44–60%) during a field experiment (Dubey et al., 2007).

In soil, water, and air quality conservation and management program, swine wastes were treated with *A. niger* resulted in 90% reduction in Cu and Zn from the wastes (Anonymous, 2001). In adsorption efficiencies experiment, the percentage uptake of lead ion by *A. niger* was observed from 6.71 to 64.95% for dead and live biomass, respectively (Awofolu et al., 2006). These characters represent broader use of *A. niger* biopesticides with pesticides.

3.6   CONCLUSION AND FUTURE PROSPECTS

Plant diseases are significant limiting factor on crop production worldwide, and their management is essential to increase food production. In view of the adverse effects of chemical pesticides, fungal biological control agents offer a potential substitute. Numerous potentially useful microorganisms are available especially plant growth promoting, phosphate solubilizer *A. niger*. The organism can be applied directly to soil or as a seed treatment or foliar spray to reduce the disease severity and also efficiently increases the plant health. Overall performance of the fungus, *A. niger* against plant diseases and nematodes is at levels that ensure their commercial exploitation. This warrents research effort toward identification of more efficacious and environmentally adaptable strains in disturbed climatic conditions, development of suitable mass production technologies, and development of efficient immobilization systems with more shelf life.

KEYWORDS

- mycoparasitism
- plant pathogens
- metabolites
- pathogenic fungi
- crop damage
- growth promoting fungi

REFERENCES

Abarca, M. L.; Accensi, F.; Cano, J.; Cabanes, F. J. Taxonomy and Significance of Black Aspergilli. *Antonie van Leeuwenhock.* **2004,** *86,* 33–49.

Abarca, M. L.; Bragulat, M. R., Castella, G.; Cabanes, F. J. Ochratoxin a Production by Strains of *Aspergillus niger* Var. Niger. *Appl. Environ. Microbiol.* **1994,** *60,* 2650–2652.

Abdel-Rahman, F. H.; Nsaif, R. H.; Massoud, S. I. In *Nematicidal Activity of Soil Microorganisms' Metabolites,* Abstracts, Society of Nematologists 43rd Annual Meeting, Estes Park, CO, Aug 7–11, 2004; p 303.

Accensi, F.; Cano, J.; Figuera, L.; Abarca, M. L;, Cabenes, F. J. New PCR Method to Differentiate Species in the *Aspergillus niger* Aggreagate. *FEMS Microbiol. Lett.* **1999,** *180,* 191–196.

Ahmad, I.; Ansari, M. I.; Aqil, F. Biosorption of Ni, Cr and Cd by Metal Tolerant *Aspergillus niger* and *Penicillium* Sp. Using Single and Multi-Metal Solution. *Indian J. Exp. Biol.* **2006,** *44,* 73–76.

Alexopoulos, C. J.; Mims, C. W.; Blackwell, M. *Introductory Mycology;* John Wiley & Sons (ASIA) Pte. Ltd: Singapore, Asia, 2002; p 869.

Al-Musallam, A. Revision of the Black *Aspergillus* Species. Ph.D. Thesis. Rijks University Utrecht, Utrecht. 1980.

Anderson, J. B.; Stasovski, E. Molecular Phylogeny of Northern Hemisphere Species of *Armillaria. Mycologia.* **1992,** *84,* 505–516.

Angappan, K.; Dureja, P.; Sen, B. In *Multipronged actions of biocontrol agent, Aspergillus niger AN27,* Second International Crop Science Congress on Crop Producting and Sustainability–shoping the future, New Delhi, India, Nov 17–24, 1996; Abstract of Poster Session, 1996, p 301.

Anonymous 2001; All India Coordinated Research Project on Chickpea, Annual Report (2000–01), IIPR: Kanpur, India, 2001.

Anwer, M. A.; Bushra, A. *Aspergillus niger*-A Novel Heavy Metal Bio-Absorbent and Pesticide Tolerant Fungus. *Res. J. Chem. Environ.* **2015,** *19*(2), 57–66.

Anwer, M. A.; Khan, M. R. *Aspergillus niger* as Tomato Fruit (*Lycopersicum Esculentum* Mill.) Quality Enhancer and Plant Health Promoter. *J. Post-harvest Technol.* **2013,** *01*(01), 36–51.

Asea, P. E. A.; Kucey, R. M. N.; Stewart, J. W. B. Inorganic Phosphate Solubilization by Two *Penicillium* Species in Solution Culture and Soil. *Soil Biol. Biochem.* **1998,** *20,* 459–464.

Awofolu, O. R.; Okonkwo, J. O.; Der Merwe, R. R. V.; Badenhorst, J.; Jordaan, E. A New Approach to Chemical Modification Protocols of *Aspergillus niger* and Sorption of Lead Ion by Fungal Species. *EJB.* **2006,** *9*(4), 341–348.

Azcon, A. C.; Gianinazzi-Pearson, V.; Fardeau, J. C.; Gianinazzi, S. Effect of Vesicular Arbuscular Mycorrhizal Fungi and Phosphate Solubilizing Bacteria on Growth and Nodulation of Soybean in a Neutral Calcareous Soil Amended with ^{32}P- ^{45}Ca Tricalcium Phosphate. *Plant Soil.* **1986,** *96,* 3–15.

Bai, Z. H.; Zhang, H. X.; Qi, H. Y.; Peng, X. W.; Li, B. J. Pectinase Production by *Aspergillus niger* Using Wastewater in Solid State Fermentation for Eliciting Plant Disease Resistance. *Bioresour. Technol.* **2004,** *99*(1), 49–52.

Bajpai, P. D.; Sundara Rao, W. V. B. Phosphate Solubilizing Bacteria. Solubilization of Phosphate in Liquid Culture by Selected Bacteria as Affected by Different pH values. *Soil Sci. Plant Nutr.* **1971a,** *17,* 41–43.

Bajpai, P. D.; Sundara Rao, W. V. B. Phosphate Solubilizing Bacteria. *Soil Sci. Plant Nutr.* **1971b,** *17,* 46–63.

Baker, K. F.; Cook, R. J. *Biological Control of Plant Pathogens*; Freeman & Co.: San Francisco, California, 1974; p 433.

Baker, K. F. Evolving Concepts of Biological Control of Plant Pathogens. *Annu. Rev. Phytopathol.* **1987,** *25,* 67–85.

Bayman, P.; Coty, P. J. Genetic Diversity in *Aspergillus flavus*: Association with Aflatoxin Production and Morphology. *Can. J. Bot.* **1993,** *71,* 23–31.

Baytak, S.; Turker, A. R.; Cevrimli, B. S. Application of Silica Gel 60 Loaded with *Aspergillus niger* as a Solid Phase Extractor for the Separation/Preconcentration of Chromium(III), Copper(II), Zinc(II), and Cadmium(II). *J. Sep. Sci.* **2005,** *28,* 2482–2488.

Benitez, T.; Rincon, A. M.; Limon, M. C.; Codon, A. C. Biocontrol Mechanisms of *Trichoderma* Strains. *Inter. Microbiol.* **2004,** *7*(4), 249–260.

Bhatnagar, H. Integrated Use of Biocontrol Agents with Fungicides to Control Wilt Incidence in Pigeon-Pea. *World J. Microbiol. Biotechnol.* **1995,** *11,* 564–566.

Bhatti, M. A.; Kraft, J M. Effect of Inoculum Density and Temperature on Root Rot and Wilt of Chickpea. *Plant Dis.* **1992,** *76,* 960–963.

Birthal, P. S.; Sharma, O. P. Biocontrol Agents Available for Various Crop Diseases. In *Integrated Pest Management in Indian Agriculture*; Mruthyunjaya, Ed.; NCAP, Chandu Press: New Delhi, India, 2004.

Blumenthal, C. Z. Production of Toxic Metabolites in *Aspergillus niger*, *Aspergillus oryzae*, and *Trichoderma reesei*: Justification of Mycotoxin Testing in Food Grade Enzyme Preparations Derived from the Three Fungi. *Regul. Toxicol. Pharmacol.* **2004,** *39,* 214–228.

Boller, R. A.; Schroeder, H. W. Influence of *Aspergillus candidus* on Production of Afla-toxin in Rice by *Aspergillus parasiticus. Phytopathology.* **1974,** *64,* 121–123.

Braud, A.; Karine, J. E.; Vieille, E.; Triller, A.; Lebeau, T. Changes in Extractability of Cr and Pb in a Polycontaminated Soil after Bioaugmentation with Microbial Producers of Biosurfactants, Organic Acids and Siderophores. *Water Air Soil Pollu. Focus.* **2006,** *6,* 261–279.

Bruckner, H.; Pryzybylski, M. Isolation and Structural Characterization of Polypetides Antibiotics of the Peptaidol Class by HPLC with Field Desorption and Fast Atom Bombardment Mass Spectrometry. *J. Chromatogr.* **1984,** *296,* 263–275.

Buchi, G.; Francisco, M. A.; Murry, W. W. Aspersitin – A New Metabolites of *Aspergillus parasiticus. Tetrahedron Lett.* **1983,** *24,* 2527–2530.

Buscot, F.; Wipf, D.; Di Battista, C.; Munch, J. C.; Botton, B.; Martin, F. DNA Polymor-phism in Morels: PCR-RFLP Analysis of the Ribosomal DNA Spacers and Microsatel-lite-Primed PCR. *Mycolo. Res.* **1996,** *100,* 63–71.

Campbell, C. K. Forms of Aspergillosis. In *The Genus, Aspergillus;* Powell, K. A., Renwick, A., Peberdy, J. F., Eds.; Plenum Press: New York, 1994; pp 313–319.

Chakraborty, B. N.; Ayana, Mahapatra, S.; Das, S.; Nath, S. P.; Ray, S. K.; Khatua, R. C.; Dasgupta, D. Management of Damping off of Chilli (*Capsicum frutescens*) through Integration with Bio-Antagonists and Botanicals Under Green House Conditions. In *National Symposium on Plant Protection-Technology Interface (December 28–29, 2007)*; Khan, M. R., Jha, S., Sen, C., Eds.; West Bengal, India, 2007; p 95.

Chand, R.; Girish, Pandey, M. K.; Chakrabarti, D. K. Investigations on Role of Bioagents in IPM of Mango Malformation Disease. In *National Symposium on Plant Protection-Technology Interface (December 28–29, 2007);* Khan, M. R., Jha, S., Sen, C., Eds.; 2007; West Bengal, India, p 95.

Chattopadhyay, C.; Sen, B. Integrated Management of *Fusarium* Wilt of Muskmelon Caused by *Fusarium oxysporum. Indian J. Mycol. Plant Pathol.* **1996,** *26,* 162–170.

Chet, I.; Inbar, J.; Hadar, I. Fungal Antagonists and Mycoparasites. In *The Mycota IV: Environmental and Microbial Relationships;* Wicklow, D. T., Soderstrom, B., Eds.; Springer-verlag, Berlin, **1997,** pp 165–184.

Cook, R. J.; Baker, K. F. *The Nature and Practice of Biological Control of Plant Patho-gens*; American Phytopathological Society: Saint Paul, MN, 1983; p 539.

Cooke, R. C.; Whipps, J. M. *Ecophysiology of Fungi;* Blackwell Scientific Publications: London, 1993; p 337.

Cotty, P. J.; Donner, M.; Atehnkeng, J.; Bandyopadhyay, R.; Sikora, R. A. Distribution of Aspergillus Section Flavi in Soils of Maize Fields in Three Agroecological Zones of Nigeria. *Soil Biol. Biochem.* **2009,** *41,* 37–44.

Creppy, E. E.; Kane, A.; Dirheimer, G.; Lafarge-Frayssinet, C.; Mousset, S.; Frayssinet, C. Genotoxicity of Ochratoxin A in Mice: DNA Single-Strand Break Evaluation in Spleen, Liver and Kidney. *Toxicol. Lett.* **1985,** *28,* 29–35.

Dahiya, J. S.; Singh, D. P. Inhibitory Effects of *Aspergillus niger* Culture Filtrate on Mortality and Hatching of Larvae of *Meloidogyne* spp. *Plant Soil.* **1985,** *86,* 145–146.

Das, I. K. *In Kalisena, a Novel Biopesticide for Disease Free Crop and Better Yield.* Proceedings, International Symposium on Development of Microbial Pesticides and Insect Pest Management, Nove12–13, 1998; BARC Mumbai and Hindustan Antibiotics Ltd.: Pune, India, 1998.

De Bach, P. The Scope of Biological Control. *In Biological Control of Insect Pests and Weeds;* DeBach, P., ed.; Chapman and Hall Ltd.: London, 1964; pp 3–20.

de Hoog, G. S.; Guarro, J.; Gene, J.; Figueras, M. J. *Atlas of Clinical Fungi*; 2nd Ed., Centraalbureau Voor Schimmelcultures: Utrecht, The Netherlands, 2000; Vol. 1.

Deshmukh, A. J.; Mehta, B. P.; Patil, V. A. *In Vitro* Evaluation of Some Known Bioagents to Control *Colletotrichum gloeosporioides* Penz. and Sacc., Causing Anthracnose of Indian Bean. *Inter. J. Pharma. Bio. Sci.* **2010,** *6*(2), 1–6.

Dhillion, S. S. Effect of *Trichoderma harzianum, Beijerinckia mobilis* and *Aspergillus niger* on Arbuscular Mycorrhizal Infection and Sporulation in Maize, Wheat, Millet, Sorghum, Barley and Oats.(Zeitschrift Fur Pflanzenkrankheiten und Pflanzenschutz). *J. Plant Dis. Protect.* **1994,** *101*(3), 272–277.

Domich, K. H.; Gams, W.; Andeson, T. *compeniun of Soil Fungi;* Academic Press: London, 1980; Vol. 1 and 2, pp 405, 859.

Dorrenhaus, A.; Flieger, A.; Golka, K.; Schulze, H.; Albrecht, M.; Degen, G. H.; Follmann, W. Induction of Unscheduled DNA Synthesis in Primary Human Urothelial Cells by the Mycotoxin Ochratoxin A. *Toxicol. Sci.* **2000,** *53,* 271–277.

Dubey, S. C.; Suresha, M.; Singh, B. Evaluation of *Trichoderma* Species against *Fusarium oxysporum* F. Sp. *Ciceris* for Integrated Management of Chickpea Wilt. *Biol. Contr.* **2007,** *40*(1), 118–127.

Eapen, S. J.; Venugopal, M. N. Field Evaluation of *Trichoderma* spp. and *Paecilomuces lilacinus* for Control of Root Knot Nematodes and Fungal Diseases of Cardamom Nurseries. *Indian J. Nematol.* **1995,** *25,* 15–16.

Eisendle, M.; Oberegger, H.; Buttinger, R.; Illmer, P.; Haas, H. Biosynthesis and Uptake of Siderophores is Controlled by the Pacc-Mediated Ambient-Ph Regulatory System in *Aspergillus nidulans. Eukaryotic Cell.* **2004,** *3,* 561–563.

El-Hasan, A.; Walker, F.; a Buchenauer, H. *Trichoderma harzianum* and its Metabolite 6-Pentyl-Alpha-Pyrone Suppress Fsaric Acid Produced by *Fusarium moniliforme. J. Phytopathol.* **2007,** *156,* 79–87.

Esteban, A.; Abarca, M. L.; Bragulat, M. R.; Cabanes, F. J. Study of the Effect of Water Activity and Temperature on Ochratoxin A Production by *Aspergillus carbonarius. Food Microbiol.* **2006,** *23,* 634–640.

Federico, G. R.; Maria, M. R.; Reynoso, M. F.; Sofia, N. C.; Adriana, M. T. Biological Control by *Trichoderma* spp. of *Fusarium solani* Causing Peanun Brown Root Rot Under Field Condition. *Crop Prot.* **2007,** *26,* 549–555.

Fujimoto, Y.; Miyagawa, H.; Tsurushima, T.; Irie, H.; Okamura, K.; Ueno, T. Structures of Antafumicins AaA and B, Novel Antifungal Substances Produced by the Fungus *Aspergillus niger* NH 401. *Biosci. Biotechnol. Biochem.* **1993,** *57,* 1222–1224.

Gaur, A. C. In *Phosphate Solubilizing Microorganisms and Their Role in Plant Growth and Crop Yields,* Proceeding, National Sympsoim on Soil Biology, Hissar, India, 1985a; pp 125–138.

Gaur, A. C. In *Phosphate Solubilizing Bacteria as Bio-Fertilizer,* Proceeding, National Seminar on Development and Use of Biofertilizers, Ministry of Agriculture: New Delhi, India, 1985b.

Gaur, A. C. *Phosphate Solubilising Microorganisms as Biofertilizers*; Omega Scientific Publishes: New Delhi, India, 1990; p 176.

Gilman, J. C. *A Manual of Soil Fungi Biotech Books*; 2nd Indian Ed., Biotech Books: Delhi, India, 2001; pp 195–196.

Gomathi, V. Gnanamanickam, S. S. Polygalacturonase-Inhibiting Proteins in Plant Defense. *Curr. Sci.* **2004,** *87*(9), 1211–1217.

Goodwin, P. H.; Annis, S. L. Rapid Identification of Genetic Variation and Assay. *Appl. Environ. Microbiol.* **1991,** *57,* 2482–2486.

Goswami, J.; Pandey, R. K.; Tewari, J. P.; Goswami, B. K. Management of Root Knot Nematode on Tomato Through Application of Fungal Antagonists, *Acremonium strictum* and *Trichoderma harzianum. J. Environ. Sci. Health B.* **2008,** *43*(3), 237–240.

Gracia de Salamone, I. E.; Hynes, R. K. I.; Nelson, L. M. Cytokinin Production by Plant Growth Promoting Rhizobacteria and Selected Mutants. *Can. J. Microbiol.* **2001,** *47,* 404–411.

Harman, G. E.; Howell, C. R.; Viterbo, A.; Chet, I.; Lorito, M. *Trichoderma* Species-Opportunistic, Avirulent Plant Symbionts. *Nature Rev.* **2004,** *2,* 43–56.

Harwing, J.; Kuiper-Goodman, T.; Scott, P. M. Microbial Food Toxicants: Ochratoxins. In *Handbook of Foodborne Diseases of Biological Origin*; Rechcigl, M., Ed.; CRC Press: Boca Raton, FL, 1983; pp 193–238.

Hayman, D. S. Phosphorus Cycling by Soil Microorganisms and Roots. In *Soil Microbiology. A Critical Review;* Walker, N., Ed.; Butterworths: London & Boston, 1975; pp 67–92.

Hell, K.; Fandohan, P.; Bandyopadhyay, R.; Kiewnick, S.; Sikora, R.; Cotty, P. J. Pre- and Post-Harvest Management of Aflatoxin In Maize: An African Perspective. In *Mycotoxins: Detection Methods, Management, Public Health, and Agricultural Trade;* Leslie, J. F., Bandyopadhyay, R., Visconti, A., Eds.; CAB International: Oxfordshire, 2008; pp 219–229.

Henis, Y.; Ghaffar, A.; Baker, R. Factor Affecting Suppression to *Rhizoctonia solani* in Soil. *Phytopahology.* **1979,** *69,* 1164–1169.

Horn, B. W.; Wicklow, D. T. Factors Influencing the Inhibition of Aflatoxin Production in Corn by *Aspergillus niger. Can. J. Microbiol.* **1983,** *29,* 1087–1091.

Howell, C. R. The Role of Antibiosis in Biocontrol. In *Trichoderma & Gliocladium*; Harman, G. E., Kubicek, C. P., Eds.; Taylor & Francis: Padstow, UK, 1998; Vol. 2, pp 173–184.

Howell, C. R. Mechanism Employed by *Trichoderma* spp. in the Biological Control of Plant Diseases; the History and Evolution of Current Concepts. *Plant Dis.* **2003,** *87,* 4–10.

IARC 1976; Some Naturally Occurring Substances, IARC Monographs on the Evaluation of Carcinogenic Risk, International Agency for Research on Cancer : Lyon, France, 1976; Vol. 10.

IARC 1983; Some Food Additives, Feed Additives and Naturally Occurring Substances, IARC Monographs on the Evaluation of Carcinogenic Risk of Chemicals to Humans, International Agency for Research on Cancer : Lyon, France, 1983; Vol. 31.

IARC 1993; Some Naturally Occurring Substances: Food Items and Constituents, Hetero-
cyclic Aromatic Amines, and Mycotoxins. IARC Monographs on the Evaluation of
Carcinogenic Risk of Chemicals to Humans, International Agency for Research on
Cancer: Lyon, France, 1993; Vol. 56, p 571.

Jadhav, R. S.; Thaker, N. V.; Desai, A. An Iron-Inefficient Variety of Peanut Plant, Grown
Hydroponically with the Catechol Siderophore. *World J. Microbiol. Biotechnol.* **1994,**
10, 360–361.

Jennings, D. H. Some Perspectives on Nitrogen and Phosphorus Metabolism in Fungi. In
Nitrogen, Phosphorus and Sulphur Utilization by Fungi; Boddy, L., Marchant, R., Read,
D. J., Eds.; Cambridge University Press: Cambridge, UK, 1989.

Khan, M. R.; Anwer, M. A. Molecular and Biochemical Characterization of Soil Isolates
of *Aspergillus niger* and Assessment of Antagonism against *Rhizoctonia Solani. Phytopathol. Mediterr.* **2007,** *46,* 304–315.

Khan, M. R.; Anwer, M. A. DNA and Some Laboratory Tests of Nematode Suppressing
Efficient Soil Isolates of *Aspergillus niger. Indian Phytopathol.* **2008,** *61*(2), 212–225.

Khan, M. R.; Anwer, M. A. Fungal Based Bioinoculants for Plant Disease Management.
In *Microbes and Microbial Technology: Agricultural and environmental Applications;*
Pichtel, J., Ahmad, I., Eds.; Springer: USA, 2011: pp. 447–489.

Khan, M. R.; Jairajpuri, M. S. *Nematode Infestations, Part I: Field Crops;* The National
Academy of Sciences: Allahabad, India, 2010; p 325.

Khan, M. R.; Khan, S. M. Biomanagement of *Fusarium* Wilt of Tomato by the Soil Application of Certain Phosphate Solubilizing Microorganism. *Int. J. Pest Manage.* **2001,**
47(3), 227–231.

Khan, M. R.; Khan, S. M. Effects of Root Dip Treatment with Certain Phosphate Solubilizing
Microorganisms on the Fusarial Wilt of Tomato. *Bioresour. Technol.* **2002,** *85,* 213–215.

Khan, M. R.; Altaf , S.; Mohidin, F. A.; Khan, U.; Anwer, A. Biological Control of Plant
Nematodes with Phosphate Solubilizing Microorganisms. In *Phosphate Solubilizing
Microbes for Crop Improvement;* Khan, M. S., Zaidi, A., Eds.; Nova Science Publisher
Inc.: New York, USA, 2009; pp 395–426

Khan, M. R. *Biological Control of Fusarial Wilt and Root-Knot of Legumes;* Department
of Biotechnology, Ministry of Science and Technology, Government of India Publication: New Delhi, India, 2005; p 50.

Khan, M. R. Prospects of Microbial Control of Root-Knot Nematodes Infecting Vegetable
Crops. In *Biotechnology: Plant Health Management;* Sharma, N., Singh, H. B., Eds.;
International Book Distributing Co.: Lucknow, India, 2007; pp 659–690.

Khan, M. W.; Khan, M. R. Fungal-Nematode Interactions. In *Nematode Post Management
an Appraisal of Eco-friendly approaches;* Swarup, G., Dasgupta, D. R., Gill, J. S., Eds.;
Nematological Society of India: New Delhi, India, 1995; pp 70–78.

Kirkpatrick , J. D.; Mai, W. F.; Parker, K. G.; Fisher, E. G. Effects of Phosphorus and
Potassium Nutrition of Sour Cherry on the Soil Population Levels of Five Plant Parasitic
Nematodes. *Phytopahtology.* **1964,** *54,* 706–712.

Kis-Papo, T.; Oren, A.; Grishkan, I.; Wasser, S. P.; Nevo, E. Survival of Filamentous Fungi
in Hypersaline Dead Sea Water. *Microb. Ecol.* **2003,** *45*(2), 183–190.

Knight, W. G.; Allen, M. F.; Jurinak, J. J.; Dudley, L. M. Elevated Carbondioxide and Solution Phosphorus in Soil with Vascular-Arbuscular Mycorrhizal Western Wheatgrass. *Soil
Sci. Am. J.* **1989,** *53,* 1075–1082.

Kumar, R.; Ahmad, S.; Saxena, S. K. Disease Complex in Chickpea Involving *Meloidogyne incognita* and *Fusarium oxysporum. Int. Nematol. Netw. Newsl.* **1998,** *5,* 12–14.

Kusters-van Someren, M. A.; Samson, R. A.; Visser , J. The Use Of RFLP Analysis in Classification of the Black *Aspergilli*: Reinterpretation of *Aspergillus niger* Aggregate. *Curr. Genet.* **1991,** *19,* 21–26.

Lal, L. *Phosphatic Biofertilizers;* Agrotech Publishing Academy: Udaipur, India, 2002; pp 48–50.

Leeman, M.; den Ouden, F. M.; van Pelt, J. A.; Dirkx, F. P. M.; Steijl, H.; Bakker, P. A. H. M.; Schipper, R. Iron Availability Affects Induction of Systemic Resistance to *Fusarium. Phytopathology.* **1996,** *86,* 149–155.

Lewis, J. A.; Papavizas, G. C. Biocontrol of Plant Diseases: The Approach of Tomorrow. *Crop Prot.* **1991,** *10,* 95–105.

Li, S. L.; George, E.; Marschner, H. Phosphorus Depletion and Ph Decrease at the Root-Soil and Hyphae-Soil Interfaces of VA Mycorrhizal White Clover Fertilized with Ammonium. *New Phytol.* **1991,** *119,* 397–404.

Lodhi, M. K. Biological Control of Different Soil Borne Fungal Diseases of Potato (*Solanum tuberosum* L.) Raised Through Tissue Culture by Using Vesicular Arbuscular Mycorrhiza and Other Antagonistic Fungi. Ph.D. Thesis, Department of Botany, University of Punjab, 2004, p 240.

Mach, R. L.; Zeilinger, S. Regulation of Gene Expression in Industrial Fungi: *Trichoderma. Appl. Microbiol. Biotechnol.* **2003,** *60,* 515–522.

Majumdar, S.; Sen, B. Kalisena, In *A Novel Biopesticide for Disease Free Crop and Better Yield,* Proceedings of National Symposium on Development of Microbial Pesticides and Insect Pest Management, Pune, India, November 12–13, 1998; BARC, Mumbai and Hindustan Antibiotics Ltd.: Pimpri, India, 1998.

Mankau, R. Toxicity of Cultures of *Aspergillus niger* to Mycophagous Nematode *Aphelenchus avenae. Phytopathology.* **1969 a,** *59,* 13.

Mankau, R. Nematicidal Activity of *Aspergillus niger* Culture Filtrates. *Phytopathology.* **1969b,** *59,* 1170.

Medina, A.; Jakobsen, I.; Vasiilev, N.; Azcon, K.; Larsen, J. Fermentation of Sugar Beet Waste by *Aspergillus niger* Facilitate Growth and P Uptake of External Mycelium of Mixed Populations of Arbuscular Mycorrhizal Fungi. *Soil Biol. Biochem.* **2007,** *39,* 485–492.

Misra, A. K. Present Status of Important Diseases of Guava in India with Special Reference to Wilt. In *International Guava Symposium;* Singh, G., Kishun, R., Chandra, R., Eds.; ISHS: Lucknow, India, 2007; p 735.

Mittal, V.; Singh, O.; Nayyar, H.; Kaur, J.; Tewari, R. Stimulatory Effect of Phosphate – Solubilizing Fungal Strains (*Aspergillus awamori* and *Penicillium citrinum*) on the Yield of Chickpea (*Cicer arietinum* L. cv. GPF2). *Soil Biol. Biochem.* **2008,** *40,* 718–727.

Molina, G. C.; Davide, R. G. Evaluation of Microbial Extracts for Nematicidal Activity against Plant Parasitic Nematodes *Meloidogyne incognita* and *Radopholus similis. Philipp. Agric.* **1986,** *69,* 173–186.

Mondal, G.; Dureja, P.; Sen. B. Fungal Metabolites from *Aspergillus niger* AN27 Related to Plant Growth Promotion. *Indian J. Exp. Biol.* **2000,** *38*(1), 84–87.

Mondal, G. 1998. *In Vitro* Evaluation of *Aspergillus niger* AN 27 against Soil Borne Fungal Plant Pathogens and Field Testing against *Macrophomina Phaseolina* on Potato, Ph.D. Thesis, I.A.R.I., New Delhi, India, p 117.

Monte, E. Understanding *Trichoderma*: Between Biotechnology and Microbial Ecology. *Int. Microbiol.* **2001**, *4*, 1–4.

Morean, R. The Biological Liberation of Soluble Phosphates from Insoluble Phosphorus Compounds in Soil. *C. R. Acad. Sci.* **1959**, *249*, 1804–1806.

Mostafa, M. A,; Yovssef, Y. A. Studies on Peritrophie Mycorrhiza between *Tropaeolum majus* L. and Two Associated Rhizospherie Fungi. III. Effect of Fungal Metabolites on Relative Growth Vigour of *Tropaeoliem mains* L. *Egypt J. Bot.* **1962**, *111*, 137–151.

Muhammad, S.; Amusa, N. A. *In Vitro* Inhibition of Growth of Some Seedling Blight Inducing Pathogens by Compost-Inhabiting Microbes. *Afr. J. Biotechnol.* **2003**, *2* (6), 161–164.

Mullenborn, C.; Steiner, U.; Ludwig, M.; Oerke, E. C. Effect of Fungicides on the Complex of Fusarium Species and Saprophytic Fungi Colonizing Wheat Kernels. *Eur. J. Plant Pathol.* **2008**, *120*, 157–166.

Nair, M. G.; Burke, B. A. A Few Fatty Acid Methyl Ester and Other Biologically Active Compounds from *Aspergillus niger*. *Phytochemistry.* **1988**, *27*(10), 3169–3173.

Nazar, R. N.; Hu, X.; Schmidt, J.; Culham, D.; Robb, J. Potential Use of PCR Amplified Ribosomal Intergenetic Sequences in the Detection and Differentiation of *Berticillium* Wilt Pathogen. *Physiol. Mol. Plant Pathol.* **1991**, *39*, 1–11.

Nielsen, K. F.; Mogensen, J. M.; Johansen, M.; Larsen, T. O.; Frisvad, J. C. Review of Secondary Metabolites and Mycotoxins from the *Aspergillus niger* Group. *Anal. Bioanal. Chem.* **2009**, *395*, 1225–1242.

NTP1989; Toxicology and Carcinogenesis Studies of Ochratoxin A (CAS No. 303-47-9) in F344/N Rats (Gavage Studies), Technical Report Series No 358. NTIS Publication No. PB90-219478/AS, National Toxicology Program. Research Triangle Park, Bethesda: NC and MD, 1989; p 142.

O' Sullivan, D. J.; O'Gara, F. Traits of Fluorescent *Pseudomonas* spp. Involved in Suppression of Plant Root Pathogens. *Microbiol. Rev.* **1992**, *56*, 662–676.

Osiewacz, H. D. *Molecular Biology of Fungal Development*; Marcel Dekker: New York, NY, 2002.

Palakshappa, M. G.; Kulkarni, S.; Hedge, R. K. Effect of Organic Amendments on the Survivals Ability of *Sclerotium rolfsii*, a Causal Agent of Foot-Rot of Betelvine. *Mysore J. Agric. Sci.* **1989**, *23*(3), 332–336.

Pandya, U.; Saraf, M. Application of Fungi as a Biocontrol Agent and Their Biofertilizer Potential in Agriculture. *J. Adv. Dev. Res.* **2010**, *1*(1), 90–99.

Pant, H.; Pandey, G. Efficacy of Biocontrol Agents for the Management of Root-Knot Nematode on Chickpea. *Ann. Plant Protect. Sci.* **2001**, *9*, 117–170.

Papavizas, G. C. Biological Control of Soil Borne Diseases. *Summa. Phytopathol.* **1985**, *11*, 173–179.

Parenicova, L.; Benen, J. A. E.; Samson, R. A.; Visser, J. Evaluation of RFLP Analysis of the Classification of Selected Black Aspergilli. *Mycol. Res.* **1997**, *101*, 810–814.

Parenicova, L.; Skouboe, P.; Frisvad, J.; Samson, R. A.; Rossen, L.; ten Hoor-Suykerbuyk, M.; Visser, J. Combined Molecular and Biochemical Approach Identifies *Aspergillus japonicus* and *Aspergillus aculeatus* as Two Species. *Appl. Envir. Microbiol.* **2001**, *67*, 521–527.

Park, J. C.; Nemoto, Y.; Homma, T.; Sato, R.; Matsuoka, H.; Ohno, H.; Takatori, K.; Kurata, H. Adaptation of *Aspergillus niger* to Several Antifungal Agents. *Microbiology.* **1994**, *40*, 2409–2414.

Perrone, G.; Susca, A.; Cozzi, G. K.; Ehrlich, K.; Varga, J.; Frisvad, J. C.; Meijer, M.; Noonim, P.; Mahakarnchanakul, W.; Samson, R. A. Biodiversity Of *Aspergillus* Species in Some Important Agricultural Products. *Stud. Mycol.* **2007,** *59,* 53–66.

Persson, Y.; Veehuis, M.; Nordbring-Hertz, B. Morphogenesis and Significance of Hypal Coiling by Nematode-Trapping Fungi in Mycoperasitic Relationships. FEMS *Microbiol. Ecol.* **1985,** *31,* 283–91.

Radhakrishna, P.; Sen, B. Efficacy of Different Methods of Inoculation of *Fusarium oxysporum* and *F. Solani* for Inducing Wilt in Muskmelon. *Indian Phytopathol.* **1986,** *38,* 70–73.

Rahe, J. E.; Utkhede, R. S. Integrated Biological and Chemical Control of Sclerotial Pathogens. In *Ecology and Management of Soilborne Plant Pathogens*; Parker, C. A., Rovira, A. D., Moore, K. J., Wong, P. T. W., Kollmorgen, J. F., Eds.; American Phytopathological Society: St Paul, US, 1985; pp 124–126.

Raijmakers, J. M.; der Sluis, I. V.; Koster, M.; Bakker, P. A. H. M.; Weisbeek, P. J.; Schippers, B. Utilization of Heterologous Siderophores and Rhizosphere Competence of Fluorescent *Pseudomonas* spp. *Can. J. Microbiol.* **1995,** *41,* 126–135.

Ramachandran, S.; Fontanille, P.; Pandey, A.; Larroche, C. Stability of Glucose Oxidase Activity of *Aspergillus niger* Spores Produced by Solid-State Fermentation and Their Role as Biocatalysts in Bioconversion Reaction. *Food Technol. Biotechnol.* **2008,** *46*(2), 190–194.

Rao, A. B. Role of Microorganisms in Plant Nutrition Under Acid Conditions. In *Biofertilizers;* Vyas, L. L., Ed.; Scientific Publication: Jaudpur, Rajasthan, India, 1990; pp 67–84.

Raper, K.; Fennell, D. *The genus Aspergillus;* Williams & Wilkins Company: Baltimore, MD, 1965.

Rekha, A.; Saxena, S. K. Influence of Certain Rhizosphere Fungi Together with *Rhizoctonia solani* and *Meloidogyne incognita* on Germination of 'Pusa Ruby' Tomato Seeds. *Indian Phytopathol.* **1999,** *52*(2), 121–126.

Sant, D.; Casanova, E.; Segarra, G.; Aviles, M.; Reis, M.; Trillas, M. I. Effect of *Trichoderma asperellum* Strain T34 on Fusarium Wilt and Water Usage in Carnation Grown on Compost-Based Growth Medium. *Biol. Control.* **2010,** *53*(3), 291–296.

Scher, F. M.; Baker, R. Effect of *Pseudomonas putida* and a Synthetic Iron Chelator on Induction of Soil Suppressiveness of *Fusarium* Wilt Pathogens. *Phytopathology.* **1982,** *72,* 1567–1573.

Sehgal, M.; Jeswani, M. D.; Kalra, N. Management of Insect, Disease and Nematode Pests of Rice-Wheat in the Indo-Gangetic Plains. *J. Crop Prod.* **2001,** *4*(1), 167–226.

Sen, B.; Kumar, S.; Majunder, S. Fifty Years towards Biological Control of Fusarial Wilts in India. In *Plant Pathology-Fifty Years of Research in India;* Verma, J. P., Verma, A., Eds.; Malhotra Publishing House: New Delhi, India, 1997; pp 675–688.

Sen, B.; Kumar, S.; Majumdar, S.; Mondal, G.; Angappan, K.; Mukherjee, K.; Chattopadhyay, L.; Das, I. K. In *Kalisena, a Novel Biopesticide for Disease Free Crop and Better Yield,* Proceedings of National on Development of Microbial Pesticides and Insect Post Management, Nove12–13, 1998; BARC, Mumbai and Hindustan Antibiotics Ltd.: Pune, India, 1998.

Sen, B.; Mukherjee, K.; Chattopadhyay, C.; Patibanda, A. K.; Sharma, J. In *Aspergillus niger, a Potential Biocontrol Agent for Soil-Borne Plant Pathogens,* Proceeding of Global Conference on Advances in Research on Plant Conference and their Management.

Society of Mycology and Plant Pathology, Rajasthan Agriculture Universoty: Udaipur, India, 1995; pp 161, XI 29.

Sen, B. *Consultation on the Application of Biotechnology in Plant Pest Management;* (Feb 25–28, 1997), IARI, FAO, RAP Publication 40: New Delhi, India, 1997; 276–278.

Sen, B. Biocontrol: A Success Story. *Indian Phytopathol.* **2000,** *53*(3), 243–249.

Shah, N. H.; Khan, M. I.; Azam, M. F. Studies on the Individual and Concomitant Effect of *Aspergillus niger, Rhizoctonia solonia*, and *M. javanica* on Plant Growth and Nematode Reproduction on Chilli (*Capsicum annuum* L). *Ann. plant Prot. Sci.* **1994.** *1*(2), 75–78.

Sharma, H. K.; Prasad, D.; Sharma, P. *Compatibility of Fungal Bioagents as Seed Dressers with Carbofuran in Okra against Meloidogyne Incognita: National Symposium On Recent Advances and Research Priorities in Indian Nematology;* (Dece 9–11, 2005), IARI: New Delhi, India, 2005; p 72.

Sharma, J.; Sen, B. Antagonistic Potentials of Soil and Dung Isolates of *Aspergillus* spp. against *Fusarium solani* (Mart) Sacc. Causing Wilt of Curbits. *Indian J. Microbiol. Geol.* **1991a,** *2,* 91–97.

Sharma, J.; Sen, B. Interaction of Soil Microflora with Cucurbit with Pathogen, *Fusarium solani. Indian Phytopathol.* **1991b,** *44,* 94–96.

Siddiqui, Z. A.; Kazuyoshi, F. Biocontrol of *Meloidogyne incognita* on Tomato Using Antagonistic Fungi, Plant-Growth-Promoting Rhizobacteria and Cattle Manure. *Pest Manag. Sci.* **2009,** *65*(9), 943–948.

Singh, R.; Singh, B. K.; Upadhyay, R. S.; Rai, B.; Lee, Y. S. Biological Control of *Fusarium* Wilt Disease of Pigeonpea. *Plant Pathol. J.* **2002,** *18*(5), 279–283.

Singh, S.; Mathur, N. *In Vitro* Studies of Antagonistic Fungi against the Root-Knot Nematode, *Meloidogyne incognita. Biocon. Sci. Technol.* **2010,** *20*(3), 275–283.

Singh, S. M.; Azam, M. F.; Khan, A. M.; Saxena, S. K. Effect of *Aspergillus niger* and *Rhizoctonia solani* on Development of *Meloidogyne incognita* on Tomato. *Curr. Nematol.* **1991,** *2,* 163–166.

Sperber, J. I. Solution of Mineral Phosphates by Soil Bacteria. *Nature.* **1958,** *180,* 994–995.

Stirling, G. R. Bioconrol of Plantpathogenic Nematode and Fungus. *Phytopathology.* **1993,** *83,* 1525–1532.

Suarez, B.; Rey, M.; Castillo, P.; Monte, E.; Llobell, A. Isolation and Characterization of PRA1, Atrypsin Like Protease from the Biocontrol Agent *Trichoderma harzianum* CECT 2413 Displaying Nematicidal Activity. *Appl. Microbiol. Biotechnol.* **2004,** *65,* 46–55.

Sutton, D. A.; Fothergill, A. W.; Rinaldi, M. G. *Guide to Clinically Significant Fungi, 1st Ed.;* Williams & Wilkins: Baltimore, MD, 1998.

Upadhyay, R. S.; Rai, B. Fungistatic Activity of Different Indian Soils against *Fusarium udum* Butler. *Plant Soil.* **1981,** *63,* 407–415.

Vaddar, U. B.; Patil, A. B. Studies on Grape Rhizosphere Microorganisms. *Karnataka J. Agr. Sci.* **2007,** *20* (4), 932.

van Tuyl, J. M. *Genetics of Fungal Resistance to Systemic Fungicides;* Wageningen University: WAU dissertation no. 679, Wageningen, Netherland, 1977.

Varga, J.; Kevei, F.; Harari, Z.; Toth, B.; Teren, J.; Croft, J. H.; Kozakiewicz, Z. Genotypic and Phenotypic Variability among Black Aspergilli. In *Integration of Modern Taxonomic Methods for Penicillium and Aspergillus Classification;* Samson, R. A., Pitt, J. I., Eds.; Harwood: Amsterdam, Netherland, 2000; pp 397–411.

Varga, J.; Kevei, F.; Vriesema, A.; Debets, F.; Kozakiewicz, Z.; Croft, J. H. Mitochondrial DNA Restriction Fragment Length Polymorphisms in Field Isolates of the *Aspergillus niger* Aggregate. *Can. J. Microbiol.* **1994,** *40,* 612–621.

Vassilev, N.; Franco, I.; Vessileva, M.; Azcon, R. Improved Plant Growth with Rock Phosphate Solubilized by *Aspergillus niger* Grown on Sugarbeet Waste. *Bioresource Technol.* **1996,** *55,* 237–241.

Vassilev, N.; Vassileva, M.; Nikolaeva, I. Simultaneous P Solubilizing and Bio-Control Activity of Microorganisms: Potentials and Future Trends. *Appl. Microbiol. Biotechnol.* **2006,** *71,* 137–144.

Vey, A.; Hoagland, R. E.; Butt, T. M. Toxic Metabolites of Fungal Biocontrol Agents. In *Fungi as Biocontrol Agents: Progress, Problems and Potential;* Butt, T. M., Jackson, C., Magan, N., Eds.; CAB International: Bristol, Eangland, 2001; pp 311–346.

Vigalys, R.; Heter, M. Rapid Genetic Identification and Mapping of Enzymatically Amplified Ribosomal DNA from *Cryptococcus* Species. *J. Bacteriol.* **1990,** *172,* 4238–4246.

White, T. J.; Bruns, T.; Lee, S.; Taylor, J. Amplification and Direct Sequencing of Fungi Ribosomal RNA Gene for Phylogenetics. In *PCR Protocols – A Guide to Methods and Applications;* Innis, M. A., Gelfad, D. H., Suinsky, J. J., White, T. J., Eds.; Academic Press: San Diego, CA, 1990; pp 315–322.

Wicklow, D. T.; Hesseltine, C. W.; Shotwell, O. L.; Adams, G. L. Interference, Competition and Aflatoxin Levels in Corn. *Phytopathology.* **1980,** *70,* 761–764.

Youssef, Y. A.; Mankarios, A. T. Production of Plant Growth Substances by Rhizosphere Mycoflora of Broad Bean and Cotton. *Biologia. Plantarum.* (PRAHA). **1975,** *17*(3), 175–181.

Yuh-Shan, H. Comment on 'Biosorption of Cadmium Using the Fungus *Aspergillus niger*' by Barros, L. M., Macedo, G. R., Duarte, M. M. L., Silva, E. P. and Lobato, A. K. C. L. *Braz. J. Chem. Eng.* **2005,** *22*(2), 319–322.

Zepnik, H.; Pahler, A.; Schauer, U.; Dekant, W. Ochratoxin A Induced Tumour Formation: Is there a Role of Reactive Ochratoxin A Metabolites? *Toxicol. Sci.* **2001,** *59,* 59–67.

Zhang, W.; Mullaney, E. J.; Lei, X. G. Adopting Selected Hydrogen Bonding and Ionic Interactions from *Aspergillus fumigatus* Phytase Structure Improves the Thermostability of *Aspergillus niger* PhyA Phytase. *Appl. Env. Microbiol.* **2007,** *73,* 3069–3076.

CHAPTER 4

FLOURESCENT PSEUDOMONADS: A POTENTIAL BIO-CONTROL AGENTS AGAINST PLANT DISEASES

KAHKASHAN ARZOO[1], ERAYYA[2*], and NISHANT PRAKASH[3]

[1]Department of Plant Pathology, GBPUA&T, Pantnagar 263145, Uttarakhand, India

[2]Department of Plant Pathology, Bihar Agricultural University, Sabour 813210, Bhagalpur, Bihar, India

[3]Department of Plant Pathology, Krishi Vigyan Kendra, Arwal, BAU, Sabour 813210, Bhagalpur, Bihar, India

*Corresponding author. E-mail: erayyapath@gmail.com

CONTENTS

ABSTRACT

The crop production and protection techniques which we all follow today have caused considerable damage to our soil and water. So, microbiologists and plant pathologists are facing an urgent challenge of the development of environment friendly alternatives for chemical pesticides for combating crop diseases. Biological control of plant diseases involving use of antagonistic microorganisms offers an excellent alternative to pest management by chemicals. It can be achieved by either promoting the native antagonists to reach a density sufficient to suppress a pathogen(s) or introducing alien antagonists. An important group of the microbial communities, existing in rhizosphere, which exert beneficial effects on plant growth upon root colonization, are termed as plant growth-promoting rhizobacteria (PGPR). These PGPR have been considered important in sustainable agriculture because of their biocontrol potentials and plant growth promotional activities. Many strains of *Pseudomonas fluorescens* and related fluorescent *Pseudomonas* spp. show potential for biological control of phytopathogens especially root pathogens. The fluorescent pseudomonads include all *Pseudomonas* sp. with the ability to produce fluorescent pyoverdine siderophore (also known as pseudobactin). The modes of action that play a role in disease suppression by these bacteria include: siderophore mediated competition for iron, antibiosis, production of lytic enzymes, and induced systemic resistance (ISR). Superior root colonization and effective functioning in the rhizosphere are the key criteria when selecting strains and researches aim at better understanding the molecular basis of these traits and the signaling process regulating the ecology of these bioagents *in situ*. Current genomic analysis for rhizosphere competence and biocontrol traits will likely lead to the development of novel tools for effective management of indigenous and inoculated fluorescent pseudomonads biocontrol agents and a better exploitation of their plant-beneficial trait for sustainable agriculture.

4.1 INTRODUCTION

Environmental concerns have led to the need of sustainable use of natural resources. The crop production and protection techniques which we all follow today have caused considerable damage to our soil and water. It has resulted in an urgent need to change our present crop-management strategies to more environmental cleaner techniques. One of the biggest ecological challenges confronting microbiologists and plant pathologists in the near future is the development of environment friendly alternatives for chemical pesticides for combating crop diseases. Biological control of plant diseases involving use of antagonistic microorganisms offers an excellent alternative to pest management by chemicals. Biological control as a strategy and philosophy for reduction of crop loss from plant disease is not new to agriculture (Cook et al., 1982). Even at earliest farmers practiced biological control by rotating their crops, burry disease infested crop residues and fertilizing with organic manures. These and many other traditional practices gave effective disease control by allowing time and opportunity for biological destruction of disease organisms. The challenge today is how to achieve effective biological control of pests in modern agriculture where farmers must specialize with one or two crops, use minimal or no tillage to save fuel and stop soil erosion and where organic manures are unavailable, inconvenient, or too expensive.

Biological control of plant pathogens can be achieved by either promoting the native antagonists to reach a density sufficient to suppress a pathogen(s) or introducing alien antagonists. Though some of the earlier works, related to promoting the native antagonists was by using organic amendments, the recent trend is to isolate, multiply, and introduce the antagonists to soil or specific court of infection to achieve a successful biological suppression of a disease. One time skepticism that biological control by introduced antagonist may not work stands invalid in the present day as evidenced by enormous literature in favor of successful biological control. Research has reached a state where we have commercial formulations of biological control agents (bioagents) sold in the market of several countries. There remains no doubt that biological control has come to stay and in near future it will be an integral part of disease management strategies for many crops. Bioagents are microorganisms that adversely affect the population of other microorganisms (e.g., target pathogen) growing in association with them. Generally bioagents have potential to interfere in the life process of plant pathogens. At present, use of beneficial microorganisms (biopesticides) is considered one of the most promising methods for rational and ecologically safe crop-management practices. A vast number of microorganisms present in rhizosphere have been considered important in sustainable agriculture because of their biocontrol potentials and plant growth promotional activities. Bioagents includes virtually all classes of organisms, for example, fungi, bacteria, nematodes, protozoa, viruses, and seed plants. The biocontrol of plant pathogens can be achieved by either promoting natural biocontrol agents to reach density enough to control a pathogen or introduced biocontrol agents.

The rhizosphere, volume of soil surrounding roots and influenced chemically, physically, and biologically by the plant root, is a highly favorable habitat for the proliferation of microorganisms and exerts a potential impact on plant health and soil fertility. Many microorganisms are attracted by nutrients exuded from plant roots and this "rhizosphere effect" was first described by Hiltner (1904). Root exudates rich in amino acids, monosaccharides, and organic acids, serve as the primary source of nutrients, and support the dynamic growth and activities of various microorganisms within the vicinity of the roots (Lugtenberg et al., 2001; Nelson, 2004). An important group of the microbial communities, existing in rhizosphere, which exert beneficial effects on plant growth upon root colonization, was first defined by Joseph Kloepperand Milton Schroth and termed as plant growth-promoting rhizobacteria (PGPR) (Kloepper & Schroth, 1978).

There are four considerations that have to be taken in order to classify a microbe as a plant growth promoter. First, these microbes cannot be endophytic like mycorrhizal fungi and *Rhizobium*. Second, after inoculation, they should exhibit a higher density population in the rhizosphere; this is important because a population that declines quickly will not be able to compete with native micro flora present in soil. Third, these microbes have to exhibit the capability of colonizing effectively the root surface and influence positively the plant growth. Finally, they should not cause harm to humans or other beneficial microbes (Mercado, 1997; Jimenez et al., 2001).

PGPR can be classified according to their beneficial effects. They can have an impact on plant growth and development in two different ways: indirectly or directly. The indirect promotion of plant growth occurs when these bacteria decrease or prevent some of the deleterious effects of a phytopathogenic organism by any one or more of several different mechanisms like, production of certain metabolites such as siderophores, antibiotics and enzymes, such as glucanase or chitinase that can function as antagonistic products with the capability of inhibiting or terminating the growth of pathogenic microbes (Jimenez et al., 2001). On the other hand, the direct promotion of plant growth by PGPR generally entails providing the plant with a compound that is synthesized by the bacterium or facilitating the uptake of nutrients from the environment (Glick et al., 1999). For instance, biofertilizers can fix nitrogen, which can subsequently be used by the plant, thereby improving plant growth when the amount of nitrogen in the soil is limiting. Phytostimulators can directly promote the growth of plants, usually by the production of phytohormones (such as auxin) and volatile growth stimulants (such as ethylene and 2, 3-butanediol).

In PGPR, which are sometimes called biopesticides; the biocontrol aspect is most conspicuous. Biocontrol PGPR are able to protect plants from infection by phytopathogenic organisms (Bloemberg & Lugtenberg, 2001; Vessey, 2003; Haas & Defago, 2005). These PGPR, which mostly belong to *Pseudomonas* and *Bacillus* spp., are antagonists of recognized root pathogens (Haas & Defago, 2005). This chapter is primarily focused on the biocontrol potential of fluorescent pseudomonads.

4.2 FLUORESCENT PSEUDOMONADS

The genus *Pseudomonas* covers one of the most diverse and ecologically significant groups of bacteria. Members of the genus are found in large numbers in a wide range of environmental niches, such as terrestrial and marine environments, as well as in association with plants and animals. This genus is heterogeneous and harbors plant, animal, and human pathogenic species. This almost universal distribution of *Pseudomonas* suggests a remarkable degree of genomic diversity and genetic adaptability. *Pseudomonas* spp. are often used as model root-colonizing bacteria (Lugtenberg et al., 2001).

Members of the genus *Pseudomonas* are rod-shaped Gram-negative bacteria that are characterized by metabolic versatility, aerobic respiration (some strains also have anaerobic respiration with nitrate as the terminal electron acceptor and/or arginine fermentation), motility owing to one or several polar flagella, and a high genomic G + C content (59–68%). The term pseudomonads (*Pseudomonas* like bacteria) is often used to describe the strains for which taxonomic affiliation has not been established in detail.

Many strains of *Pseudomonas fluorescens* show potential for biological control of phytopathogens especially root pathogens. In taxonomic terms, several of them are indeed *P. fluorescens* sensu stricto, while others belong in fact to neighboring species of the "*P. fluorescens*" complex or to ill-defined related species within the fluorescent *Pseudomonas* spp. The fluorescent pseudomonads include all *Pseudomonas* sp. with the ability to produce fluorescent pyoverdine siderophore(s) (also known as pseudobactin). This large and heterogeneous group comprises, most notably, *P. aeruginosa*, *P. putida*, *P. fluorescens*, and *P. syringae* (Bossis et al., 2000; Haas & Defago, 2005). These bacteria have become prominent models for rhizosphere ecological studies and analysis of bacterial secondary metabolism. In recent years, knowledge on their plant-beneficial traits has been considerably enhanced by widening the focus beyond the case of phytopathogen directed antagonism (Couillerot et al., 2009).

4.3   MODE OF ACTION OF FLUORESCENT PSEUDOMONADS IN BIOLOGICAL CONTROL

The modes of action that play a role in disease suppression by these bacteria include: siderophore mediated competition for iron, antibiosis, production of lytic enzymes, and induced systemic resistance (ISR).

4.3.1   SIDEROPHORE MEDIATED COMPETITION FOR IRON

Although siderophores are part of primary metabolism (because iron is an essential element), on occasion they also behave as antibiotics (which are commonly considered to be secondary metabolites). Siderophores are low-molecular-weight molecules that are secreted by microorganisms to take up iron from the environment and their modes of action in suppression of disease were thought to be solely based on competition for iron with the pathogen (Visca, 2004).

   Competition for iron between pathogens and siderophores of fluorescent pseudomonads has been implicated in the biocontrol of wilt diseases caused by *Fusarium oxysporum* (Kloepper et al., 1980; Scher & Baker, 1982), damping-off cotton caused by *Pythium ultimum* (Loper, 1988), and *Pythium* root rot of wheat (Baker & Cook, 1988). The role of siderophores produced by fluorescent pseudomonads in plant growth promotion was first reported by Kloepper et al. (1981). Suryakala et al. (2004) have reported that siderophores exerted maximum impact on *F. oxysporum* than on *Alternaria* sp. and *Colletotrichum capsici*. The common siderophores produced by fluorescent pseudomonads are pyoverdin (Pvd) or pseudobactin and pyochelin.

4.3.1.1   SIDEROPHORES PRODUCED BY BIOCONTROL FLUORESCENT PSEUDOMONADS

a.   Pyoverdin

Fluorescent *Pseudomonas* spp. are characterized by the production of yellow-green pigments termed Pvds which fluoresce under UV light and function as siderophores (Demange et al., 1987). This extracellular diffusible pigment called Pvd or pseudobactin has high affinity for Fe^{3+} ions. Ferripyoverdin (Pvd complexed with Fe^{3+}) interacts with a specific outer-membrane receptor, which is present in the producer but might also occur in some non-producers. Subsequently, Fe^{3+} is transported into the cytoplasm and reduced to Fe^{2+} (Visca, 2004).

In iron depleted media *in vitro*, Pvd-producing *Pseudomonas* spp. inhibit the growth of bacteria and fungi with less potent siderophores (Kloepper et al., 1980). On low-iron agar plates, Pvd that is deposited on a filter disc can produce a halo of inhibition on a susceptible microorganism. Therefore, under certain conditions, Pvd functions as a diffusible, bacteriostatic, or fungistatic antibiotic, whereas ferripyoverdin does not (Scher & Baker, 1982).

b.   Pyochelin

Another pseudomonad siderophore, pyochelin, has been identified as an antifungal antibiotic in a screening program (Phoebe et al., 2001). However, it has not yet been investigated whether iron deprivation is the antibiotic mechanism that is involved. As pyochelin is a relatively weak Fe^{3+}chelator, but a good Cu^{2+} and Zn^{2+}chelator (Cuppels et al., 1987; Visca et al., 1992), it may be able to deprive some fungi of copper and/or zinc. This example shows that the distinction between siderophores and typical antibiotics is blurred.

4.3.2   ANTIBIOSIS

PGPR can be utilized in low-input sustainable agricultural applications, such as biocontrol, on account of their ability to synthesize secondary metabolites with antibiotic properties (Franks et al., 2006). An intuitive, simple explanation of how the biological control of soil-borne pathogens could work was discussed at the 1963 international symposium entitled "Ecology of soil-borne plant pathogens - prelude to biological control" (Haas & Defago, 2005). The idea was that antagonistic microorganisms could compete with pathogens, particularly by producing antibiotic compounds. In soil, these antibiotics could interfere with pathogen development, for example, during spore germination and the onset of root infection. This symposium shaped early research on the biocontrol mechanisms that operate in the rhizosphere. This explanation was appealing, as many soil microorganisms (e.g., *Streptomyces*, *Bacillus*, and *Pseudomonas* spp.) were known to be excellent antibiotic producers *in vitro*.

Comprehensive lists of antibiotics that are involved in biocontrol, producer strains, target pathogens, and host plants have been compiled by Raaijmakers et al. (2002) and Morrissey et al. (2004). There are six classes of antibiotic compounds for which the experimental evidence most clearly supports a function in the biocontrol of root diseases: phenazines (Phzs), phloroglucinols (2,4-diacetylphloroglucinol (DAPG)), pyoluteorin (PLT), pyrrolnitrin (PRN), cyclic lipopeptides (all of which are diffusible like oomycinA, viscosinamide), and hydrogen cyanide (HCN; which is volatile).

4.3.2.1 ANTIBIOTICS MADE BY FLUORESCENT PSEUDOMONADS

a. Phzs

Phz is a low-molecular-weight secondary metabolite, nitrogen containing heterocyclic antimicrobial compound consisting of brightly colored pigment produced by the bacterial genera pertaining to *Pseudomonas*, *Burkholderia*, *Brevibacterium*, and *Streptomyces*. More than 50 naturally occurring Phz compounds have been described. A few strains of PGPR produce 10 different Phz derivatives at the same time. Commonly identified derivatives of Phz produced by *Pseudomonas* spp. are pyocyanin, PCA, PCN, and hydroxyphenazines (Turner & Messenger, 1986; Tambong & Hofte, 2001).

The Phzs, which are analogues of flavin coenzymes, inhibit electron transport and are known to have various pharmacological effects on animal cells. In the presence of ferripyochelin, Phzs catalyze the formation of hydroxyl radicals, which damage lipids and other macromolecules (Britigan et al., 1992). Interestingly, reduced Phz-1-carboxamide can release soluble Fe^{2+} ions from insoluble Fe^{3+} $(OH^-)_3$ at neutral pH, which raises the possibility that Phzs might contribute to iron mobilization in soils (Hernandez et al., 2004).

b. DAPG (PHL)

The polyketide antibiotic DAPG is a phenolic molecule synthesized by the condensation of three molecules of acetyl coenzyme A with one molecule of malonyl coenzyme A to produce the precursor monoacetylphloroglucinol, which is subsequently trans-acetylated to generate PHL utilizing a CHS-type enzyme (Shanahan et al., 1992). Biosynthetic locus of DAPG is highly conserved. It comprises the biosynthetic genes phlACBD (Keel et al., 2000).

Among the secondary metabolites, DAPG has received the particular attention because of its production by a wide range of pseudomonads used for the biological control. The implication of Phl in biocontrol was evidenced for a few strains in plant experiments where wild-type pseudomonads protected better than mutant derivatives in which Phl production was inactivated (Keel et al., 1992; Duffy et al., 2004). In addition, acquisition of the ability to produce Phl conferred biocontrol potential to Phl-pseudomonads (Vincent et al., 1991; Fenton et al., 1992; Baker et al., 2007).

DAPG has been found to be involved in protection of plants by a number of root diseases like take-all of wheat caused by *Gaeumannomyces graminis* var. *tritici* (Keel et al., 1992), *Fusarium* wilt of tomato (Tamietti et al., 1993), black root of tobacco caused by *Thielaviopsis basicola* (Stutz et al., 1986) and damping-off of sugar beet caused by *P. ultimum* (Fenton et al., 1992; Shanahan et al., 1992). Phl causes membrane damage to *Pythium* spp. and is particularly inhibitory to zoospores of this *Oomycete* (de Souza et al., 2003). At high concentrations, Phl is phytotoxic (Keel et al., 1992). In addition to its antifungal activity, DAPG has some antiviral properties and also inhibits the growth of soft-rotting bacteria and cyst nematodes of potato (Cronin et al., 1997). Phl can also induce a plant systemic response resulting in resistance to pathogens (Iavicoli et al., 2003).

c. PRN

PRN is a chlorinated phenylpyrrole antibiotic produced by several fluorescent and non-fluorescent pseudomonads. It was first isolated from *Burkholderia pyrrocinia* (Arima et al., 1964). Pseudomonad species such as *P. fluorescens*, *P. chlororaphis*, *P. aureofaciens*, *B. cepacia*, *Enterobacter agglomerans*, *Myxococcus fulvus*, and *Serratia* sp. also produce PRN antibiotics (Hammer et al., 1999). PRN has been described as an inhibitor of fungal respiratory chains (Tripathi et al., 1969). It also has been used as an antimycotic topical antibiotic in human medicine and its synthetic analogues, a phenyl pyrrol derivative, have been developed for use as agricultural fungicides (Ligon et al., 2000). PRN persists actively in the soil for at least 30 days.

d. PLT

PLT is an aromatic polyketide antibiotic consisting of a resorcinol ring derived through polyketide biosynthesis. PLT is produced from several *Pseudomonas* spp. including strains that suppress plant diseases caused by phytopathogenic fungi (Maurhofer et al., 1994). PLT mainly inhibits the oomycetous fungi including *P. ultimum* against which it is strongly active when applied to seeds. PLT-producing pseudomonads decrease the severity of *Pythium* damping-off (Nowak-Thompson et al., 1999).

e. Cyclic lipopeptides

Cyclic lipopeptides, which include biocontrol-active substances as well as toxins of phytopathogenic pseudomonads, have surfactant properties, and are able to insert into membranes and perturb their function, which results in broad antibacterial and antifungal activities. Some lipopeptides of *Bacillus* spp. chelate cations (e.g., Ca^{2+}), but this property has not been investigated for lipopeptides from *Pseudomonas* spp. (Nybroe & Sørensen, 2004).

f. Aldehydes, alcohols, ketones, and sulfides

P. chlororaphis (PA23) isolated from soybean roots produced antifungal volatiles belonging to aldehydes, alcohols, ketones, and sulfides. It was inhibitory to all the stages of *Sclerotinia sclerotiorum*. Effective antifungal volatiles were benzothiazole, cyclohexanol, n-decanal, dimethyl trisulfide, 2-ethyl 1-hexanol, and nonanal. These substances completely inhibited the growth of mycelium, germination of ascospores and the survival of sclerotia. These volatiles would come in direct contact with the overwintering structures and destruct the sclerotial bodies leading to the reduction in inoculum potential and thereby prevents the disease occurrence (Fernando et al., 2004). Bacterial volatiles also promote growth of plants. The 2, 3-butadienol enhanced the growth of *Arabidopsis thaliana* and inhibited the pathogen *Erwinia carotovora* (Ryu et al., 2003). Production of inhibitory volatiles may increase the survival rate of bacteria in soil, by eliminating potential competitors for nutrients (Mackie & Wheatley, 1999).

g. HCN

The cyanide ion derived from HCN is a potent inhibitor of many metalloenzymes, especially copper-containing cytochrome c oxidases (Blumer & Haas, 2000). Production of HCN by certain strains of fluorescent pseudomonads has been involved in the suppression of soil-borne pathogens like black root rot of tobacco (Stutz et al., 1986) and take-all of wheat (Defago et al., 1990). HCN from *P. fluorescens* strain CHAO not repressed by fusaric acid played a significant role in disease suppression of *F. oxysporum* f. sp. *radicis-lycopersici* in tomato (Duffy et al., 2003). *P. fluorescens* HCN inhibited the mycelial growth of *Pythium in vitro* (Westsleijn, 1990). The cyanide producing strain CHAO stimulated root hair formation, indicating that the strain induced and altered plant physiological activities (Voisard et al., 1989). Four of the six PGPR strains that ISR in cucumber against *C. orbiculare* produced HCN (Wei et al., 1991). Fluorescent *Pseudomonas* strain RRS1 isolated from rajanigandha (tuberose) produced HCN and the strain improved seed germination and root length (Saxena et al., 1996).

Ramette et al. (2003) reported that HCN is a broad-spectrum antimicrobial compound involved in biological control of root disease by many plant associated fluorescent pseudomonads. Further, they noted that the enzyme HCN synthase is encoded by three biosynthetic genes (henA, henB, and henC). Pessi and Haas (2000) reported that low oxygen concentration is a prerequisite for the activity of the transcription factor ANR which positively regulates HCN biosynthesis.

4.3.3   PRODUCTION OF LYTIC ENZYMES

Production of fungal cell wall-degrading enzymes excreted by microorganisms is frequently involved in the attack of phytopathogenic fungi (Picard et al., 2000). Lysis by cell wall-degrading enzymes excreted by microorganisms is a well-known feature of mycoparasitism. Chitinase, β-1, 3 glucanase and cellulase are especially important fungus controlling enzymes due to their ability to degrade the fungal cell wall components such as chitin, β-1, 3 glucan, and glycosidic bonds.

Chitinases are PR-proteins which hydrolyze chitin, major cell wall component constituents for 3 to 10% of higher fungi and cuticle of peritrophic membrane in insects. Chitinase cleaves a bond between C1 and C4 of two consecutive N-acetyl glucosamine (GlcNAc) by either endolytic mechanism or exolytic mechanism. Nielsen et al. (1998) reported that in the sugar beet rhizosphere fluorescent pseudomonads inhibit plant pathogenic fungi *Rhizoctonia solani* by production of cell wall-degrading endochitinase. Biological control of *F. solani*, mainly via laminarinase and chitinase activities of *P. stutzeri* YPL-1, has been reported (Lim et al., 1991). Fridlender et al. (1993) reported that β-1, 3 glucanase producing *P. cepacia* decreased the incidence of diseases caused by *R. solani*, *S. rolfsii*, and *P. ultimurn*.

4.3.4   INDUCED SYSTEMIC RESISTANCE (ISR)

Induced resistance is a state of enhanced defensive capacity developed by a plant when appropriately stimulated (Kuc, 1982). Induced resistance is not the creation of resistance where there is none, but the activation of latent resistance mechanisms that are expressed upon subsequent, so-called "challenge" inoculation with a pathogen (Van Loon, 1997). Induced resistance can be triggered by certain chemicals, non-pathogens, avirulent forms of pathogens, incompatible races of pathogens, or by virulent pathogens under circumstances where infection is stalled owing to environmental conditions (Van Loon et al., 1998).

In the early 1990s, three independent research groups reported a PGPR-elicited plant defense response in cucumber, carnation, and bean when PGPR was inoculated into plants at a separate site than the site of pathogen challenge to avoid direct contact between the two microorganisms. This response was termed ISR (Alstrom, 1991; Van Peer et al., 1991; Wei et al., 1991). The involvement of ISR is typically studied in systems in which the *Pseudomonas* bacteria and the pathogen are inoculated and remain spatially separated on the plant, for example, the bacteria on the root and the pathogen on the leaf or stem, or by use of split root systems. Since no direct interactions are possible between the two populations, suppression of disease development has to be plant-mediated (Baker et al., 2007).

Rhizobacteria mediated ISR has been demonstrated against fungi, bacteria, nematodes, and viruses in Arabidopsis, bean, brinjal, carnation, chili, cucumber, potato, radish, sugarcane tobacco, and tomato. In the last decade, it has become clear that elicitation of ISR is a widespread phenomenon, not only for fluorescent pseudomonads but for a variety of nonpathogenic microorganisms and bioagents.

A large number of defense enzymes have been found to be associated with ISR. These are phenylalanine ammonia lyase (PAL), chitinase, β-1,3-glucanase, peroxidase (PO), polyphenol oxidase (PPO), superoxide dismutase (SOD), catalase (CAT), lipoxygenase (LOX), ascorbate peroxidase (APX), and proteinase inhibitors (Koch et al., 1992; Schneider & Ullrich, 1994; Van Loon, 1997). These enzymes also bring about liberation of molecules that elicit the initial steps in induction of resistance, phytoalexins, and phenolic compounds (Keen &Yoshikawa, 1993; Van Loon et al., 1998).

4.3.4.1.1   Fluorescent Pseudomonads as Inducers of Defense

Maize plants rose from *P. fluorescens* treated seeds showed higher activity of PO, PPO, and PAL, when leaf sheaths were inoculated with the pathogen, *R. solani*. The bacterized seeds with *P. fluorescens* lead to accumulation of higher phenolic compounds and higher activity of PO, PPO, and PAL that may play a role in defense mechanism in plants against pathogen (Sivakumar & Sharma, 2003). Kloepper et al. (2004) also observed, control of nematode diseases in tomato and bell pepper by treatments with PGPR strains through induction of systemic resistance.

Pieterse et al. (2003) proposed that *P. fluorescens* WCS417r mediates ISR without the activation of PR-protein genes, whereas Saravanakumara et al. (2007) reported that the protection of tea against *Exobasidium vexans* induced by a bacterial suspension of *P. fluorescens* Pf1 was followed by the induction of chitinase, β-1,3-glucanase, PPO, and PO. De-Vleess-chauwer et al. (2008) demonstrate the ability of *P. fluorescens* WCS374r to trigger ISR in rice (*Oryza sativa*) against the leaf blast pathogen *Magnaporthe oryzae* using salicylic acid (SA) non-accumulating NahG gene containing rice, an ethylene-insensitive OsEIN2 antisense line, and the jasmonate deficient mutant hebiba. The results revealed that WCS374r-induced resistance is regulated by an SA-independent but jasmonic acid/ethylene-modulated signal transduction pathway. Bacterial mutant analysis uncovered a pseudobactin-type siderophore as the crucial determinant responsible for ISR elicitation.

Sari et al. (2008) reported that wheat seedlings treated with *P. fluorescence* CHA0 before inoculation with *G. graminis* var. *tritici*, a pathogen of take-all of wheat acquired induced resistance against it. The treatment with CHA0 enhanced activities of soluble peroxidases (SPOX), cell wall bound peroxidases (CWPOX), β1, 3-glucanases, β1, 4-glucanases, and phenolic compounds in wheat roots. The activities of SPOX and β1, 4-glucanases were maximum of 4th day and that of CWPOX, β1, 3-glucanases, and total phenolics were maximum on 6th day after inoculation.

4.4 BIOCONTROL BY FLUORESCENT PSEUDOMONADS IN PRACTICE

The use of fluorescent *Pseudomonas* spp. as bioagents as inoculants of soil or plants has been successfully implemented in agronomic field trials. Their use have limited ecological impact on indigenous saprophytic population and there is no side effect on rhizosphere functioning. Many commercial inoculation products are available and researches for their long-term storage and effective antagonistic activity are going on. The plants susceptible for diseases caused by fungal soil pathogens *G. graminis* var. *tritici*, *F. oxysporum*, *R. solani*, *T. basicola*, *Phytophthora cinnamomi*, for nematodes *Meloidogyne incognita* and for bacteria *Streptomyces scabies*, and *Ralstonia solanaciarum* have been found to do well in disease suppressive soil in contrast to conducive soils. In these suppressive soils *P. fluorescence* and related fluorescent pseudomonads have been found to play an important role.

4.5   CONCLUSION AND FUTURE PROSPECTS

Superior root colonization and effective functioning in the rhizosphere are the key criteria when selecting strains and researches aim at better understanding the molecular basis of these traits and the signaling process regulating the ecology of these bioagents *in situ*. Successful root colonization is influenced by many factors such as genetic factors, abundance of growth substances, indigenous bacteria as well as abiotic factors such as soil humidity, pH, and temperature (Smith et al., 1999; Garbeva et al., 2004; Verma et al., 2004). The requirement of bacteria for different nutrients might explain why bacteria most often colonize plants in a species-specific manner (Dunn et al., 2003). It was previously demonstrated that an effective biological control strain isolated from one region may not perform as effectively in other soils or plants (Kiely et al., 2006). Promising results are obtained with the development of genetically improved strains with higher plant protection ability by either reprogramming the regulation of existing biocontrol traits (Mark et al., 2006) or by the introduction of novel mechanisms such as degradation of pathogen quorum sensing molecules (Molina et al., 2003) or by 1-aminocyclopropane-1-carboxylic acid (ACC) deaminase activity (Wang et al., 2000). Current genomic analysis for rhizosphere competence and biocontrol traits will likely lead to the development of novel tools for effective management of indigenous and inoculated fluorescent pseudomonads biocontrol agents and a better exploitation of their plant-beneficial trait for sustainable agriculture.

KEYWORDS

- plant pathogens
- bioagents
- PGPR
- fluorescent pseudomonads

- rhizosphere
- antibiosis
- siderophores
- lytic enzymes
- ISR

REFERENCES

Alstrom, S. Induction of Disease Resistance in Common Bean Susceptible to Halo Blight Bacterial Pathogen after Seed Bacterization with Rhizosphere Pseudomonads. *J. Genet. Appl. Microbiol.* **1991,** *37,* 495–501.

Arima, K.; Imanaka, H.; Kausaka, M.; Fukuda, A.; Tameera, C. Pyrrolinitrin, A New Antibiotic Substance, Produced by *Pseudomonas. Agr. Biol. Chem.* **1964,** *28,* 575–576.

Baker, J. O.; Cook, R. J. Role of Siderophores in Suppression of *Pythium* Species and Production of Increased Growth Response of Wheat by Fluorescent Pseudomonads. *Phytopathology.* **1988,** *78,* 778–782.

Baker, P. A. H. M.; Pieterse, C. M. J.; van Loon, L. C. Induced Systemic Resistance by Fluorescent *Pseudomonas* Spp. *Phytopathology.* **2007,** *97,* 239–243.

Bloemberg, G. V.; Lugtenberg, B. J. J. Molecular Basis of Plant Growth Promotion and Biocontrol by Rhizobacteria. *Curr. Opin. Plant Biol.* **2001,** *4,* 343–350.

Blumer, C.; Haas, D. Mechanism, Regulation, and Ecological Role of Bacterial Cyanide Biosynthesis. *Arch. Microbiol.* **2000,** *173,* 170–177.

Bossis, E.; Lemanceau, P.; Latour, X.; Gardan, L. The Taxonomy of *Pseudomonas fluorescens* and *Pseudomonas putida*: Current Status and Need for Revision. *Agronomie.* **2000,** *20,* 51–63.

Britigan, B. E.; Roeder, T. L.; Rasmussen, G. T.; Shasby, D. M.; McCormick, M. L.; Cox, C. D. Interaction of the *Pseudomonas aeruginosa* Secretory Products Pyocyanin and Pyochelin Generates Hydroxyl Radical and Causes Synergistic Damage to Endothelial Cells. Implications for *Pseudomonas* Associated Tissue Injury. *J. Clin. Invest.* **1992,** *90,* 2187–2196.

Cook, R. J.; Baker, K. F. *The Nature and Practices of Biological Control of Plant Pathogens;* APS Books: St. Paul, MN, 1982; p 599.

Couillerot, O.; Prigent-Combaret, C.; Caballero-Mellado, J.; Moënne-Loccoz, Y. *Pseudomonas fluorescens* and Closely Related Fluorescent Pseudomonads as Biocontrol Agents of Soil-Borne Phytopathogens. *Lett. Appl. Microbiol.* **2009,** *48,* 505–512.

Cronin, D.; Moënne-Loccoz, Y.; Fenton, A.; Dunne, C.; Dowling, D. N.; O'Gara, F. Role of 2, 4-Diacetylphloroglucinol in the Interactions of the Biocontrol Pseudomonad Strain F113 with the Potato Cyst Nematode *Globodera rostochiensis. Appl. Environ. Microbiol.* **1997,** *63,* 1357–1361.

Cuppels, D. A.; Stipanovic, R. D.; Stoessl, A.; Stothers, J. B. The Constitution and Properties of a Pyochelin–Zinc Complex. *Can. J. Chem.* **1987,** *65,* 2126–2130.

de Souza, J. T.; Arnould, C.; Deulvot, C.; Lemanceau, P.; Gianinazzi-Pearson, V.; Raaijmakers, J. M. Effect of 2, 4-Diacetylphloroglucinol on Pythium: Cellular Responses and Variation in Sensitivity among Propagules and Species. *Phytopathology.* **2003,** *93,* 966–975.

De Vleesschauwer, D.; Djavaheri, M.; Baker, P. A.; Höfte, M. *Pseudomonas fluorescens* WCS 374r-Induced Systemic Resistance in Rice against *Magnaporthe oryzae* is Based on Pseudobactin-Mediated Priming for a Salicylic Acid-Repressible Multifaceted Defense Response. *Plant Physiol.* **2008,** *148,* 1996–2012.

Defago, G.; Berling, C. H.; Burger, U.; Haas, D.; Kahr, G.; Keel, C.; Voisard, C.; Wirthner, P.; Wuthrich, B. Suppression of Black Root Rot of Tobacco and Other Root Diseases by Strains of *Pseudomonas fluorescens*: Potential Applications and Mechanisms. In *Biological Control of Soil Borne Plant Pathogens;* Hornby, D., Ed.; CAB International: Wallingford, Oxon, UK, 1990; pp 93–108.

Demange, P.; Wenderbaum, S.; Bateman, A.; Dell, A.; Abdallah, M. A. Bacteiral Siderophores: Structure and Physicochemical Properties of Pyoverdins and Related Compounds. In *Iron Transport in Microbes, Plants and Aniamls;* Winkleman, G., Van Der Helm, D., Neilands, J. B., Eds.; VCH Chemie: Weinheim, Germany, 1987; pp 167–187.

Duffy, B.; Keel, C.; De´fago, G. Potential Role of Pathogen Signaling in Multitrophic Plant-Microbe Interactions Involved in Disease Protection. *Appl. Environ. Microbiol.* **2004,** 70, 1836–1842.

Duffy, B.; Schouten, A.; Raajimakers, J. Pathogen Self-Defense: Mechanisms to Counteract Microbial Antagonism. *Annu. Rev. Phytopathol.* **2003,** *45,* 501–538.

Dunn, J. D.; Johnson, B. J.; Kayser, J. P.; Waylan, A. T.; Sissom, E. K.; Drouillard, J. S. Effects of Flax Supplementation and a Combined Trenbolone Acetate and Estradiol Implant on Circulating Insulin-Like Growth Factor-I and Muscle Insulin-Like Growth Factor-I Messenger RNA Levels in Beef Cattle. *J. Anim. Sci.* **2003,** *81,* 3028–3034.

Fenton, A. M.; Stephans, P. M.; Crowley, J.; O'Callaghan, M.; Gara, F. O. Exploitation of Gene(s) Involved in 2, 4-Diacetylphloroglucinol Biosynthesis to Confer a New Biocontrol Capacity to *Pseudomonas* Strains. *Appl. Environ. Microbiol.* **1992,** *58,* 3873–3878.

Fernando, W. G. D.; Ramarathnam, R.; Krishnamoorthy, A. S.; Savchuk, S. Identification and Use of Bacterial Organic Volatiles in Biological Control of *Sclerotinia Sclerotiorum. Soil Biol. Biochem.* **2004,** *36.*

Franks, A.; Ryan, R. P.; Abbas, A.; Mark, G. L.; O'Gara, F. Molecular Tools for Studying Plant Growth-Promoting Rhizobacteria (PGPR). In *Molecular Approaches to Soil, Rhizosphere and Plant Microorganism Analysis;* Cooper, J. E., Rao, J. R., Eds.; CAB International: Wallingford, US, 2006; pp 116–131.

Fridlender, M.; Inbar, J.; Chet, I. Biological Control of Soil-Borne Plant Pathogens by β-1,3-Glucanase-Producing *Pseudomonas cepacia. Soil Biol. Biochem.* **1993,** *25,* 1211–1221.

Garbeva, P.; van Veen, J. A.; van Elsas, J. D. Assessment of the Diversity and Antagonism toward *Rhizoctonia solani* AG3 of *Pseudomonas* Species in Soil from Different Agricultural Regimes. *FEMS Microbiol. Ecol.* **2004,** *47,* 51–64.

Glick, B. R.; Patten, C. L.; Holguin, G.; Penrose, D. M. *Biochemical and Genetic Mechanisms Used by Plant Growth Promoting Bacteria;* Imperial College Press: London, UK, 1999.

Haas, D.; Défago, G. Biological Control of Soil-Borne Pathogens by Fluorescent Pseudomonads. *Nat. Rev. Microbiol.* **2005**, *3*, 307–319.

Hammer, P. E.; Burd, W.; Hill, D. S.; Ligon, J. M.; van Pee, K. H. Conservation of the Pyrrolnitrin Gene Cluster among Six Pyrrolnitrin-Producing Strains. *FEMS Microbiol. Lett.* **1999**, *180*, 39–44.

Hernandez, M. E.; Kappler, A.; Newman, D. K. Phenazines and Other Redox-Active Antibiotics Promote Microbial Mineral Reduction. *Appl. Environ. Microbiol.* **2004**, *70*, 921–928.

Hiltner, L. Überneuere Erfahrungen und Probleme auf DemGebiete der Bodenbakteriologie Unterbesonderer Berücksichtigung der Gründüngung und Brache. *Arb. Dtsch. Landwirtsch. Ges.* **1904**, *98*, 59–78.

Iavicoli, A.; Boutet, E.; Buchala, A.; Métraux, J. P. Induced Systemic Resistance in *Arabidopsis thaliana* in Response to Root Inoculation with *Pseudomonas fluorescens* CHA0. *MPMI.* **2003**, *16*, 851–858.

Jimenez, D.; Virgen, G.; Tabares, F.; Olalde, V. Bacterias Promoter as Delcrecimiento De Plantas: *Agrobiotechnologia. Avancey Perspectiva.* **2001**, *20*, 395–400.

Keel, C.; Schnider, U.; Maurhofr, M.; Voisard, C.; Laville, J.; Burger, U.; Wirthner, P.; Haas, D.; Défago, G. Suppression of Root Diseases by *Pseudomonas fluorescens* CHA0: Importance of the Bacterial Secondary Metabolite 2, 4-Diacetylphloroglucinol. *MPMI.* **1992**, *5*, 4–13.

Keel, U. S.; Seematter, A.; Maurhofer, M.; Blumer, C.; Duffy, B.; Bonnefoy, C. G.; Reimmann, C.; Notz, R.; Défago, G.; Haas, D.; Keel, C. Autoinduction of 2, 4-Diacetylphloroglucinol Biosynthesis in the Biocontrol Agent *Pseudomonas fluorescens* CHA0 and Repression by the Bacterial Metabolites Salicylate and Pyoluteorin. *J. Bacteriol.* **2000**, *182*, 1215–1225.

Keen, N. T.; Yoshikawa, M. β-1, 3-Endoglucanase from Soybean Releases Elicitor Active Carbohydrates from Fungus Cell Wall. *Pl. Physiol.* **1993**, *71*, 460–465.

Kiely, G.; Morgan, G.; Moles, R.,; Byrne, P.; Jordan, P.; Daly, K.; Doody, D.; Tunney, H.; Kurz, I.; Bourke, D.; O'Reilly, C.; Ryan, D.; Jennings, E.; Irvine, K.; Carton, O. *Pathways for Nutrient Loss to Water with Emphasis on Phosphorus;* Final Integrated Synthesis Report: Eutrophication from Agriculture Sources (2000-LS-S-MS2). EPA, Ireland, 2006. Available at: http://www.epa.ie.

Kloepper, J. W.; Leong, J.; Teintze, M.; Schroth, M. N. Pseudomonas Siderophores: A Mechanism Explaining Disease-Suppressive Soils. *Curr. Microbiol.* **1980**, *4*, 317–320.

Kloepper, J. W.; Schroth, M. N. In *Station de Pathologie Végétale et Phytobactériologie*, Proceedings of the IVth International Conference on Plant Pathogenic Bacteria, INRA, Angers, France, Aug. 27–Sept. 2, 1978; *2*, 879–882.

Kloepper, J. W.; Schroth, M. N. Development of a Powder Formulation of Rhizobacteria for Inoculation of Potato Seed Pieces. *Phytopathol.* **1981**, *71*, 590–592.

Kloepper, J. W.; Reddy, M. S.; Kenney, D. S. In *Application of Rhizobacteria in Transplant Production and Yield Enhancement,* XXVI International Horticultural Congress, ISHS, Acta Horticulture, *631*, 2004.

Koch, E.; Meier, B. M.; Eiben, H. G.; Slusarenko, A. A Lipoxygenase from Leaves of Tomato (*Lycopersicon esculentum* Mill.) is Induced in Response to Plant Pathogenic Pseudomonads. *Pl. Physiol.* **1992**, *99*, 571–576.

Kuc, J. Induced Immunity to Plant Disease. *Bioscience.* **1982**, *32*, 854–860.

Ligon, J. M.; Hill, D. S.; Hammer, P. E.; Torkewitz, N. R.; Hofmann, D.; Kempf, H. J.; Van Pee, K. H. Natural Products with Antifungal Activity from Pseudomonas Biocontrol Bacteria. *Pest Manag. Sci.* **2000,** *56,* 688–695.

Lim, H. S.; Kim, Y. S.; Kim, S. D. *Pseudomonas stutzeri* YPL-1 Genetic Transformation and Antifungal Mechanism against *Fusarium solani*, an Agent of Plant Root Rot. *Appl. Environ. Microbiol.* **1991,** *57,* 510–516.

Loper, J. E. Role of Fluorescent Siderophore Production in Biological Control of *Pythium ultimum* by a *Pseudomonas fluorescens* Strain. *Phytopathol.* **1988,** *78,* 166–172.

Lugtenberg, B. J. J.; Dekkers, L.; Bloemberg, G. V. Molecular Determinants of Rhizosphere Colonization by Pseudomonas. *Ann. Rev. Phytopathol.* **2001,** *39,* 461–490.

Mackie, A. E.; Wheatley, R. E. Effects of the Incidence of Volatile Organic Compound Interactions between Soil Bacterial and Fungal Isolates. *Soil Biol. Biochem.* **1999,** *31,* 375–385.

Mark, G. L.; Morrissey, J. P.; Higgins, P.; O'Gara, F. Molecular Base Strategies to Exploit Pseudomonas Biocontrol Strains for Environmental Biotechnological Applications. FEMS. *Microbiol. Eco.* **2006,** *56,* 167–177.

Maurhofer, M.; Hase, C.; Meuwly, P.; Metraux, J. P.; Defago, G. Induction of Systemic Resistance of Tobacco to Necrosis Virus by the Root-Colonizing *Pseudomonas fluorescens* Strain CHA0: Influence of the GacA Gene and of Pyoverdine Production. *Phytopatholog.* **1994,** *84,* 139–146.

Mercado, W. Alternativaspara el Control de *Rhizoctonia solani, Myrothecium roridum* Causantes de Cancros y Pudricion de la Raiz en Cafetos de Viveros. Tesis SB608, C6M47, Universidad de Puerto Rico Recinto Universitario de Mayaguez, 1997.

Molina, L.; Constantinescu, F.; Michel, L.; Reimmann, C.; Duffy, B.; Defago, G. Degradation of Pathogen Quorum Sensing Molecules by Soil Bacteria: A Preventive and Curative Biological Control Mechanism. *FEMS. Microbiol. Eco.* **2003,** *45,* 71–81.

Morrissey, J. P.; Abbas, A.; Mark, L.; Cullinane, M.; O'Gara, F. Biosynthesis of Antifungal Metabolites by Biocontrol Strains of *Pseudomonas*. In *Pseudomonas: Biosynthesis of Macromolecules and Molecular Metabolism*; Ramos, J. L., Ed.; Kluwer Academic/Plenum Publishers 3: New York, NY, 2004; pp 635–670.

Nelson, E. B. Microbial Dynamics and Interactions in the Spermosphere. *Annu. Rev. Phytopathol.* **2004,** *42,* 271–309.

Nielsen, M. N.; Sorensen, J.; Fels, J.; Pedersen, H. C. Secondary Metabolite and Endochitinase Dependent Antagonism toward Plant-Pathogenic Micro-Fungi of *Pseudomonas fluorescens* Isolates from Sugar Beet Rhizosphere. *Appl. Environ. Microbiol.* **1998,** *64,* 3563–3569.

Nowak-Thompson, B.; Chaney, N.; Wing, J. S.; Gould, S. J.; Loper, J. E. Characterization of the Pyoluteorin Biosynthetic Gene Cluster of *Pseudomonas fluorescens* Pf-5. *J. Bacteriol.* **1999,** *181,* 2166–2174.

Nybroe, O.; Sørensen, J. Production of Cyclic Lipopeptides by Fluorescent Pseudomonads. In *Pseudomonas, Biosynthesis of Macromolecules and Molecular Metabolism;* Ramos, J. L. Ed.; Kluwer Academic/Plenum Publishers: New York, NY, 2004; Vol. 3, pp 147–172.

Pessi, G.; Haas, D. Transcriptional Control of the Hydrogen Cyanide Biosynthetic Genes HCN ABC by the Anaerobic Regular ANR and the Quorum-Sensing Regulators LasR and RhlR in *Pseudomonas aeruginosa. J. Bacteriol.* **2000,** *182,* 6940–6949.

Phoebe, Jr. C. H.; Combie, J.; Albert, F. G.; Van Tran, K.; Cabrera, J.; Correira, H. J. "Extremophilic Organisms as an Unexplored Source for Antifungal Compounds". *J. Antibiot.* **2001,** *54,* 56–65.

Picard, C.; Di Cello, F.; Ventura, M.; Fani, R.; Gluckert. A. Frequency and Biodiversity of 2, 4-Diacetylphloroglucinol-Producing Bacteria Isolated from the Maize Rhizosphere at Different Stages of Plant Growth. *Appl. Environ. Microbiol.* **2000,** *66,* 948–955.

Pieterse, C. M. J.; Van Pelt, J. A.; Verhagen, B. W. M.; Ton, J.; Van Wees, S. C. M.; Léon-Kloosterziel, K. M.; Van Loon, L. C. Induced Systemic Resistance by Plant Growth-Promoting Rhizobacteria. *Symbiosis.* **2003,** *35,* 39–54.

Raaijmakers, J. M., Vlami, M.; de Souza, J. T. Antibiotic Production by Bacterial Biocontrol Agents. *Antonie van Leeuwenhoek.* **2002,** *81,* 537–547.

Ramette, A.; Frapolli, M.; Defago, G.; Moenne-Loccoz, Y. Phylogeny of HCN Synthase-Encoding Hcnbc Genes in Biocontrol Fluorescent Pseudomonads and its Relationship with Host Plant Species and HCN Synthesis Ability. *Molecular Biol. Pl. Microbe Interaction.* **2003,** *16,* 525–535.

Ryu, C. M.; Farag, M. A.; Hu, C. H.; Reddy, M. S.; Wei, H. X.; Pare, P. W.; Kloepper, J. W. Bacterial Volatiles Promote Growth in Arabidopsis. *Proc. Nation. Acad. Sci. USA.* **2003,** *100,* 4927–4932.

Sari, E.; Etebarian, H. R.; Aminian, H. Effects of *Pseudomonas fluorescence* CHA0 on the Resistance of Wheat Seedling Roots to the Take all Fungus *Gaeumannomyces graminis* var. *tritici. Plant Prodn. Sci.* **2008,** *11*(3), 298–306.

Saravanakumara, D.; Charles, V.; Kumarb, N.; Samiyappan, R. PGPR-Induced Defense Responses in the Tea Plant against Blister Blight Disease. *Crop Protect.* **2007,** *26,* 556–565.

Saxena, A.; Sharma, A.; Goel, R.; Johri, B. N. In *Functional Characterization of a Growth Promoting Fluorescent Pseudomonads from Rajnigandha Rhizosphere,* 37th Annual Conference of the Association of Microbiologists of India, IIT: Chennai, India, December 4–6, 1996; p 135.

Scher, F. M.; Baker, R. Effect of *Pseudomonas putida* and a Synthetic Iron Chelator on Induction of Soil Suppressiveness to Fusarium Wilt Pathogens. *Phytopathology.* **1982,** *72,* 1567–1573.

Schneider, S.; Ullrich, W. R. Differential Induction of Resistance and Enhanced Enzyme Activities in Cucumber and Tobacco Caused by Treatment with Various Abiotic and Biotic Inducers. *Physiol. Mol. Pl. Pathol.* **1994,** *45,* 291–304.

Shanahan, P.; O'Sullivan, D. J.; Simpson, P.; Glennon, J. D.; O'Gara, F. Isolation of 2, 4-Diacetylphloroglucinol from a Fluorescent Pseudomonad and Investigation of Physiological Parameters Influencing its Production. *Appl. Environ. Microbiol.* **1992,** *58,* 353–358.

Sivakumar, G.; Sharma, R. C. Induced Biochemical Changes Due to Seed Bacterization by *Pseudomonas fluorescens* in Maize Plants. *Indian Phytopath.* **2003,** *56,* 134–137.

Smith, E.; Naidu, R.; Alston, A. N. Chemistry of Arsenic in Soil. I. Adsorbtion of Arsenite and Arsenate by Selected Soil. *J. Environ. Qual.* **1999,** *28,* 1719–1726.

Stutz, E.; Défago, G.; Kern, H. Naturally Occurring Fluorescent Pseudomonads Involved in Suppression of Black Rot of Tobacco. *Phytopathology.* **1986,** *76,* 181–195.

Suryakala, D.; Aheshwaridevi, P. V.; Lakshmi, K. V. Chemical Characterization and *in Vitro* Antibiosis of Siderophores of Rhizosphere Fluorescent Pseudomonads. *Indian J. Microbiol.* **2004,** *44,* 105–108.

Tambong, J. T.; Höfte, M. Phenazines are Involved in Biocontrol of *Pythium myriotylum* on Cocoyam by *Pseudomonas aeruginosa* PNA1. *Eur. J. Plant Pathol.* **2001**, *107*, 511–521.

Tamietti, G.; Ferraris, L., Matta, A.; Abbattista Gentile, I. Physiological Responses of Tomato Plants Grown in Fusarium Suppressive Soil. *Phytopathology.* **1993**, *138*, 66 –76.

Tripathi, R. K.; Gottlieb, D. Mechanism of Action of the Antifungal Antibiotic Pyrrolnitrin. *J. Bacteriol.* **1969**, *100*, 310–318.

Turner, J. M.; Messenger, A. J. Occurrence, Biochemistry and Physiology of Phenazine Pigment Production. *Adv. Microb. Physiol.* **1986**, *27*, 211–275.

Van Loon, L. C.; Baker, P. A. H. M.; Pieterse, C. M. J. Systemic Resistance Induced by Rhizosphere Bacteria. *Ann. Rev. Phytopathol.* **1998**, *36*, 453–483.

Van Loon, L. C. Induced Resistance in Plants and the Role of Pathogenesis-Related Proteins. *Eur. J. Plant Pathol.* **1997**, *103*, 753–65.

Van Peer, R.; Niemann, G. J.; Schippers, B. Induced Resistance and Phytoalexin Accumulation in Biological Control of Fusarium Wilt of Carnation by *Pseudomonas* Sp. Strain WCS417r. *Phytopathology.* **1991**, *81*, 728–734.

Verma, A.; Abbot, L.; Werner, D.; Hampp, R. *Soil Surface Microbiolgy;* Springer: Berlin, 2004.

Vessey, J. Plant Growth Promoting Rhizobacteria as Biofertilizers. *Plant Soil.* **2003**, *255*(2), 571–86.

Vincent, M. N.; Harrison, L. A.; Mukherji, P. Genetic Analysis of the Antifungal Activity of a Soil Borne *Pseudomonas aureofaciens* Strain. *Appl. Environ. Micribiol.* **1991**, *57*, 2928–2934.

Visca, P. In *Pseudomonas*; Ramos, J. L., Ed.; Kluwer Academic/Plenum Publishers: New York, NY, 2004; Vol. 2, pp 69–123.

Visca, P.; Colotti, G.; Serino, L.; Verzili, D.; Orsi, N.; Chiancone, E. Metal Regulation of Siderophore Synthesis in *Pseudomonas aeruginosa* and Functional Effects of Siderophore-Metal Complexes. *Appl. Environ. Microbiol.* **1992**, *58*, 2886–2893.

Voisard, C.; Keel, C.; Haas, D.; Defago, G. Cyanide Production by *Pseudomonas fluorescens* Helps Suppress Black Root Rot of Tobacco under Gnotobiotic Conditions. *EMBO J.* **1989**, *8*, 351–358.

Wang, C.; Knill, E.; Glick, B. R.; Defago, G. Effect of Transferring 1-Aminocyclopropane-1-Carboxylic Acid (ACC) Deaminase Genes into *Pseudomonas fluorescence* Strain CHAO and its Derivative CHA96 on their Growth Promoting and Disease Suppressive Capacities. *Can. J. Microbiol.* **2000**, *46*, 1–10.

Wei, G.; Kloepper, J. W.; Tuzun, S. Induction of Systemic Resistance of Cucumber to *Colletotrichum orbiculare* by Select Strains of Plant Growth-Promoting Rhizobacteria. *Phytopathology.* **1991**, *81*, 1508–1512.

Weststeijn, W. A. Fluorescent Pseudomonads Isolate E11-2 as Biological Agent for *Pythium* Root Rot in Tulips. *Netherlands J. Pl. Pathol.* **1990**, *96*, 262–272.

PART II
Entomopathogens

CHAPTER 5

ENTOMOPATHOGENIC FUNGI: INTRODUCTION, HISTORY, CLASSIFICATION, INFECTION MECHANISM, ENZYMES, AND TOXINS

WAHEED ANWAR, AHMAD ALI SHAHID*, and
MUHAMMAD SALEEM HAIDER

*Institute of Agricultural Sciences, University of the Punjab,
New Campus, Lahore 54000, Pakistan*

Corresponding author. E-mail: ahmadali.shahid@gmail.com

CONTENTS

ABSTRACT

Entomopathogenic fungi attacks on insect body, cause infection in series, and eventually kill them. These fungi are considered to play a vital role as biological control agent of insect populations and a large number of fungal species infect insects. These insect pathogenic species are found in a wide range of adaptations and infecting capacities including obligate and facultative pathogens. Entomopathogenic fungi are reported from the divisions of Deuteromycota, Ascomycota as well as Zygomycota, Oomycota, and Chytridiomycota. Entomopathogenic fungus varies in specificity widely within genera, between genera and among species. Entomopathogenic fungal infection cycle involves several steps: conidial attachment, germination, penetration through the insect cuticle, vegetative growth within the host, fungal protrusion outside the insect, and conidiogenesis. Mechanical force, some metabolic acids, and enzymatic processes are involved in initial interaction in pathogenesis and enzymes responsible for pathogenesis of insects can be grouped into lipases, chitinases, peptidases, and proteases. Different metabolites are produced by entomopathogenic fungi act as toxins and play a vital role in the infection development, and result in a number of symptoms in the insect. The characteristic of secondary metabolites to act toxin against the insect cellular lines can be the possibility in the development of new bio-insecticides. Many insect pathogenic fungus based bio-insecticides have been produced and commercially manufactured so far. The use of microbial insecticides should be a contribution toward all fields of agriculture, sustainable agriculture, forestry, and horticulture. Just like the genes of *Bacillus thuringiensis*, which are inserted in different crops through biotechnology to get resistance against different pathogens and insect pests, the genes (e.g., chitinases, proteases, etc.) of entomopathogenic fungus can be incorporated in different crops to get resistance against other pathogens and insect pests in future.

5.1   INTRODUCTION

Entomopathogenic fungi attacks on insect body, cause infection in series, and eventually kill them. Singkaravanit in 2010 defined entomopathogenic fungi as "A group of fungi that kills an insect by attacking and infecting its insect host." These fungi are natural enemies of arachnids, insects, and play important role for the regulation of their host populations (Eilenberg et al., 2001). These fungi are considered to play a vital role as biological control agent of insect populations and a large number of fungal species infect insects. These insect pathogenic species are found in a wide range of adaptations and infecting capacities including obligate and facultative pathogens. Spreading of fungal diseases is common in many insect species while some species may not be affected (Steinhaus, 1975). Mode of action of these fungi is mainly by contact and penetration (Nadeau et al., 1996). Entomopathogenic fungi are responsible for mortality in pest populations and some fungal species are evaluated for their role as a biocontrol agent in agriculture. In biological control, traditional approach is useful to cropping system as a fungal material, using an inoculative or inundative biological control strategy (Eilenberg et al., 2001). According to Khachatourians and Sohail (2008), over 700 species of entomopathogenic fungi from 90 genera are pathogenic to insects.

5.2   HISTORY

Even before the Christian era, man has been worried about unhealthy situation of his cultivated insect species, *Bombyx mori* (*B. mori*), silkworm, honey bee, and *Apis mellifera*. In 1956, Steinhaus reported that diseases of insects were observed in Orient for the first time. He quoted the muscardine silkworms infected with *Beauveria bassiana* (*B. bassiana*) was used for the treatment of paralysis. The diseased larvae were used as a remedy against many diseases like abscesses, sore throat, and toothache. Steinhaus suggested the first report of entomopathogenic fungi was published in 1726 and depicts the "Chinese plant worm." In 1858 Gray and in 1892 Cooke highlighted the depiction of this species, which was held in high esteem by the Emperor's physicians in China for several centuries, which was known as summer herb-winter worm. The preparation of the drug, the effects of which are compared to those of ginseng, was suitably exotic, being stuffed into a duck, and then slowly roasted. The fungal products were absorbed by the duck meat was eaten. Gray described that the *Cordyceps robertsii* infects the Lepidoptera larvae and native people of New Zealand used the fungal-infected insects; eating the fruit-bodies and extracting a pigment for tattooing purposes. He described that by chafing the extract into the wounds was effective as it act as antiseptic. Gray wrote the book in which entomopathogenic fungi were discussed in detail.

The fruiting bodies of *Cordyceps* were of curiosity for the early naturalists but the relationship between the insect and fungus was puzzled. Gray and Cooke quoted that in early eighteenth century descriptions of *Cordyceps* from the West Indies: "By the latter end of July the tree arrives at its full growth, and resembles a coral branch…and bears several little pods, which, dropping off, become worms, and from them flies, like the English caterpillar." A modern French collector took this as evidence of transmutation: "He sees in these plant-animals a proof of the passage and mutation of animal species into the vegetable, and reciprocally from the vegetable to animal" (Cooke, 1892). Torrubia reported dead wasps in Cuba in 1749 and described the disease in the trees, *Polistes* infected with *Cordyceps sphecocephala*, is given in full by Steinhaus (1956) who included the original illustration.

In 1956, Steinhaus cites eighteenth century European accounts of diseased insects, mainly flies, in which the causal agent of the disease belongs to Entomophthorales. At that time, the theory of spontaneous generation was common and insects as causal agent of diseases in animals was not a strong belief. The germ theory was established as a result of investigation of white muscardine disease of silkworm. E. Metchnikoff was also an eminent scientist not only in entomology but also in microbial diseases. In 1870s, he worked on the green muscardine (*Metarhizium anisopliae*). In 1858, Gray stated about *Cordyceps robertsii* that the parasite becomes connected with the caterpillar by means of the seed being taken in with the food and thus passing into the internal insect. It is now known that most entomopathogenic fungi invade through the cuticle of the insect and infection through ingestion of spores is rare (Roberts & Humber, 1984).

In latter part of nineteenth century, much importance was given to the taxonomy of the entomopathogenic fungi and famous mycologists include; A. Giard, P. A. Saccardo, M. J. Berkeley, C. L. Spegazzini, and P. Veillemin but the major contribution was from R. Thaxter in the Entomophthorales (Thaxter, 1888). During 1895–1925, most research was done on the use of entomopathogenic fungi as biocontrol agent in combating pests of agricultural importance. During 1925–1960, the interest in this field was waned and then a lull period predominated. Some major advances were made in taxonomy by Petch (1938) in Old World and E. B. Mains in the New World. In 1941, the monographic treatment of *Cordyceps* by Kobayasi was a significant work. Now the interest in entomopathogenic fungi has developed again and has resulted in detailed study of arthropod population and their habitat.

5.3 CLASSIFICATION

Entomopathogenic fungi are reported from the divisions of Deuteromycota, Ascomycota as well as Zygomycota, Oomycota, and Chytridiomycota (Samson et al., 1988). Entomopathogenic fungi of different genera, such as Zygomycota, Deuteromycota, and Hyphomycetes are presently under research. Infection of fungus is reported in other arthropods, insects, and other species that are not pests of cultivated crops. Spiders are infected by *Gibellula* species and different species of *Erynia, Cordyceps* attack ants. In addition, knowledge about the ecology and biology of entomopathogenic fungi can be found from Steinhaus (1964), Bałazy (1993), Evans (1989), and Eilenberg (2002).

5.3.1 PHYLUM OOMYCOTA

Fungi that have cellulose in their coenocytic hyphae, without chitin and biflagellate zoospores are considered as oomycetes. Sexual reproduction can occur either on the same hyphae or on different hyphae between archegonia and antheridia (collectively called as gametangia). Numerous species are parasites and saprophytes of plants and animals while species of two genera are pathogenic to mosquito larvae. A mosquito larvae pathogen, *Lagenidium giganteum* is studied best (Glare, 2010; Scholte et al., 2004). Some other *Lagenidium* species are pathogenic to crabs and other aquatic crustaceans (Hatai et al., 2000). Zattau and McInnis (1987) thought that these species infect the insect through secondary zoospore.

5.3.2 PHYLUM CHYTRIDIOMYCOTA

The fungal groups that are without cellulose and contain chitin walls are included in this phylum. Single flagellum zoospores and gametes are settled and a thallus is grown which then converts into a coenocytic hyphae or a resting spore. When the comparisons are based on SSU rRNA phylogenetic, this fungal group is considered as basal. Most of the usual insect infecting Chytridiomycetes is contained in Blastocladiales genus Coelomomyces. In Coelomomyces, above 70 insect pathogenic species are described (Barr, 2001). Mostly the Hemipterans and Dipteranas were recognized as the source of these fungi. Sporangia of these fungi like copepods, is thick walled and resistant and the zoospores are flagellated. Insect pathogenic *Myriophagus* (Chytridiales) are found on pupae of Diptera and Coelomycidium (Blastocladiales) found on mosquitoes and blackflies contain some other species (Samson et al., 1988).

5.3.3   PHYLUM ZYGOMYCOTA

Hyphae that are multicellular, non-septate, and zygospores by the joining of gametangia, are considered as phylum Zygomycota conventionally. But, Zygomycota has not been found as monophyletic on the basis of molecular analyses (Tanabe et al., 2000). The class Trichomycetes, within the phylum Zygomycota, consists of species that are mostly related with insects. The species *Smittium morbosum* (Trichomycetes) was reported by Cooper and Sweeney (1986), as mosquito pathogen. Trichomycetes has mostly weak or symbiotic associations unlike true infectious agents (Beard & Adler, 2002). Insect death is mostly related to some species of *Mucor* (Mucorales). The order Entomophthorales, contain above 200 insect infecting species within phylum Zygomycota. Some species are competent of producing secondary spores as of primary spores and a few produce long-lasting resting spores.

5.3.4   ASCOMYCOTA AND DEUTEROMYCOTA

Ascomycota includes the fungus with haploid and mycelia with septation and the ascospores are produced in a sac called ascus on the ascomata, the fruiting body. Normally, in every ascus, the number of ascospores produced is eight. Zoospores are not produced in ascomycetes (Shah & Pell, 2003). The taxonomy of sexual (teleomorphic) and asexual (anamorphic) conidial stages always remained in confused classification when their production is not on single occurrence. More than 300 insect infecting species are present in *Cordyceps* that is best-known ascomycetes. Chalkbrood disease in bees occurs by the genus *Ascosphaera*, which has dimorphism in sexuality. Klich (2007) described that the spores of *Ascosphaera*, in an unusual manner of insect infecting species are ingested by the bee larvae that germinate in gut causing infection. The most important insect infecting species occur in *Aspergillus, Metarhizium, Hirsutella, Beauveria, Aschersonia, Culicinomyces, Lecanicillium, Paecilomyces, Tolypocladium*, and *Sorosporella*. Mostly, these genera have a linkage with one or many genera that can be verified with biological studies or by the molecular studies presenting the genetic relationship between telemorphs and anamorphs (Huang et al., 2010; Bischoff et al., 2009).

5.3.5 BASIDIOMYCOTA

A very few Basidiomycetes have been noticed that act as insect pathogen. It has been described by some researchers that *Uredinella* and *Septobasidiales* genera *Septobasidium* are considered infection-causing agents in insects but most of them have a symbiotic relationship with insects like scales (Table 5.1; Samson et al., 1988).

TABLE 5.1   Entomopathogenic Fungi Classification (Adapted from Roy et al., 2006).

Division: A. **Zygomycota**		
Class: Zygomycetes		
Order: Entomophthorales		
Family: i. Entomophthoraceae		ii. Neozygitaceae
	Genus: *Entomophthora*	**Genus:** *Neozygites*
	Entomophaga	
	Erynia	
	Eryniopsis	
	Furia	
	Massospora	
	Pandora	
	Strongwellsea	
	Tarichium	
	Zoophthora	
Division: B. **Ascomycota**		
Class: Sordariomycetes		
Order: Hypocreales		
Family: Clavicipitaceae		
Genus:	*Beauveria*	
	Cordyceps	
	Cordycepioideus	
	Lacanicillium	
	Metarhizium	
	Nomuraea	

5.4   IMPORTANCE IN NATURE

Fungal entomopathogens have played many unpredicted roles, such as plant disease antagonists, plant growth promoting fungi, rhizosphere colonizers, and their presence as fungal endophytes (Vega et al., 2009).

5.4.1   FUNGAL ENDOPHYTES

Endophytes are capable of infecting above ground internal plant tissues and unable to cause symptoms. They are receiving great attraction due to their huge diversity and presence in all areas of the world (Arnold & Lutzoni, 2007; Saikkonen et al., 2006). Although, many fungal endophytes provide defense to their host plants against different pathogens (Rudgers et al., 2007; Arnold & Lewis, 2005). Various fungi conventionally are recognized as insect pathogens that have been isolated as endophytes, which include different species of *Isaria, Acremonium, Cladosporium, Beauveria, and Clonostachys* (Vega et al., 2008).

5.4.2   PLANT GROWTH-PROMOTING FUNGI AND RHIZOSPHERE COLONIZERS

Soil microbiota of Hypocreales members are important entomopathogenic fungi and mostly isolated in temperate regions from soils are of genera *Metarhizium, Isaria,* and *Beauveria* (Meyling & Eilenberg, 2007). In natural environment, soil provides many challenges and opportunities to entomopathogenic fungi. It reduces the effects of destructive solar radiation help in water availability and protect against harsh environmental conditions (Inglis et al., 2001; Rangel et al., 2005; Roberts & Campbell, 1977; Gaugler et al., 1989). In addition, soil is a home for various hosts of insect and stability in closeness to possible hosts is a major feature in the development of fungal entomopathogenicity (Humber, 2008). Conversely, many antimicrobial metabolites are secreted by microorganisms in soil that helps entomopathogenic fungi to infect their hosts. Groden and Lockwood (1991) recognized an important tendency of low-mortality rate of the Colorado potato beetle by *B. bassiana* with enlarged soil fungistasis levels and opportunistic soil microorganisms are capable of getting energy from dead insects that are infected by entomopathogenic fungi. Various hypocrealean entomopathogen species secrete secondary metabolites inside their insect hosts that are thought to assist fungus during saprotrophic phase of insect utilization (Strasser et al., 2000). Many insects that are infected and killed by different species of *Metarhizium* and *Beauveria* in soil produce restricted somatic growth. Entomopathogenic fungi rely mainly on the insect for carbon source (Inglis et al., 2001; Pereira et al., 1993). Free carbon is abundant in rhizosphere because entomopathogenic fungi interact with plant roots for their survival and growth (Leger, 2008). Majority of the carbon is assimilated by plant is transferred into the soil in the form of lysates, sloughed root cells, exudates, and mucilage (Bardgett, 2005; Andrews & Harris, 2000). In the rhizosphere, carbon is utilized by a large number of saprotrophic microorganisms (Whipps, 2001). For the most part, it is still unclear whether this is merely in one direction contact benefit only microbial saprotrophs, or whether a mutualistic interaction that has originated in plants to obtain mineral nutrients or defense against herbivores and parasites (Singh et al., 2004).

5.5   HOSTS OF ENTOMOPATHOGENIC FUNGI

Entomopathogenic fungus varies in specificity widely within genera, between genera and among species. *B. bassiana* (muscardine fungus) and *Metarhizium anisopliae* (*M. anisopliae*) are best studied and have a wide host range including hundreds of insect species. Many other hosts have been reported in the recent years including species of Coleoptera, Lepidoptera, Diptera, Homoptera, and Hymenoptera. The prevalent hosts of *M. anisopliae* are mostly coleopteran species that include more than 70 scarab species. The isolate of any fungus is more important than species for bio-insecticides (Vestergaard et al., 2003). The strains of insect infecting fungus are specific to hosts and very rarely infect both pest and beneficial species. The insect pathogenic fungus, *Zoophthora radicans* has been reported from above 80 species of insects of Homoptera, Lepidoptera, Diptera, and Coleoptera.

With many exceptions, the specific strains are usually more infection causing to the insects mostly to the original host that is closely related. While other groups of insect infecting fungi are restricted to host range. In rare cases, insect infecting fungus can kill some species in specific situations, such as *M. anisopliae* is hardly reported as mosquito pathogen, but in controlled laboratory conditions, several isolates are pathogenic to the larvae of mosquito (Daoust & Roberts, 1982). In certain cases, the pathogenic fungus can kill only the vulnerable hosts. The laboratory and field vulnerability of different insects to different entomopathogenic fungi and even to different isolates is different. Similarly, the virulence of different life stages of insect is different toward the entomopathogenic fungi. Mostly, the fungi opposite to viral, bacterial and protozoan pathogens, penetrates the insect directly without digestion, and due to this reason non-feeding stages of insect, such as pupae are also susceptible to pathogenic fungi. *B. bassiana* and *M. anisopliae* both can infect all the life stages of insects. Usually, fungi are called as allergens as they produce toxin and have the capacity to infect vertebrates. Therefore, fungi are an important source of allergic reactions in humans. But, the fungi that cause infections in insects are not allergic to human beings. Above all, vertebrates are not susceptible to the insect infecting fungi (Lacey et al., 2001). The exception comes in certain cases, such as *Aspergillus flavus* (*A. flavus*) is famous pathogen of vertebrates. This is why; they are not used for commercial production (Table 5.2).

5.6   ECOLOGY OF ENTOMOPATHOGENIC FUNGI

Entomopathogenic fungi are vital and well-known factor of most terrestrial ecosystems because individual species of entomopathogenic fungi are distributed in a different way. Yet, many fungal species can be found basically all over the world. Some species of *B. bassiana* are reported from tropical rainforest and north of the Arctic Circle (Aung et al., 2008). *Tolypocladium cylindrosporum, B. bassiana, and M. anisopliae* have been reported from Norway (Klingen et al., 2002), and *Isaria farinosa, M. anisopliae, B. bassiana* are reported from Finland (Vanninen, 1995). Insect associated fungi have also been isolated from Arctic Greenland and Antarctica (Eilenberg et al., 2007). Many cosmopolitan fungi belonging to the genus *Neozygites, Lecanicillium, Beauveria*, and *Conidiobolus* are frequently present on Antarctic sites (Bridge et al., 2005). Diversity of entomopathogenic fungi is not affected by altitude in range up to 1608 m (Quesada-Moraga et al., 2007). Conversely, Sun and Liu (2008) evaluated the significance of altitudes on the diversity of entomopathogenic fungi. Several groups of insect associated fungi were found in different habitats, such as many insect pathogenic myco-floras could be found in the

TABLE 5.2 Occurrence and Pathogenicity of Various Entomopathogenic Fungi from Different Insect Hosts.

Sr. no.	Insect species	Fungi	Insect order	Crop plant	Country	References
1.	*Adoryphorus couloni* (Burm.)	*Metarhizium*	Coleoptera	–	Australia	Rath et al. (1995)
2.	*Ancognatha scarabaecide*	*Metarhizium*	Coleoptera	–	Columbia	Rath et al. (1995)
3.	*Aphodius tasmaniae* (Postuare)	*Metarhizium*	Coleoptera	–	Australia	Rath et al. (1995)
4.	*Coleomegilla maculata*	*Metarhizium*	Coleoptera	–	Australia	Rath et al. (1995)
5.	*Phyllophaga anxia* (LeConte)	*Metarhizium*	Coleoptera	–	Canada	Butt et al. (1994)
6.	*Papilla japonica* Newn.	*Metarhizium*	Coleopteran	–	Japan	Butt et al. (1994)
7.	*Pterostichus* sp. n.					
8.	*Forficula auricularia* L.	*Metarhizium*	Orthoptera	–	UK	Butt et al. (1994)
9.	*Acyrthosiphon pisum* (Maskell)	*Metarhizium*	Orthoptera	–	UK	Butt et al. (1994)
10.	*Nilaparvata lugens* (Stal.)	*Metarhizium*	Orthoptera	Rice	India, Sri Lanka	Ambethgar (1997)
11.	*Agrotis segetum* (D. & S.)	*Metarhizium*	Orthoptera	–	UK	Butt et al. (1994)
12.	*Austracris guttulosa* (Walker)	*Metarhizium*	Orthoptera	Cruciferous vegetables	Australia	Butt et al. (1994)
13.	*Locusta migratoria* (L)	*Metarhizium*	Orthoptera	–	Brazil	Kleepies and Zimmerman (1992)
14.	*Mahanarva postica* (Stal.)	*Metarhizium*	Orthoptera	–	Brazil	Kleepies and Zimmerman (1992)
15.	*Orinebius kanetataki* (Finot)	*Metarhizium*	Orthoptera	–	Brazil	Klelepies and Zimmerman (1992)
16.	*Cosmopolites sordidus* (Ger.)	*Beauveria* spp.	Coleopteran	Banana	Kenya	Kaaya et al. (1991)
17.	*Leptinotarsa decemlineata* (Say)	*Beauveria* spp.	Coleopteran	Rice	India	Puzari et al. (1997)

TABLE 5.2 *(Continued)*

Sr. no.	Insect species	Fungi	Insect order	Crop plant	Country	References
18.	*Plocaederus ferrugineus* (Linn.)	*Beauveria* spp.	Coleopteran	Cashew	India	Ambethagar et al. (1994)
19.	*Aphis gossypii* (Glov.)	*Beauveria* spp.	Homoptera	Beans	Columbia	Landa (1984)
20.	*Bemisia argentifolii* (Geog)	*Beauveria* spp.	Homoptera	Beans	Columbia	Landa (1984)
21.	*Nilaparvata lugens* (Stal.)	*Beauveria* spp.	Homoptera	Rice	India	Ambethgar (1996)
22.	*Agrotis segetum*	*Beauveria*	Lepidoptera	Crucifers	UK	Butt et al. (1994)
23.	*Cnaphalocrocis*	*Beauveria*	Lepidoptera	Rice	India	Ambethgar (1991)
24.	*Helicoverpa armigera* (Hub.)	*Beauveria*	Lepidoptera	Chickpea pigeonpea	India	Gopalkrishnan and Narayan an (1988)
25.	*Achaea janata* Linn.	*N. rileyi*	Lepidoptera	Castor bean	India	Vimia Devi (1994)
26.	*Anticarsia gemmatalis* (Hb.)	*N. rileyi*	Lepidoptera	Velvet bean and soybean	Brazil Ecuador	Stansly et al. (1990); Gilreath et al. (1986)
27.	*Helicoverpa armigera* (Hub)	*N. rileyi*	Lepidoptera	Potato and tomato	Indonesia and Korea	Hugar and Hegde (1996)
28.	*Heliothis subflexa* (Guene)	*N. rileyi*	Lepidoptera	–	USA	Ignoffo (1981)
29.	*Spodoptera exigua* (Hbst.)	*N. rileyi*	Lepidoptera	Soybean	India	Phadke et al. (1978)
30.	*Spodoptera litura* (F.)	*N. rileyi*	Lepidoptera	Tobacco, chili	India, Japan	Rao and Phadke (1977)
31.	*Stenachroia elongella* Hmps.	*N. rileyi*	Lepidoptera	Sorghum	India	Phadke et al. (1978)
32.	*Myzus persicae*	*Verticillium lecanii*	Homoptera	Salad crops	–	Chandler (1992)

over ground environment and in soil. In the Białowiea forest of Poland, soil surface layer and soil litter are dominated by *Hypocreales* members but Entomophthorales members are found in understory trees. *Gibellula* is a spider pathogenic fungi found in rush communities and meadows (Sosnowska et al., 2004). In temperate forest habitats, pathogens of forest pests are frequently reported as members of Entomophthorales (Burges, 1981a). A diverse insect pathogenic fungal species of the genus *Cordyceps* were present in humid tropical forests (Evans, 1982; Aung et al., 2008). While other species of *Hypocreales*, such as *Metarhizium, Beauveria*, and *Isaria* were the prevailing fungi originate from insect in soil (Keller & Zimmerman, 1989; Samson et al., 1988). In spite of the fact that both *B. bassiana* and *M. anisopliae* are widespread but *B. bassiana* appears to be extremely susceptible to disturbed environmental conditions and so limited to natural habitats. Many reports illustrate that incidence of entomopathogenic fungi is high in forest soils as compared to intensively cultivated soils (Vanninen et al., 1989; Bałazy, 2004; Vanninen, 1995). Entomopathogenic fungi are normally present in leaf litter and soil of intercontinental forests but occurrence of these fungi in temperate forests is comparatively small in comparison with tropical forests (Aung et al., 2008; Evans, 1982). Range of entomopathogenic fungi is extremely high in temperate forests as compared to agricultural areas (Sosnowska et al., 2004) and these fungi may differ in their occurrence among the diverse forests (Chandler et al., 1997; Mietkiewski et al., 1991).

5.7 INFECTION MECHANISM OF ENTOMOPATHOGENIC FUNGI

The mode of infection of various entomopathogenic fungi is basically quite similar. A typical infection cycle involves several steps: conidial attachment, germination, penetration through the insect cuticle, vegetative growth within the host, fungal protrusion outside the insect, and conidiogenesis (McCoy et al., 1988). During attachment to the host's body, mucilage associated with conidia and enzymes play an important role. For the subsequent germination of spores a high humidity is mandatory. The next step, penetration of the insect cuticle, which rarely occurs via wounds, is thought to be a combination of mechanical and enzymatic means, in which proteases, lipases, and chitinases are the most important enzymes (Butt, 1990). After penetration fungi proliferate within the body of the host. The insect may be killed by some combination of mechanical damage produced by the fungus, nutrient exhaustion, and the action of fungal toxins (Gillespie & Claydon, 1989; Butt & Goettel, 2000).

5.7.1 CONIDIAL ATTACHMENT WITH THE CUTICLE

Host location is a casual incident for entomopathogenic fungi and attachment is an inactive process with the assist of water or wind. In the establishment of mycosis, initial step is acquired by the adhesion of fungal spores to cuticle of susceptible host. It was evaluated that dry spores of *B. bassiana* have an outer layer of interwoven fascicles of hydrophobic rodlets and this rodlet layer is thought to be particular to the conidial stage. The attachment of dry spores to the cuticle was recommended to be due to nonspecific hydrophobic forces (Boucias et al., 1988). Lectins, a kind of carbohydrate binding glycoprotein is present on the conidial surface of *B. bassiana* and these lectins involved in binding between insect cuticle and conidia. Interaction mechanism between cuticle and fungal spores continue to be recognized (Latge et al., 1988).

5.7.2 CONIDIAL GERMINATION ON THE CUTICLE

Conidial attachment to host surface promotes the germination and growth, which are extremely affected by the accessibility of nutrients, oxygen, water, pH, temperature, and by the influence of lethal host-surface compound. Broad host range of fungi germinate in reaction to a variety of nonspecific nitrogen and carbon sources (Sandhu, 1995) and fungi with limited host range emerge to have more specific supplies for germination (Leger et al., 1989). Conidia of *B. bassiana* germinate on the host surface and formulate a distinction infection structure called appressorium. Adaptations for chemical energy and physical concentration over small area may be represented by appressorium. Therefore, appressorium development contributes an essential part in establishing a pathogenic interaction with the host. Development of appressorium might be inclined by the topography of host surface and biochemical investigation specify the participation of cyclic AMP (cAMP) and intracellular second messengers Ca2+ in appressorium formation (Figure 5.1; Leger et al., 1991).

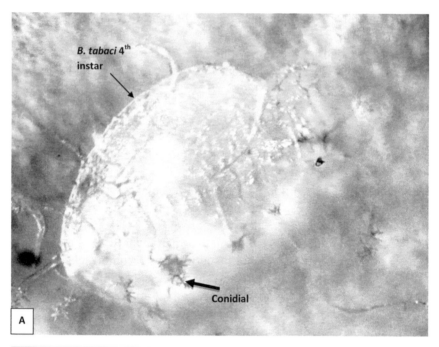

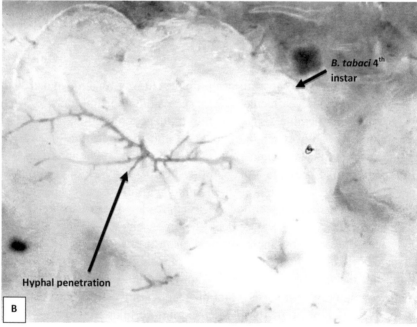

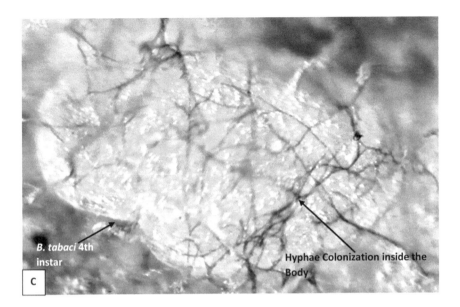

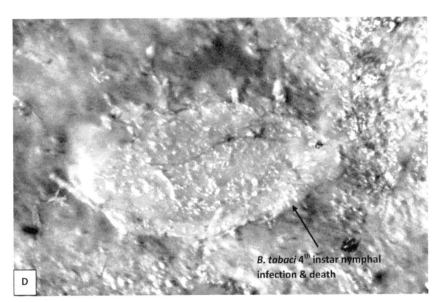

FIGURE 5.1   Infection of entomopathogenic fungi on nymphal stage of *Bemisia tabaci*: (A) conidial germination on nymph cuticle, (B) hyphal penetration inside the body, (C) colonization, and (D) infection on whole body.

5.7.3 PENETRATION INTO THE HOST CUTICLE

Entomopathogenic fungi obtain nutrients for their reproduction and growth by penetration throughout the cuticle into the insect body. Enzymatic degradation and mechanical pressure are responsible for entry into host as evidenced by the physical separation of lamellae by penetrated hyphae. The cuticle is composed of two layers: the outer epicuticle and the procuticle. Chitin is absent in epicuticle and consists of extremely complex thin structure but contains phenol-stabilized proteins and is enclosed by a waxy layer contain lipids, sterols, and fatty acids (Hackman, 1984). The procuticle is responsible for the greater part of the cuticle and consists of chitin fibrils rooted into a protein matrix collectively with quinines, proteins, and lipids (Neville, 1984). Chitin is structured helically to form a laminate structure in many areas of cuticle. It have been suggested that extracellular enzymes, such as lipases, proteases, and chitinases promote the degradation of insect cuticle in fungal pathogenesis. Synergistic action of several different enzymes plays a key role in penetration of insect cuticle (Leger et al., 1989).

5.7.4 COLONIZATION INTO THE BODY CAVITY

After the penetration of germ tube through the cuticle and insect epidermis, the fungus colonizes, or multiplies into the body cavity of insect. In some Entomophthorales, this multiplication may be by the protoplasts while in some Hyphomycete (e.g., *M. anisopliae*); blastospores are involved in the initial proliferation (Bidochka & Hajek, 1998). With the presence of cuticle, some other humoral and cellular defense methods are used by the insect against invasion of fungus. Toxins are produced by some insect pathogenic fungi and more of them aid to increase pathogenesis, and play an insecticidal role. While some other such fungi produce antimicrobial metabolites. Destruxins (DTXs) of *Metarhizium* species are among such fungal metabolites that increase the pathogenicity of fungus (Roberts, 1992).

5.7.5 INFECTION IN THE BODY

In the initial infection stages, no considerable behavioral symptoms are observed. But, some days before death, symptoms start to appear, such as reduced co-ordination, feeding activity (e.g., grasshopper and locust infected with *M. anisopliae*). Some other behavioral responses include behavioral fever, increased feeding (e.g., Colorado potato beetles infected with *B. bassiana*), positive or negative phototropism or geotropism, and altered mating (Noma & Strickler, 2000). "Behavioral fever," is another response of fungal infection in which the insect changes its body temperature by basking in the sun or using warm surfaces for their positioning, for example, response of locusts and grasshoppers in the infection of *M. anisopliae* and *B. bassiana* (Figure 5.2; Blanford & Thomas, 2001).

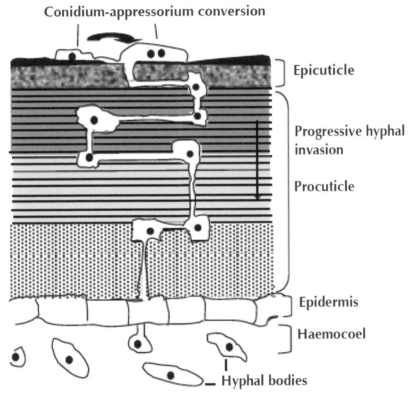

FIGURE 5.2 Infection mechanism of *Beauveria bassiana*: structure of the insect cuticle and penetration mode of fungal hyphae. Formation of the appressorium from the conidia helps in invasion of cuticle and hyphal penetration to the hemocoel subsequently. (Reprinted from Sandhu, S. S.; Sharma, A. K.; Beniwal, V.; Goel, G.; Batra, P.; Kumar, A.; Jaglan, S.; Sharma, A. K.; Malhotra, S. "Myco-Biocontrol of Insect Pests: Factors Involved, Mechanism, and Regulation". J. Pathog. 2012, 2012, Article ID 126819, 10. doi:10.1155/2012/126819. https://creativecommons.org/licenses/by/3.0/)

5.8   METHODS FOR ISOLATION OF ENTOMOPATHOGENIC FUNGI

5.8.1   SELECTIVE MEDIA

Different fungi are present in soil and they perform numerous functions. Many fungi and bacteria can grow on artificial culture media under lab conditions. This ability of the fungi to grow on artificial media has been exploited for their isolation from soil. Some microorganisms are isolated on specific media. Entomopathogenic fungi have been isolated on selective media. Bacterial growth is inhibited by adding broad-spectrum antibiotics, such as streptomycin, tetracycline, or chloramphenicol (Goettel & Inglis, 1997). The major impedance in using this method of isolation is that the hypocrealean entomopathogenic fungi grow relatively slowly in comparison to the ubiquitous opportunistic saprotrophic fungi found in the soil environment. So the media is modified with such contents, which restrict the growth of these fungi and prevent them overgrowing the required species. The most studied species are *B. bassiana*, *B. brongniartii*, and *M. anisopliae*.

A list of selective media for *Beauveria* was suggested by Goettel and Inglis (1997). Veens semi selective medium is usually used for the isolation of *Metarhizium* species (Hu & St Leger, 2002) and it is referred to Veen and Ferron (1966) for first time described by them. The medium is composed of chloramphenicol used as antibiotics and cycloheximide and dodine as the fungicides (Goettel & Inglis, 1997).

5.8.2   INSECT BAIT METHOD

The saprotrophic abilities have been exploited by the use of selective media for hypocrealean entomopathogenic fungi. However, insect bait method is used to evaluate the infection capability of the fungi. Originally, the insect bait method was used for the isolation of entomopathogenic nematodes from soil, but along with nematodes, sometimes fungi were also isolated additionally (Zimmermann, 1986). So, it was suggested by Zimmermann (1986) that this method can used for the isolation of entomopathogenic fungi. Insects which can easily reared and susceptible to fungi are used to make the method feasible. The highly susceptible larvae of the wax moth, *Galleria mellonella* (Lepidoptera: Pyralidae) are used traditionally as bait insect but larvae of mealworm, *Tenebrio molitor* (Coleoptera: Tenebrionidae) are also good. The indigenous species of many entomopathogenic fungi are isolated using larvae of *G. mellonella* in soil samples (Keller et al., 2003).

5.9 ENZYMES PRODUCED BY ENTOMOPATHOGENIC FUNGI

Mechanical force, some metabolic acids, and enzymatic processes are involved in initial interaction in pathogenesis. The enzymes responsible for pathogenesis of insects can be grouped into lipases, chitinases, peptidases, and proteases.

5.9.1 LIPASES

Protein and chitin are the major component of cuticle of insects, however, the epicuticular, which is external outermost surface layer, is formed by a complex mixture of non-polar lipids (Blomquist & Vogt, 2003). The epicuticle is waterproof and it acts against microbial attack as the first barrier. It is composed of heterogeneous mixture of fatty acids, esters, long-chain alkenes, and lipids. The importance of lipases in order to hydrolyze the ester bounds of lipoproteins, fats, and waxes at the interior of the insect integument is well known (Ali et al., 2009). Epicuticular lipids are involved where the chemical communication events occur (Blomquist & Vogt, 2003). Epicuticular layer keeps the cuticular surface dry as a result the penetration of chemicals and insecticide is affected (Blomquist et al., 1987). Large amount of protease (Pr1) is produced by fungus after the break down of epicuticle. The protease then degrades the proteinaceous material in the procuticle. Da Silva et al. (2010) demonstrated the role of lipases in the penetration of tegument and break down process. Lipases also help in the attachment of the spores to the epicuticle, after which alkenes and fatty acids break down in the cuticle surface initiates (Supakdamrongkul et al., 2010). From the initial infection stages, the *Bbcyp52x1* gene cluster of *B. bassiana* is involved in the epicuticle degradation. However, this degradation is no longer important when the cuticle has already broken down. For this reason, break down of the lipid substrates by the entomopathogenic fungi is the initial stage of the pathogenesis and spontaneously the cuticle penetration is performed (Zhang et al., 2012).

5.9.2 PROTEASES AND PEPTIDASES

Main constituent of insect cuticle is protein and chitin; thus peptidases and proteases of EPF act as virulence factor and cause the degradation of the insect cuticle. They are important for the growth of saprophytic fungi, and involved in activation of the prophenol oxidase of the hemolymph (Sheng et al., 2006). Pathogenicity and virulence are important features of entomopathogenic fungi. The proteases production by entomopathogenic fungi is related to pathogenicity and virulence. The proteases are most important in the infection process (Mustafa & Kaur, 2009). After the break down of epicuticle by the lipases, large amount of Pr1 protease is produced by the fungus, and the proteins are degraded by this enzyme. Chymoleastases collagenases, proteases, and protein degrading enzymes have been identified and characterized from *Verticillium lecanii* (*V. lecanii*), *B. bassiana*, *E. coronate, Lagenidium giganteum, B. brongniartii*, and *A. aleyrodis* (Sheng et al., 2006). Aminopeptidases and exopeptidases degrade the solubilized proteins until amino acids are produced, which act as nutrients for entomopathogenic fungi (Wang et al., 2002). The trypsin-like protease Pr2 and subtilisin-like serine-protease Pr1 are the most studied proteolytic enzymes (Leger et al., 1994). The activities of Pr1 and Pr2 have been determined in *Nomuraea rileyi* (*N. rileyi*), *Metarhizium flavoviride*, and *M. anisopliae* (Bidochka & Meltzer, 2000). During the initial stage of cuticle break down, these proteases are released and involved in a signal transduction mechanism, which activates the protein kinase A (PKA) mediated by AMPc (Fang et al., 2009). Mustafa and Kaur (2009) demonstrated that, protease Pr1 is the most important enzyme in infection process and the key extracellular enzyme for cuticle penetration and that in the absence of this enzyme the infection cannot be achieved.

5.9.3 CHITINASES

Chitin is a polymer of β-1, 4 N-acetyl glucosamine, and naturally found polymer after cellulose. The exoskeleton of invertebrates and cellular walls of the fungi are mainly composed of chitin (Seidl, 2008). Chitinases hydrolyze the β-1, 4 bonds of chitin polymer producing a predominant N, N'-diacetylchitobiose. Some bacteria, fungi, plants, vertebrates, and insects produce chitinases (Lu et al., 2005). Chitinases act synergistically with proteases to break down the cuticle of insect (Leger et al., 1991). The genome of filamentous fungi have between 10 and 25 chitinases, which are involved in diverse physiologic functions include: (a) degradation of chitin in exoskeletons of arthropods and cellular walls of the fungi; (b) remodeling of cellular wall during the branching, autolysis, hyphae growth, competence, and fusion of hyphae; and (c) defense against other fungi present in the same environment (Yang et al., 2007).

5.10 METABOLITES OR TOXINS OF DIFFERENT ENTOMOPATHOGENIC FUNGI

Different metabolites are produced by entomopathogenic fungi. These metabolites act as toxins and play a vital role in the infection development, and result in a number of symptoms in the insect. The characteristic of secondary metabolites to act toxin against the insect cellular lines can be the possibility in the development of new bio-insecticides. New bio-insecticides can be developed by using biotechnology and molecular biology techniques for the control of pest (Arboleda et al., 2011). Death caused by toxin is the result of a severe damage of the tissues, loss of nutrient intake, dehydration of cell, and toxicosis (Tellez et al., 2009) finally the hyphae emerge from the insect body, sporulates, and start a new infection cycle. Degradation by enzymatic action provides nutrients to the fungus, which facilitates in its proliferation inside the insect (Hasan et al., 2013). A large amount of toxins are produced by *M. anisopliae* and *B. bassiana* inside their hosts. Many toxins, such as beauverolides, bassianolide, beauvericin, and isarolides have been isolated from the hosts infected by *B. bassiana* (Hamill et al., 1969a). DTXs and cytochalasins are isolated from *M. anisopliae* and various insect tissues are affected by the toxins.

Following is the brief discussion of some of the secondary metabolites produced by the entomopathogenic fungi and their biological activity.

5.10.1   EFRAPEPTINS

Efrapeptins are peptide antibiotics, which form a complex mixture. They are produced by a soil Hyphomycete fungus named *Tolypocladium niveum* (syn. *Tolypocladium inflatum*, *Beauveria nivea*) (Jackson et al., 1979).

Efrapeptins C–G depicted toxicity against Colorado potato beetle and *Leptinotarsa decemlineata* (Krasnoff et al., 1991). All peptides showed that mitochondrial ATPase inhibitory activity when tested against preparations from entomopathogenic fungi and efrapeptins also inhibits photophosphorylation in chloroplasts (Lucero et al., 1976). The peptides are probably catalytic-site competitive inhibitors, which bind to the soluble F^1 part of the mitochondrial ATPase (Lardy et al., 1975). It has been proposed that efrapeptins bind at the catalytic site and block accessibility of an important arginine residue on the enzyme at the adenine nucleotide-binding site (Kohlbrenner & Cross, 1978). In some cases, it has also been shown that ATPases and certain mutant cell lines are resistant to efrapeptins (Wise et al., 1984; Dean et al., 1986).

5.10.2   DESTRUXINS

DTXs, the insecticidal cyclodepsipeptides were originally isolated from the *entomopathogenic* fungus, *M. anisopliae* (Kodaira, 1962; Roberts, 1966, 1969). Recently, DTXs were isolated as the active principle from the chlorosis-causing plant pathogenic fungus, *Alternaria brassicae* (*A. brassicae*; Bains & Tewari, 1987). Roseotoxin B, the trans-3-methyl proline analog of DTX A is known to be produced by *Trichothecium roseum* (*T. roseum*; Engstrom et al., 1975; Engstrom, 1978).

DTXs are insecticidal compounds, which depict the activity against insects. DTX B is reported to cause chlorosis of the pathogenic fungus, *A. brassicae* (Ayer & Pena-Rodriguez, 1987; Bains & Tewari, 1987). DTXs are also known to form homodestruxin B and desmethyldestruxin B. Roseotoxin B, trans-3-methylproline analog of DTX A was reported as the poisonous metabolite of *T. roseum* (Engstrom et al., 1975; Engstrom, 1978). Along with reported insecticidal activity, DTXs acquire immunodepressant potential in model systems of insects (Cerenius et al., 1990).

This characteristic is important for a group of cyclic peptides derivatives of fungi that are clinically fundamental in immunodepressant drugs. It was reported that DTX E was exhibit cytotoxic and cytostatic activity on mouse leukemia cells (Odier et al., 1987). Calcium channels were also activated by DTXs in insect muscles (Samuels et al., 1988c). DTXs do not have ionophoric properties, not like beauvericin, a cyclodepsipeptide formed by the entomopathogenic fungus *B. bassiana* (Abalis, 1981; Samuels et al., 1988c). The DTXs production and pathogenicity of the fungus suggested to be correlated and the role of the pathogen in the host

(Samuels et al., 1988a). These insect toxins apparently can be the cause of mortality of insect after the attack of the fungi (Samuels et al., 1988b).

5.10.3 BEAUVERICIN

Beauvericin has been isolated from the following microorganisms: *Paecilomyces fumosoroseus* (Bernardini et al., 1975), entomopathogenic deuteromycetous fungi *B. bassiana* (Hamill et al., 1969a). Beauvericin has been reporetd to probably act as an ionophore, which has the ability to make complexes with divalent cations (Prince et al., 1974; Dorschner & Lardy, 1968). It is cytotoxic (Vey et al., 1973) and also reported to possess insecticidal sctivity against mosquito larvae and blowflies (Grove & Pople, 1980). Originally, beauvericin was reported like a toxin against brine shrimp (Hamill et al., 1969b). No data available on the activity of beauvericin as a mammalian toxin.

Cyclodepsipeptides is an additional structurally-related group of enniatins A–C and antibiotics that was isolated from *Fusarium* sp. (Ovichinnikov et al., 1971). Here, enniatins A–C, the aromatic amino acid N-methyl phenylalanine of beauvericin is also replaced by N-methyl derived from alphatic amnio acids isoleucine, valine, and leucine, respectively. Althoguh beauvericin and enniatins are structurally similar but do not co-occur in nature may be because the multi enzymes are responsible for biosynthesis (Zocher et al., 1982; Peeters et al., 1983). Beauvericin acts as toxin against the insect, Colorado potato beetle, with LC_{50} of 633 ppm (95% fiducial limits, 530–748 ppm), and LC^{90} of 1196 ppm (95% fiducial limits, 954–1863 ppm).

5.10.4 BASSIANOLIDE

Bassianolide is a toxic metabolite of *V. lecanii* and *B. bassiana* (Suzuki et al., 1977). The toxin was originally isolated from strains of *B. bassiana* and *V. lacanii*, which were entomogenous on the cadavars of *B. mori* pupae (Murakoshi et al., 1978).

Bassianolide, like certain other depsipeptides, probably acts as an ionophore. The molecular conformation of such depsiptides results in a structure where the molecule has a relatively hydrophilic interior (because of the amide carbonyls) and the hydrophobic exterior (because of the hydrophobic side chains). Inorganic cations can be ligated in the hydrophilic center of the molecule. Ions can then be transported across lopophilic membranes because of the overall hydrophobic nature of the complex and bassianolide is toxic to insects (Kanaoka et al., 1978). Fifth instar larvae of *B. mori* showed mortality when fed upon an artificial diet containing bassianolide, suspended in water to *B. mori* larvae caused ationic symptoms at a dose of 2 μg/larva. The toxin was lethal at a dose of > 5 μg/larva.

5.10.5   LEUCINOSTATINS

Leucinostatins have been isolated from the submerged cultures of *Paecilomyces lilacinus* (*P. lilacinus*; Arai et al., 1973), *Paecilomyces marquandi* (Rossi et al., 1983), and *Paecilomyces farinosus*.

Leucinostatins have antimicrobial properties against many fungi and gram-positive bacteria (Fukushima et al., 1983a, b). They have also been shown to have anticancer activity against Ehrlich solid carcinoma (Arai et al., 1973). Leucinostatins have intraperitoneal and oral toxicity in mice. Intraperitoneal LD_{50} for leucinostatins A and B hydrochlorides is 1.8 mg/kg body weight. The oral LD_{50} values for leucinostatins A and B hydrochlorides have been reported to be 5.4 and 6.3 mg/kg body weight, respectively. These antibiotics act as uncouplers of oxidative phosphorylation in mitochondria (Fukushima et al., 1983a, b) and there are several reports of use of leucinostatins as tools to probe mitochondrial activity.

5.11   FACTORS INFLUENCE THE NATURAL PATHOGENICITY OF ENTOMOPATHOGENIC FUNGI

In a general view, the life cycle in insect pathogenic fungi, an infective spore stage is required generally that germinates on the host cuticle (Akbar et al., 2012). The infecting fungal spores then increase in number, causing toxin production that ultimately kills the insect. The fungus comes out of the insect cadaver in suitable temperature and humidity conditions and disseminates in the environment. Several species form the resting spores that become capable of infecting at the time of favorable environmental conditions. The fungus needs a strategy for the dissemination to infect new hosts (Shahid et al., 2012). In general, the vital factors for the survival and reproduction in fungus infection are host and suitable environmental conditions.

The accomplishment of biological control using entomopathogenic fungi depends on suitable temperature and humidity conditions. Humidity is often cited as the key abiotic factor influencing the potential of fungi. A high-relative humidity or free water is important for various parts of the infection cycle, that is, germination, infection, and conidiogenesis (McCoy, 1990). The lower limit of relative humidity for spore germination is in the region of 92–93% (Hall & Papierok, 1982). It is thought that the microclimate in the phyllosphere largely determines the performance of fungi (Ravensberg et al., 1990). The obstacle of high-humidity requirement can be overcome with improvements of formulation and application strategies (Bateman et al., 1993). For fungal growth inside the host and subsequent killing of the insect, the fungus is independent of external conditions. Compared with the relative humidity, the temperature seems to be less importance in glasshouses. Many entomopathogenic fungal species grow and sporulate at temperatures between 15 and 30 °C (Osborne et al., 1990a). In general, the optimum growth and germination rates on artificial media varied around 25 °C for *A. aleyrodis* and *V. lecanii*, and above 30 °C germination and growth decline rapidly or are impaired (Hall, 1981; Fransen, 1987). The optimum temperature for growth often coincides with optimum temperature for infection. However, temperatures below optimum not necessarily mean that infection is less successful. Although germination and growth of fungi are retarded, the development rate of the host insect may be affected likewise. Thus, the number of insects escaping fungal infection by molting will not increase as a result of a slower development of the fungus (Hall, 1981).

Wind and sunlight are the other abiotic conditions influencing pathogen and host encounter. Spores dispersal can be assisted with wind but in some cases, it decreases the humidity by removing free water. In most fungi, sunlight and ultraviolet (UV) rays can detriment to the infection perseverance but, a substantial variation in vulnerability to sunlight and UV is found between diverse strains and species (Fargues et al., 1996). For aquatic species of entomopathogens, humidity is not a limiting factor but the temperature, chemical factors like organic pollutants, and salinity are important factors (Gilbert & Gill, 2010).

In addition to abiotic factors, also biotic factors, such as host plant and insect, take part in the efficacy of entomopathogenic fungi. Fungal processes take place in the boundary layer surrounding the leaf or insect in which the air is undisturbed. The humidity in this layer is partly determined by the host-plant characteristics, such as leaf size and shape, leaf hairiness, position of the leaf on a plant (Ferro & Southwick, 1984). The host-plant chemistry may also influence the fungus directly, for instance, by inhibiting or stimulating fungal germination (Blakeman, 1971) or via the insect, influencing its susceptibility, or its development rate (Elliot et al., 2000; Poprawski et al., 2000).

5.12   ENTOMOPATHOGENIC FUNGI AS BIOPESTICIDE

Many studies on entomopathogenic fungi have been done for their development as bio-control agents against ticks, mites, and insects (Goettel et al., 2005; Vincent et al., 2007). This strategy involved in the development of entomopathogenic fungi as biocontrol agent is that the pest control depends on the activity of the released compound but not on successive generations of the fungus. Using this strategy, more than 171 products have been developed against at least 12 different species of fungi (Faria & Wraight, 2007).

Since the 1960s, worldwide a considerable number of myco-acaricides and myco-insecticides have been developed. The most common products are based on *M. anisopliae* (33.9%), *I. fumosorosea* (5.8%), and *B. brongniartii* (4.1%) among the 171 products (Faria & Wraight, 2001). The target of most of the products is the insects belonging to the orders Coleoptera, Thysanoptera, Orthoptera, Lepidoptera, and Hemiptera. The commercial products based on *B. bassiana* as myco-insecticides are more stable as compared with other insecticides against Lepidoptera (Thakur et al., 2011). It has been reported by many researches that fungi are a potential biological control agent. The characteristics which make them favorable biocontrol agent are that they are target specific, their resting stage or saprobic phase-producing capability, their fast reproductive capabilities, they have short generation time. The main condition for the development of an entomogenous fungus as a myco-biocontrol agent includes virulence of the fungus and the susceptibility of the insect. A broad host range is possessed by deuteromycetes fungi, particularly *Metarhizium* and *Beauveria* depict promising agent as myco-biocontrol products and presently used as myco-insecticides. New opportunities have been provided by molecular techniques for genetic engineering for the study of fungi used in myco-biocontrol of insect pests. The gene encoding for virulence and pathogenesis can be isolated and its role for pathogenesis can tested, which offer a rational basis for the enhancement of strain (Sandhu et al., 2012). Some important myco-biocontrol agents are discussed here.

5.12.1 BEAUVERIA SP.

B. bassiana, also known as imperfect fungus is a filamentous fungus, which belongs to a class of insect pathogenic deuteromycetes. Various *B. bassiana* spp. have been isolated from a different insect all over the world, which are important of medicinally or agriculturally. *Beauveria* sp. is host specific and includes hosts as the codling moth, Colorado potato beetle and some genera of termites, *Eutectona machaeralis* (*E. machaeralis*), and *Helicoverpa armigera* (Thakur & Sandhu, 2010). This ever-present fungus has long been recognized to be the most widespread causal means of disease connected with deceased and declining insects in nature and has been scrutinized worldwide as a microbial control agent of hypogeous species (Ferron, 1981; Mcleod, 1954). As an insecticide, the spores are sprayed on affected crops as an emulsified suspension or wettable powder. *B. bassiana* causes infection on a large variety of arthropod hosts and as a result is considered as a nonselective biological insecticide. *B. bassiana* is also useful against the *Dendrolimus* spp., *Ostrinia mubilalis*, and *Nephotettix* spp.

5.12.2 VERTICILLIUM LECANII

V. lecanii is a widely distributed entomopathogenic fungus and experiment conducted in South Korean greenhouses showed that *V. lecanii* can be used as bio-control agent against *Trialeurodes vaporariorum* (Kim et al., 2002). This fungus attaches to the underside of leaf through a filamentous mycelium and attacks nymphs (Nunez et al., 2008). In 1970s, various aphids' species (like *Myzus persicae*) and whitefly were controlled by *V. lecanii* in the greenhouse chrysanthemums (Hamlen, 1979). *V. lecanii* was used as main parasite of cereal-cyst nematode populations causing its massive decline in monocultures of susceptible crops (Kerry et al., 1982). Only some isolates *V. chlamydosporium* can act as commercial biological control agents as they show variable activity.

5.12.3 METARHIZIUM SPP.

M. anisopliae has been used as a successful candidate for biocontrol of various economically important insect pests as it is very a potential pathogen on insect pests (Sandhu et al., 1993; Sandhu & Mishra, 1994). Teak skeletonizer, *E. machaeralis* has been controlled successfully through *M. anisopliae* and total bioactivity of *M. anisopliae* has been experienced against teak pest as a possible myco-biocontrol agent (Sandhu et al., 2000). Solid-state fermentation technique has been used by Hasan et al. (2002) to test spore production of *M. anisopliae*.

5.12.4 NOMURAEA SP.

N. rileyi is a dimorphic Hyphomycete, which acts as a potential ento-mopathogenic fungi, which causes epizootic death in various insects. Ignoffo (1981) showed that *N. rileyi* parasites numerous insect species belong to Lepidoptera together with *Spodoptera litura* and *Coleoptera*. *N. rileyi* is host specific and ecofriendly and these characteristics make it a good biocontrol agent to be used in insect pest management. Though, their modes of infection and development have been reported for several insect hosts, such as *Anticarsia gemmatalis*, *Heliothis zea*, *Pseudoplusia*, *B. mori*, and *Plathypena scabra*. *N. rileyi* attacking *Spilosoma* was studied in detail for its myco-biocontrol mechanism (Mathew et al., 1998). That is, *Junonia orithya* eaten by hedge plant was controlled by an epizootic of *N. rileyi*, which was proved to be the most excellent substitute to manage *Junonia orithya* (Rajak, 1991).

5.12.5 PAECILOMYCES SP.

Paecilomyces is a genus of nematophagous fungus, which causes disease in notorious nematodes resulting in their killing by pathogenesis. Thus, the fungus can be applied to the soil as a biocontrol agent against nema-todes. Eggs of cyst and root-knot nematodes are infected and assimilated by *P. lilacinus*. In 1979, after the discovery of the fungi as a biological control agent, most research has been focused on it. *Paecilomyces fumoso-roseus* causes the disease called "Yellow Muscardine" in whiteflies and is important parasite of whiteflies worldwide (Nunez et al., 2008). *Bemisia* and *Trialeurodes* spp. have been controlled successfully in both open field environments and greenhouse through strong epizootic potential of *P. lilacinus*. *P. lilacinus* can grow broadly above the leaf surface under moist environment, which increases its capability to spread quickly throughout whitefly populations and is applied in tropical and subtropical agricultural soils (Wraight et al., 2000). Mosquito sp. *Culex pipiens* has been controlled by *P. furiosus* (Table 5.3; Sandhu & Mishra, 1994).

5.13   ADVANTAGES OF USING FUNGI AS BIOCONTROL AGENT

They are highly host specific for the management of insect pests, so beneficial predators of insect are not affected. The chemicals pesticides are hazardous to mammals and environment, which are usually not affected by bio-pesticides. As the fungi have different modes of actions so the resistance development is minimized and has genes for secretion of insect toxins; therefore, they have high potentials for additional improvement by biotechnological research. Endophytic capability of fungi activates the immune system (Khan et al., 2012) and high perseverance in the environment provides long-term inhibition effects of entomopathogenic fungi on pest.

5.14   DISADVANTAGES OF THE FUNGI AS BIOCONTROL AGENTS

The main disadvantage is that they are very slow killing: normally take 2–3 weeks to destroy the insects, however, chemicals are very fast and possibly will require only 2–3 h. Pathogenesis caused by the fungi is a biological process, which is carried out at specific conditions, such as specific humidity, light period, and temperature (Khan et al., 2012). They are host specific, cannot be commercialized and their manufacture is comparatively costly and the shelf life of spores is short so frosty storeroom is necessary. Specific techniques for application are needed to be applied for retaining the long-term impacts as the efficacy and persistence of entomopathogenic fungi in the host population may differ in various insect species, also cause possible risks to immunodepressive people.

TABLE 5.3 Commercial Products of Entomopathogenic Fungi (Butt et al., 2001; Wraight et al., 2001; Copping, 2004; Zimmermann, 2007; Khachatourians, 1986).

Fungus	Product/trade name	Company/producer	Country/origin	Target pests
Culicinomyces			Austria, Belgium	Mosquito larvae
Clavisporus			Czech	
Hirsutella thompsonii	Mycar		Austria, Belgium	Citrus rust mite
Metarhizium Anisopliae	Meta-sin			Spittle bug; sugarcane frog Hopper
Verticillium lecanii	Vertalec			Aphids; coffee green bug; greenhouse whitefly trips
Beauveria bassiana	Bio-power	Stanes	India	Mite; coffee green bug
	Boverol	Fytovita	Czech Republic	
	Conidia	LST	Columbia	
	Mycotrol ES; Mycotrol-O	Laverlam International (formerly Emerald BioAgriculture)	USA	
	Natrualis	Intrachem	Italy	Aphids spittle bug; sugarcane
	Ostrinil	Arista (formerly NPP, Calliope)	France	
	Proecol	Probioagro	Venezuela	
	Racer BB	SOM Phytopharma	India	
B. brongniartii (tenella)	Beauveria Schweizer	Lbu (formerly Eric Schweizer Seeds)	Switzerland	Greenhouse whitefly thrips; mosquito larvae
	Betal	Arista (formerly NPP, Calliope)	France	
	Biolisa Kamikiri	Nitto Denko	Japan	
	Engerlingspilz	Andermatt Biocontrol AG	Switzerland	
	Melocont Pilzgerste	Agrifutur–Kwizda	Italy, Austria	

5.15   COMMERCIAL MYCO-INSECTICIDES APPLICATIONS

5.15.1   PROTECTED CROPS

Although greenhouses are prone to extremely strong human interference factors but a number of factors together with the potential for stabilizing environmental conditions, opportunity of avoiding pest invasions, or the chances for regular sanitation are accountable for the achievement of biological control programs (Dowell, 1990). In biological control, acceptance for pests in such a commercially important crop is imposing a tricky challenge but troubles associated to insecticide resistance and different factors especially, the requirement of consumers for healthy foods have produced good situation for implementation of biological control in confined crop systems worldwide. A large number of myco-insecticides are accessible for greenhouse use. Mycotal product based on *V. lecanii* is commercialized in Europe for control of *T. vaporariorum*, even though it also has a little efficacy to control *B. tabaci*. Saito (1993) found that *V. lecanii* was as efficient as buprofezin to manage the nymphs of *B. tabaci* infesting tomatoes. In several European countries, *P. fumosoroseus* (strain Apopka-97) is also being used to manage the *T. vaporariorum* and *B. tabaci*. This fungus is also registered, but not so far commercialized, in the USA and products containing this fungus (tradenamed PFR-97 in the USA and PreFeRal in Europe) are based on granular-formulated blastospores. These spores have inferior shelf life compared to conidia, but can be produced with better effectiveness. A lot of the work published for the efficacy data of these products relate to control of *T. vaporariorum*, while PreFeRal was very efficient (> 90% mortality) to control whitefly on greenhouse cucumbers and tomatoes with improved efficacy on cucumbers (Bolckmans et al., 1995). On the contrary, Vidal et al. (1998b) evaluated the same effectiveness of PFR-97 was used against whitefly infesting greenhouse cucumbers, cabbage, and three cultivars of tomato. Biological control of whitefly in both field crops and greenhouses can be achieved by different conidial-based products that are commercially available in China and Latin America (Fang et al., 1986). In greenhouses and nurseries different pests such as, thrips, whiteflies, mealy bugs, and aphids can be controlled through different formulations including oil-based emulsions, a wet table powder based on GHA strain available in market as named BotaniGard in Mexico, USA and other Central Americans countries. *B. bassiana* has a massive potential as a bio-insecticide, although this fungus rarely caused natural infection of whiteflies.

5.15.2   OPEN FIELD CROPS

Under field conditions, efficiency of myco-insecticides is normally not effective in greenhouses to elaborate the lesser number of products for these agro-ecosystems as compared to protected agriculture and this problem is the result of economic constraints to harsh climatic conditions. Most of the field crops cannot sustain the other well-organized application methods and high rates of application suggested for the crops in greenhouse. Poor effectiveness against adults of whiteflies under usual weather conditions is a new difficult problem. Crops are mostly susceptible to mass migrations of whiteflies from harvested fields or adjacent vegetation in open field environment. In a low-value field crops, consequences at present do not permit a suggestion for common wide-scale use of myco-insecticides for the control of *B. tabaci* in large hectarage. Most of the research is related on field trials with the *B. bassiana* strain and registered under the name Mycotrol. In experiments by Wraight et al. (2000), *B. bassiana* and *P. fumosoroseus* were used against nymphs of whitefly infesting honeydew melons, zucchini squash, cantaloupe and cucumbers, and 86–98% of management achieved with both pathogens subsequent 2–6 applications of low to high rates of conidia ($1.25–5.0 \times 10^{13}$/ha) at 4–7 days intervals by using a portable air-blast sprayer. Beauveria applications completed with high-pressure hydraulic sprayers and tractor-mounted airblast are effective to some extent (Wraight & Carruthers, 1999; Jaronski & Lord, 1996). Poprawski (1999) observed 65% control of *B. tabaci* on collards after weekly five applications of *B. bassiana* strain GHA. Jaronski et al. (1998) evaluated 80% control of whitefly nymphs by a single high-rate application in irrigated cotton in the Imperial Valley of California. Up till now, in compare to the above results, applications of Mycotrol by other researchers working in southern Texas and in the Imperial Valley find no efficient control of whiteflies in cotton (Liu et al., 1999; Wraight et al., 1996).

5.15.3   RECOMMENDATIONS FOR THE USE OF MYCO-INSECTICIDE

Myco-insecticides are efficient against small populations of first-instar nymphs, to stop populations from building to unmanageable levels. Different fungal pathogens are not consistent to control large populations of adults or late-instar nymphs, or any population under harsh outburst environment. Hyphomycetes such as *P. fumosoroseus* and *B. bassiana* are normally well matched with a wide collection of chemical insecticides required for management of migrating adults that can quickly overcome young seedlings and transmit plant viruses. Benefit must be taken of optimistic environmental circumstances whenever possible, for example, by selecting spring or fall crops, or by timely spray programs to match with predicted intervals of high humidity and reasonable temperature in green house or field. High efficiency of *P. fumosoroseus* and *B. bassiana* was recorded in cucurbit field experiments during which average daily highest temperatures did not exceed 32 °C (Wraight et al., 2000). Incompatible fungicides and myco-insecticide should be applied asynchronously and users must choose mechanically sprayers competent of targeting leaf undersides. Whenever possible, users must utilize such methods as increasing spray pressure, reducing the speed of ground, and quantity to maximize spray coverage. Applications must be banded on the crop line to take advantage of dose at least cost. Banding can be particularly beneficial in different crops such as cucurbits grown on extensive areas. In such cases, banded applications applied to small plants can utilize significantly better rate of application at the same cost as much lower broadcast rate (Vandenberg et al., 2000). Considerations should be given to match myco-insecticide use for compatible crops and crop cultivars and it is also easy to manage whiteflies infesting plant varieties with restricted vegetative growth. For example, crops that have a very short-cycle, such as cucumbers, pickle are easily protected than long-cycle crops. Myco-insecticide should be stored under moderate temperatures or refrigeration whenever possible. In recent times, conidia-based formulations of various entomopathogenic Hyphomycetes can be stored at room temperature (25–30 °C) for at least eight months (Alves, 1999).

5.16 COMPATIBILITY OF ENTOMOPATHOGENIC FUNGI WITH PESTICIDE

Entomopathogenic fungi that are used for biological control are very less vulnerable to the chemical action of insecticides than parasitic and predatory insects. Myco-insecticides are compatible with a wide variety of insecticides for the establishment of integrated insect pest management practices. Mostly, the compatibility of entomopathogenic fungi with pesticides used in commercial crop protection systems is critical, if these fungi are to be utilized for insect control. Different methods are available to evaluate the fungicidal activity and many studies of entomopathogenic fungi have used inhibition of growth in liquid or on solid media (Roberts & Campbell, 1977). Soper et al. (1974) experienced the efficacy of fungicides on growth of some species of entomopathogens in agar media and observed that all were inhibitory. Growth of *B. bassiana* and pathogen of numerous insect pests was inhibited by fungicide in solid and liquid media. Four fungicides used commercially for control of foliar diseases of potato were evaluated *in vitro* and under field conditions on survival of spores of *B. bassiana*, a pathogen of the Colorado potato beetle. Mancozeb, the most detrimental of the fungicides, substantially reduced the survival under all available conditions (Loria et al., 1983). Most importantly, entomopathogenic *V. lecanii* strains have been found non-pathogenic to plants and humans (Burges, 1981b).

Recommended concentrations of insecticides viz., fenitrothion, monocrotophos and phosphamidon, and the fungicides like ziram, foltaf, dithane Z-78, chlorothalonil, captan, and wettable sulfur were found safe to the fungus *N. rileyi*. Influence of nine insecticides on the natural infection of *A. gemmatalis* by *N. rileyi* was studied by Barbosa et al. (1997). The effects of trichlorfon and chlorpyrifos did not differ from the untreated control. *Baculovirus anticarsia,* diflubenzuron, endosulfan, methamidophos, monocrotophos, methyl parathion, and thiodicarb showed similar performance and caused significant decrease in the percentage of mycosed larvae. Tang and Hou (1998) evaluated five fungicides, eight insecticides, and nine herbicides commonly used in maize fields, for their inhibition to conidial germination of *N. rileyi* by paper disc method. Among them, propineb and maneb (fungicides) were extremely inhibitory to the fungus though, herbicides and insecticides examined did not affect the conidial germination significantly.

5.17   TRANSGENIC ENTOMOPATHOGENIC FUNGI

Extensive exploit of fungi for biocontrol will rely on improvements of wild-type strains by mixing character of diverse mutants and strains. Improvements may be considered of two types: (a) effectiveness of the insecticide will be improved by dropping the dose required to eradicate the insects and decrease crop damage by the pest or minimize the time to kill the pest, and (b) hypervirulent strain development is an absolute considerate of the outstanding pathology of fungi infection and improved by increasing the host range. The basic tools for understanding the pathogenesis mechanism and recombinant organisms with new characteristics are provided by molecular biology. Early development of these goals has occurred with *M. anisopliae* and to a large degree with *B. bassiana* (Hegedus, 1991). The fundamental element of recent fungal research is the genetic transformation system, which is required for the tentative exploitation of virulence genes *in vitro* and *in vivo* has been recognized (Hasan et al., 2002; Goettel & St Leger, 1990; Sandhu & Vikrant, 2006). The achievement of using these events based on the accessibility of selectable transformation markers (Sandhu et al., 2000). Different transformation techniques have been used to target particular pathogenic genes, examine virulence factors of *M. anisopliae*, and to develop a strain with improved virulence. Unraveling the molecular mechanisms of fungal pathogenesis in insects will give the foundation for the genetic engineering of entomopathogenic fungi (Sandhu et al., 2012).

5.18 FUTURE ASPECTS OF ENTOMOPATHOGENIC FUNGI

Many insect pathogenic fungus based bio-insecticides have been produced and commercially manufactured so far. Though, a number of studies have been done for the improvements in production, pesticide formulation and practical application, even then many improvements are required to search and study, and implementation. Regarding myco-insecticides, additional establishment for stabilization, formulation, mass production, and delivery are of serious implication. Natural enemies and agrochemicals compatibility testing is desired to maintain integration of these control agents into new and existing IPM systems. Growers and extension agents should be educated for identification of natural enemies, especially in developing countries. Therefore, microbial control action thresholds and sampling methods must be developed for different pests. Improvement in strains by the use of guides and selections will be a best strategy in the future. The use of microbial insecticides should be a contribution toward all fields of agriculture, sustainable agriculture, forestry, and horticulture. It should be cared that the entomopathogenic fungus should not destroy beneficial natural fauna in the environment. Strategies should be made at small and large levels for the mass production of conidia. Use of insect pathogenic fungus is unavoidable as it is an essential part of integrated pest management programs in many ecological zones (Gul et al., 2014; Faria & Wraight, 2001). Just like the genes of *Bacillus thuringiensis* (*B. thuringiensis*), which are inserted in different crops through biotechnology to get resistance against different pathogens and insect pests, the genes (e.g., chitinases, proteases, etc.) of entomopathogenic fungus can be incorporated in different crops to get resistance against other pathogens and insect pests in future. So, these entomopathogenic fungus genes can be a useful tool of GM crops in future.

KEYWORDS

- *Beauveria bassiana*
- entomophthorales
- *Bemisia tabaci*
- bassianolide
- chitinases
- lipases

REFERENCES

Abalis, I. M. Biochemical and Pharmacological Studies of the Insecticidal Cyciodepsipeptides Destruxins and Bassianolide Produced by Entomopathogenic Fungy. Ph.D. Thesis, Cornell University, Ithaca, New York, 1981, p 198.

Akbar, S.; Freed, S.; Hameed, A.; Gul, H. T.; Akmal, M.; Malik, M. N.; Khan, M. B. Compatibility of *Metarhizium anisopliae* with Different Insecticides and Fungicides. *African J. Microbiol. Res.* **2012**, *6*(17), 3956–3962.

Ali, S.; Huang, Z.; Ren, S. X. Production and Extraction of Extracellular Lipase from the Entomopathogenic Fungus *Isaria fumosoroseus* (Cordycipitaceae: Hypocreales). *Biocontrol Sci. Technol.* **2009**, *19*, 81–89. http://dx.doi.org/10.1080/09583150802588524

Alves, R. T. Development of Mycoinsecticide Formulations and Application Techniques Appropriate for Pest Control. Ph.D. Thesis, University of London, UK, 1999, p 139.

Ambethgar, V. Biological Control of Brown Plant Hopper *Nilaparvata lugens* with Entomogenous Fungi. *Madras Agric. J.* **1996**, *8*, 203–204.

Ambethgar, V. Record of White Muscardius Fungus, *Beauveria bassiana* (Bals.) Vuill. on Rice Leaf Folder Complex from Karaikol. Pondicherry Union Territory (India). *J. Ento. Res.* **1997**, *21*, 197–199.

Andrews, J. H.; Harris, R. F. The Ecology and Biogeography of Microorganisms on Plant Surfaces. *Ann. Rev. Phytopathol.* **2000**, *38*, 145–180.

Arai, T.; Mikami, Y.; Fujushima, K.; Utsumi, T.; Yarazawa, K. A New Antibotic, Leucinostatin, Derived from *Penicillium lilacinum. J. Antibiot.* **1973**, *26*(3), 157–161.

Arboleda-Valencia, J. W.; Gaitan-Bustamante, A. L.; Valencia-Jiménez, A.; Grossi-de-Sa, M. F. Cytotoxic Activity of Fungal Metabolites from the Pathogenic Fungus *Beauveria bassiana*: An Intraspecific Evaluation of Beauvericin Production. *Curr. Microbiol.* **2011**, *63*, 306–312. http://dx.doi.org/10.1007/s00284-011-9977-2

Arnold, A. E.; Lewis, L. C. Ecology and Evolution of Fungal Endophytes, and their Roles against Insects. In *Insect-Fungal Associations: Ecology and Evolution;* Vega, F. E., Blackwell, M., Eds.; Oxford University Press: New York, 2005; pp 74–96.

Arnold, A. E.; Lutzoni, F. Diversity and Host Range of Foliar Fungal Endophytes: Are Tropical Leaves Biodiversity Hotspots? *Ecology.* **2007**, *88*, 541–549.

Aung, O. M.; Soytong, K.; Hyde, K. D. Diversity of Entomopathogenic Fungi in Rainforests of Chiang Mai Province, Thailand. *Fungal Divers.* **2008**, *30*, 15–22, ISSN 1560–2745.

Ayer, W. A.; Pena-Rodriguez, L. M. Metabolities Products by *Alternaria brassicae*, the Black Spot Pathogen of Canola. Part. 1, the Phytoxic Components. *J. Nat. Prod.* **1987**, *50*(3), 400–407.

Bains, P. S.; Tewari, J. P. Purification, Chemical Characterization and Host-Specificity of the Toxin Produced by *Alternaria brassicae. Physiol. Mol. Plant Pathol.* **1987**, *30*, 259–271.

Bałazy, S. Entomophthorales. In *Flora of Poland, Fungi (Mycota)*; Instytut Botaniki: Krakow, Poland, 1993; Vol. 24.

Bałazy, S. Znaczenie Obszarów Chronionych Dla Zachowania Zasobów Grzybów Entomopatogenicznych [Significance of Protected Areas for the Preservation of Entomopathogenic Fungi]. *Kosmos.* **2004**, *53*(1), 5–16, ISSN 0023–4249 (in Polish with English summary).

Barbosa, J. V.; Daemon, E.; Bittencourt, V. R. E. P. Partial Purification and Characterization of Two Extracellular N-Acetyl-D Glucosaminidases on Larval Moulting and Nymphal Survival of *Rhipicephalus sanguineus. J. Invertebr. Pathol.* **1997**, *66*, 264–282.

Bardgett, R. D. *The Biology of Soil: A Community and Ecosystem Approach*; Oxford University Press: New York, 2005.

Barr, D. J. S. Chytridiomycota. In *Systematics and Evolution*; Springer: Berlin, Heidelberg, 2001; pp 93–112.

Bateman, R. P.; Carey, M.; Moore, D.; Prior, C. The Enhanced Infectivity of *Metarhizium flavoviride* in Oil Formulations to Desert Locusts at Low Humidities. *Ann. Appl. Biol.* **1993**, *122*, 145–152.

Beard, C. E.; Adler, P. H. Seasonality of Trichomycetes in Larval Black Flies from South Carolina, USA. *Mycologia.* **2002**, *94*(2), 200–209.

Bernardini, M.; Carilli, A.; Pacioni, G.; Santurbano, B. Isolation of Beauvericin from *Paecilomyces fumosoroseus. Phytochemistry.* **1975**, *14*, 1865.

Bidochka, M. J.; Hajek, A. E. A no Permissive Entomophthoralean Fungal Infection Increases Activation of Insect Prophenoloxidase. *J. Invertebr. Pathol.* **1998**, *72*(3), 231–238.

Bidochka, M. J.; Meltzer, M. J. Genetic Polymorphisms in Three Subtilisin-Like Protease Isoforms (*Pr*1A, *Pr*1B, and *Pr*1C) from *Metarhizium* Strains. *Can. J. Microbiol.* **2000**, *46*, 1138–1144. http://dx.doi.org/10.1139/w00-112

Bischoff, J. F.; Rehner, S. A.; Humber, R. A. A Multilocus Phylogeny of the *Metarhizium anisopliae* Lineage. *Mycologia.* **2009,** *101*(4), 512–530.

Blakeman, J. P. The Chemical Environment of the Leaf Surface in Relation to Growth of Pathogenic Fungi. In *Ecology of Leaf Surface Micro-Organisms;* Preece, T. F., Dickinson, C. H., Eds.; Academic Press: London, UK, 1971; pp 255–268.

Blanford, S.; Thomas, M. B. Adult Survival, Maturation, and Reproduction of the Desert Locust *Schistocerca gregaria* Infected with the Fungus *Metarhizium anisopliae* Var *Acridum. J. Invertebr. Pathol.* **2001,** *78*(1), 1–8.

Blomquist G. J.; Nelson D. R.; de Renobales M. Chemistry, Biochemistry and Physiology of Insect Cuticular Lipids. *Arch. Insect. Biochem. Physiol.* **1987,** *6*, 22–265. http://dx.doi.org/10.1002/arch.94006040

Blomquist G. J.; Vogt R. G. Biosynthesis and Detection of Pheromones and Plant Volatiles—Introduction and Overview. In *Insect Pheromone Biochemistry and Molecular Biology*; Blomquist, G. J., Vogt R. G., Eds.; Elsevier Academic Press: London, 2003; pp 137–200.

Bolckmans, G.; Sterk, J.; Eyal, J.; Sels, B.; Stepman, W. Preferal (Paecilomyces Fumosoroseus Strain Apopka 97), a New Microbial Insecticide for the Biological Control of Whiteflies in Greenhouses. *Med. Fac. Landbouww. Univ. Gent.* **1995,** *60*, 707–711.

Boucias, D. G.; Pendland, J. C.; Latge, J. P. "Nonspecific Factors Involved in the Attachment of Entomopathogenic Deuteromycetes to Host Insect Cuticle". *Appl. Environ. Microbiol.* **1988,** *54*(7), 1795–1805.

Bridge, P. D.; Clark, M. S.; Pearce, D. A. A New Species of *Paecilomyces* Isolated from the Antarctic Springtail *Cryptopygus antarcticus. Mycotaxon.* **2005,** *92*, 213–222, ISSN 0093-4666.

Burges, H. D. Progress in the Microbial Control of Pests. In *Microbial Control of Pests and Plant Diseases 1970–1980*; Burges, H. D., Ed.; Academic Press: London, 1981b; pp 1–6.

Burges, H. D. Strategy for the Microbial Control of Pests in 1980 and Beyond. In *Microbial Control of Pests and Plant Diseases 1970–1980*; Burges, H. D., Ed.; Academic Press, ISBN 0121433609: London and New York, 1981a; pp 797–836.

Butt T. M.; Jackson C.; Magan. N. Introduction Fungal Biological Control Agents: Progress, Problems and Potential. In *Fungi as Biocontrol Agents: Progress, Problems and Potential*; Butt, T. M., Jackson, C., Magan N., Eds.; CAB International: Wallingford, 2001; pp 1–8.

Butt, T. M. Fungal Infection Process–A Mini-Review. In *Proceedings and Abstracts, Vth International Colloquium on Invertebrate Pathology and Microbial Control;* Adelaide: Australia, 1990; pp 121–124.

Butt, T. M.; Goettel, M. S. Bioassays of Entomogenous Fungi. In *Bioassays of Entomopathogenic Microbes and Nematodes;* Navon, A., Ascher, K. R. S., Eds.; CAB International: Wallingford, UK, 2000; pp 141–191.

Butt, T. M.; Ibrahim, L.; Ball, B. V.; Clark, S. J. Pathogenicity of the Entomogenous Fungi *Metarhizium anisopliae* and *Beauveria bassiana* against Crucifer Pests and the Honeybee. *Biocontrol Serv. Technol.* **1994,** *4*, 207–217.

Cerenius, L.; Thornquist, P. O.; Vey, A.; Johansson, M. W.; Soderhall, K. The Effect of the Fungal Toxin Destruxin E on Isolated Crayfish Haemocytes. *J. Insect Physiol.* **1990,** *36*, 785–789.

Chandler, D. The Potential of Entomopathogenic Fungi to Control the Lettuce Root Aphid *Pemphigus bursarius. Phytoparasitica.* **1992,** *20*, 115–155.

Chandler, D.; Hay, D.; Reid, A. P. Sampling and Occurrence of Entomopathogenic Fungi and Nematodes in UK Soils. *Appl. Soil Ecol.* **1997,** *5*(2), 133–141, ISSN 0929–1393.

Cooke, M. C. Vegetable Wasps and Plant Worms. Society for Promoting Christian Knowledge: London, 1892, p 364.

Cooper, R. D.; Sweeney, A. W. Laboratory Studies on the Recycling Potential of the Mosquito Pathogenic Fungus *Culicinomyces clavisporus. J. Invertebr. Pathol.* **1986,** *48*(2), 152–158.

Copping L. G. The Manual of Biocontrol Agents, British Crop Protection Council, Crop Protection. *Crop Prot.* **2004,** *23*, 275–285.

Da Silva, W. O. B.; Santi, L.; Scharank, A.; Vainstein, M. H. *Metarhizium anisopliae* Lipolytic Activity Plays a Pivotal Role in *Rhipicepthalus* (*Boophilus*) *Microplus* Infection. *Fungal Biol.* 2010, *144*, 10–15. http://dx.doi.org/10.1016/j.mycres.2009.08.003

Daoust, R. A.; Roberts, D. W. Virulence of Natural and Insect-Passaged Strains of *Metarhizium anisopliae* to Mosquito Larvae. *J. Invertebr. Pathol.* **1982,** *40*(1), 107–117.

Dean, G. E.; Nelson, P. J.; Rudnick, G. Characterization of Native and Reconstituted of Chromaffin Granules. *Biochemistry.* **1986,** *25*, 4918–4925.

Dorschner, E.; Lardy, H. Specificity of Ionic Transport Induced by Beauvericin. *Antimicrob. Agents Chemother.* **1968,** *8*, 11–14.

Dowell, R. V. Integrating Biological Control of Whiteflies into Crop Management Systems. In *Whiteflies, their Bionomics, Pest Status and Management;* Gerling, D., Ed.; Intercept: Andover, UK, 1990; pp 315–335.

Eilenberg, J. Biology of Fungi from the Order Entomophthorales, with Emphasis on the Genera *Entomophthora, Strongwellsea* and *Eryniopsis*. The Royal Veterinary and Agricultural University: Copenhagen, Denmark, 2002.

Eilenberg, J.; Hajek, A.; Lomer, C. Suggestions for Unifying the Terminology in Biological Control. *BioControl.* **2001,** *46*, 387–400.

Eilenberg, J.; Schmidt, N. M.; Meyling, N.; Wolsted, C. Preliminary Survey for Insect Pathogenic Fungi in Arctic Greenland. *IOBC/WPRS Bull.* **2007,** *30*(1), 12, ISSN 1027–3115.

Elliot, S. L.; Sabelis, M. W.; Janssen, A.; van der Geest, L. P. S.; Beerling, E. A. M.; Fransen, J. Can Plants Use Entomopathogens as Bodyguards? *Ecol. Lett.* **2000,** *3*, 228–235.

Engstrom, G. W,; Delance, J. V.; Richard, J. L.; Baetz, A. L. Purification and Characterization of Roseotxin B, a Toxic Cyclodepsipeptide from *Trichothecium roseum. J. Agric. Food Chem.* **1975,** *23*(2), 244–253.

Engstrom, G. W. Amino Acid Sequence of Roseotoxin B. *J. Agric. Food Chem.* **1978,** *26*, 1403.

Evans, H. C. Entomogenous Fungi in Tropical Forest Ecosystems: An Appraisal. *Ecol. Entomol.* **1982,** *7*(1), 47–60, ISSN 1365–2311.

Evans, H. C. Mycopathogens of Insects of Epigeal and Aerial Habitats. In *Insect–Fungus Interactions*; Wilding, N., Collins, N. M., Hammond, P. M., Weber, J. F., Eds.; Academic Press: London, 1989; pp 205–238.

Fang, Q. X.; Yang, S. F.; Hu, Y. M.; Zhou, Y. Y.; Gong, Y. X.; Kang, Z. J. Control of Whitefly, *Trialeurodes vaporariorum* by *Paecilomyces fumosoroseus* Var. Beijingensis Fang Et Q. T. Chen, in Greenhouses. *Chin. J. Biol. Control.* **1986,** *2*, 129–131.

Fang, W.; Pava-Ripoli, M.; Wang, S.; St. Leger, R. J. Protein Kinase a Regulates Production of Virulence Determinants by the Entomopathogenic Fungus, *Metarhizium anisopliae. Fungal Genet. Biol.* **2009,** *46*, 277–285. http://dx.doi.org/10.1016/j.fgb.2008.12.001

Fargues, J.; Goettel, M. S.; Smits, N.; Ouedraogo, A.; Vidal, C.; Lacey, L. A.; Rougier, M. Variability in Susceptibility to Simulated Sunlight of Conidia among Isolates of Entomopathogenic Hyphomycetes. *Mycopathologia.* **1996,** *135*(3), 171–181.

Faria, M. R.; Wraight, S. P. Mycoinsecticides and Mycoacaricides: A Comprehensive List with Worldwide Coverage and International Classification of Formulation Types. *Biol. Control.* **2007,** *43,* 237–256.

Faria, M.; Wraight, P. S. Biological Control of Bemisia Tabaci with Fungi. *Crop Prot.* **2001,** *20,* 767–778.

Ferro, D. N.; Southwick, E. E. Microclimates of Small Arthropods: Estimating Humidity within the Leaf Boundary Layer. *Environ. Entomol.* **1984,** *13,* 926–929.

Ferron, P. "Pest Control by the Fungi *Beauveria* and *Metarhizium*". In *Microbial Control of Insects and Mites*; Burgess, H. D., Ed.; Academic Press: New York, NY, USA, 1981; pp 465–482.

Fransen, J. J.; Winkelman, K.; van Lenteren, J. C. The Different Mortality at Various Life Stages of the Greenhouse Whitefly, *Trialeurodes vaporariorum* (Hymenoptera: Aleyrodidae), by Infection with the Fungus *Aschersonia aleyrodis* (Deuteromycotina: Coelomycetes). *J. Invertebr. Pathol.* **1987,** *50,* 158–165.

Fukushima, K.; Arai, T.; Mori, Y.; Tsuboi, M.; Suzuki, M. Studies on Peptide Antibiotics, Leucinostatins. I. Separation, Physicochemical Properties and Biological Activities of Leucinostatins. A and B. *J. Antibiot.* **1983a,** *36*(12), 1606–1612.

Fukushima, K.; Arai, T.; Mori, Y.; Tsuboi, M.; Suzuki, M. Studies on Peptide Antibiotics, Leucinostatins. II. The Structures of Leucinostatins A and B. *J. Antibiot.* **1983b,** *36*(12), 1613–1630.

Gaugler, R.; Costa, S. D.; Lashomb, J. Stability and Efficacy of *Beauveria bassiana* Soil Inoculations. *Environ. Entomol.* **1989,** *18,* 412–417.

Gilbert, L. I.; Gill, S. S. *Insect Control: Biological and Synthetic Agents*; Gilbert, L. I., Gill, S. S., Eds.; Academic Press: London, 2010.

Gillespie, A. T.; Claydon, N. The Use of Entomogenous Fungi for Pest Control and the Role of Toxins in Pathogenesis. *Pest. Sci.* **1989,** *27,* 203–215.

Gilreath, M. E.; Mccutcheon, G. S.; Carner, G. R.; Turnipseed, S. G. Pathogen Incidence in Noctuid Larvae from Selected Soybean Genotypes. *J. Agric. Entomol.* **1986,** *3,* 213–226.

Glare, T. 11 Entomopathogenic Fungi and their Role in Regulation of Insect Populations. In *Insect Control: Biological and Synthetic Agents;* Gilbert, L. I., Gill, S. S., Eds.; Academic Press: London, 2010.

Goettel, M. S.; Eilenberg, J.; Glare, T. R. Entomopathogenic Fungi and Their Role in Regulation of Insect Populations. In *Comprehensive Molecular Insect Science;* Gilbert, L., Iatrou. K., Gill. S., Eds.; Elsevier: Boston, 2005; Vol. 6, pp 361–406.

Goettel, M. S.; Inglis, G. D. Fungi: Hyphomycetes. In *Manual of Techniques in Insect Pathology;* Lacey, L., Ed.; Academic press: London, 1997; pp 213–249.

Goettel, M. S.; Leger, R. J. S.; Bhairi, S.; Jung, M. K.; Oakley, B. R.; Roberts, D. W.; Staples, R. C. "Pathogenicity and Growth of *Metarhizium anisopliae* Stably Transformed to Benomyl Resistance," *Curr. Genet.* **1990,** *17*(2), 129–132.

Gopalkrishnan, C.; Narayanan, K. Occurrence of Two Entomofungal Pathogens, *Metarhizium anisopliae* (Metschinkoff) Sorokin Var. Minor Tulloch and *Nomuraea rileyi* on *Heliothis armigera. Curr. Sci.* **1988,** *57,* 867–868.

Gray, R. C. *Notices of Insects That are Known to Form the Bases of Fungoid Parasites*; Privately printed: London, 1858, p 22.

Groden, E.; Lockwood, J. L. Effects of Soil Fungistasis on *Beauveria bassiana* and its Relationship to Disease Incidence in the Colorado Potato Beetle, *Leptinotarsa decemlineata*, in Michigan and Rhode Island Soils. *J. Invertebr. Pathol.* **1991,** *57,* 7–16.

Grove, J. F.; Pople, M. The Insecticidal Activity of Beauvericin and the Enniatin Complex. *Mycopathologia.* **1980,** *70*(2), 103–105.

Gul, T. H.; Saeed, S.; Khan, A. Z. F. Entomopathogenic Fungi as Effective Insect Pest Management, *Appl. Sci. Bus. Econ..* **2014,** *1*(1), 10–18.

Hackman, R. H. "Cuticle: Biochemistry". In *Biology of the Integument*; Bereiter-Hahn, J., Matolts, A. G., Richards, K. S., Eds.; Springer: Berlin, Germany, 1984; pp 626–637.

Hall, R. A. The Fungus *Verticillium lecanii* as a Microbial Insecticide against Aphids and Scales. In *Microbial Control of Pests and Plant Diseases 1970–1980;* Burges, H. D., Ed.; Academic Press: London, UK, 1981, pp 483–498.

Hall, R. A.; Papierok, B. Fungi as Biological Control Agents of Arthropods of Agricultural and Medical Importance. *Parasitology.* **1982,** *84,* 205–240.

Hamill R. L.; Higgens, C. E.; Boaz, J. E.; Gorman, M. The Structure of Beauvericin, a New Depsipeptide Antibiotic Toxic to *Alternaria salina*. *Tetrahedron Lett.* **1969a,** *49,* 4255–4258.

Hamill, R. L.; Sullivan, H. R.; Gorman, M. "Determination of Pyrrolnitrin and Derivatives by Gas-Liquid Chromatography". *Appl. Microbiol.* **1969b,** *18*(3), 310–312.

Hamlen, R. A. "Biological Control of Insects and Mites on European Greenhouse Crops: Research and Commercial Implementation". *Proc. Fla. State Hortic. Soc.* **1979,** *92,* 367–368.

Hasan, S.; Ahmad, A.; Purwar, A.; Khan, N.; Kundan, R.; Gupta, G. Production of Extracellular Enzymes in the Entomopathogenic Fungus *Verticillium lecanii*. *Bioinformation.* **2013,** *9,* 238–242. http://dx.doi.org/10.6026/97320630009238

Hasan, S.; Bhamra, A. K.; Sil, K.; Rajak, R. C.; Sandhu, S. S. "Spore Production of *Metarhizium anisopliae* (ENT-12) by Solid State Fermentation". *J. Ind. Bot. Soc.* **2002,** *8,* 85–88.

Hatai, K.; Roza, D.; Nakayama, T. Identification of Lower Fungi Isolated from Larvae of Mangrove Crab, *Scylla serrata*, in Indonesia. *Mycoscience.* **2000,** *41*(6), 565–572.

Hegedus, D.; Pfeifer, T. A.; MacPherson, J. M.; Khachatourians, G. G. "Cloning and Analysis of Five Mitochondrial tRNA-Encoding Genes from the Fungus *Beauveria bassiana*". *Gene.* **1991,** *109*(1), 149–154.

Hu, G.; St. Leger, J. Field Studies Using a Recombinant Mycoinsecticide (*Metarhizium anisopliae*) Reveal That it is Rhizosphere Competent. *Appl. Environ. Microbiol.* **2002,** *68,* 6383–6387.

Huang, B.; Li, C.; Fan, M.; Lin, Y.; Li, Z. *Metacordyceps Guniujiangensis* and its *Metarhizium anamorph*: a New Pathogen on Cicada Nymphs. *Mycotaxon.* **2010,** *111*(1), 221–231.

Hugar, P. S.; Hegde, M. Occurrence of *Nomuraea rileyi* on *Helicoverpa armigera* Infesting Sorghum. *Insect Environ.,* **1996,** *2,* 44–45.

Ignoffo, C. M. The Fungus *Nomuraea rileyi* as a Microbial Insecticide. In *Microbial Control of Pests and Plant Diseases 1970–1980*; Burgus, H., Eds.; Academic Press: New York, 1981; pp 513–536.

Inglis, G. D.; Goettel, M. S.; Butt, T. M.; H. Strasser. Use of Hyphomycetous Fungi for Managing Insect Pests. In *Fungi as Biocontrol Agents: Progress, Problems and Potential;* Butt, T. M., Jackson, C. W., Magan, N., Eds.; CABI International/AAFC: Wallingford, UK, 2001, pp 23–69.

Jackson, C. G.; Linnett, P. E.; Beechey, R. B.; Henderson, P. J. F. Purification and Preliminary Structure Analysis of the Efrapeptins, a Group of Antibiotics that Inhibit the Mitochondrial Adenosine Triphosphates'. *Biochem. Soc. Transact.* **1979,** *7,* 224–226.

Jaronski, S. T.; Lord, J.; Rosinska, J.; Bradley, C.; Hoelmer, K.; Simmons, G.; Antilla, L. *In Effect of a Beauveria bassiana-based Mycoinsecticide on Beneficial Insects Under Field Conditions,* The 1998 Brighton Conference Pest and Disease, Brighton, Nov. 16,1998; British Crop Protection Council Brighton, UK, 1998; Vol. 2, pp 651–657.

Jaronski, T. J.; Lord, J. C. Evaluation of *Beauveria bassiana* (Mycotrol WP) for Control of Whitefly in Spring Cantaloupes, 1995. *Arthropod Manage. Tests.* **1996,** *21,* 103.

Kaaya, G. P.; Kokwaro, E. D.; Murithi, J. Mortalities in Adult *Glossina morsitans* Experimentally Infected with the Entomogenous Fungi, *Beauveria bassiana* and *Metarhizium anisopliae. Discov. Innov.* **1991,** *3,* 55–60.

Kanaoka, M.; Isogai, A.; Murakoshi, S.; Ichinoe, M.; Suzuki, A.; Tamura, S. Bassianolide, a New Insecticidal Cyclodepsipeptide from *Beauveria bassiana* and *Vertecillium lecanii. Agric. Biol. Chem.* **1978,** *42*(3), 629–635.

Keller, S.; Kessler, P.; Schweizer, C. Distribution of Insect Pathogenic Soil Fungi Switzerland with Special Reference to *Beauveria brongniartii a*nd *Metharhizium anisopliae. Biocontrol.* **2003,** *48,* 307–319.

Keller, S.; Zimmerman, G. Mycopathogens of Soil Insects. In *Insect-Fungus Interactions*; Wilding, N., Collins, N. M., Hammond, P. M., Webber, J. F., Eds.; Academic Press, ISBN 0127518002: London, 1989; pp 240–270.

Kerry, R.; Crumph, D.; Mullen, A. "Studies of the Cerealcyst Nematode, *Heterodera avenue* under Continuous Cereals, 1975–1978. Ii. Fungal Parasitism of Nematode Females and Eggs". *Ann. Appl. Biol.* **1982,** *100,* 489–499.

Khachatourians G. G. Production and Use of Biological Pest Control Agents. *Trends Biotechnol.* **1986,** *12,* 120–124. http://dx.doi.org/10.1016/0167-7799 (86)90144-7

Khachatourians G. G.; Sohail S. Q. Entomopathogenic Fungi. In *Biochemistry and Molecular Biology, Human and Animal Relationships,* 2nd ed.; Brakhage, A. A., Zipfel, P. F., Eds.; The Mycota VI, Springer-Verlag: Berlin, Heidelberg, 2008.

Khan, S.; Guo, L.; Maimaiti, Y.; Mijit, M.; Qiu, D. Entomopathogenic Fungi as Microbial Biocontrol Agent. *Mol. Plant Breed.* **2012,** *3*(7), 63–79.

Kim, J. J.; Lee, M. H.; Yoon, C. S.; Kim, H. S.; Yoo, J. K.; Kim, K. C. *"Control of Cotton Aphid and Greenhouse Whitefly with a Fungal Pathogen"*; Journal of National Institute of Agricultural Science and Technology: Korea, 2002, pp 7–14.

Klich, M. A. *Aspergillus flavus*: The Major Producer of Aflatoxin. *Mol. Plant Pathol.* **2007,** *8*(6), 713–722.

Klingen, I.; Eilenberg, J.; Meadow, R. Effects of Farming System, Field Margins and Bait Insect on the Occurrence of Insect Pathogenic Fungi in Soils. *Agric. Ecosyst. Environ.* **2002,** *91*(2–3), 191–198, ISSN 0167–8809.

Kodaira, Y. Studies on the New Toxic Substances to Insects, Destruxin A and B, Produced by *Oospora destructor.* Part I. Isolation and Purification of Destruxin A and B. *Agric. Biol. Chem.* **1962,** *26*(1), 36–42.

Kohlbrenner, W. E.; Cross, R. L. Efrapeptin Prevents Modification by Phenylglyoxal of an Essential Arginyl Residue in Mitochondrial *Adenosine triphosphatase*. *J. Biol. Chem.* **1978,** *253*(21), 7609–7611.

Krasnoff, S. B.; Gupta, S.; St. Leger, R. J.; Renwick, J. A. A.; Roberts, D. W. Antifungal and Insecticidal Properties of the Efrapeptins: Metabolites of the Fungus *Tolypocladium niveum*. *J. Invertebr. Pathol.* **1991,** *58*, 180–188.

Lacey, L. A.; Frutos, R.; Kaya, H. K.; Vail, P. Insect Pathogens as Biological Control Agents: Do They have a Future? *Biol. Control.* **2001,** *21*(3), 230–248.

Landa, Z. Protection against Glasshouse Whitefly, *Trialeurodicus vaporariorum* in Integrated Protection Programmes for Glasshouse Chambers. *Sbor. Ur. Zahor.* **1984,** *11*, 215–218.

Lardy, H.; Reed, P.; Lin, C. H. Antibiotics Inhibitors of Mitochondrial ATP Synthesis. *Fed. Proc.* **1975,** *34*(8), 1707–1710.

Latge, J. P.; Monsigny, M.; Prevost, M. C. "Visualization of Exocellular Lectins in the Entomopathogenic Fungus *Conidiobolus obscurus*". *J. Histochem. Cytochem.* **1988,** *36*(11), 1419–1424.

Leger, R. J. S.; Butt, T. M.; Goettel, M. S.; Staples, R. C.; Roberts, D. W. "Production *in Vitro* of Appressoria by the Entomopathogenic Fungus *Metarhizium anisopliae*". *Exp. Mycol.* **1989,** *13*(3), 274–288.

Leger, R. J. S.; Roberts, D. W.; Staples, R. C. "A Model to Explain Differentiation of Appressoria by Germlings of *Metarhizium anisopliae*". *J. Invertebr. Pathol.* **1991,** *57*(3), 299–310.

Liu, T. X.; Stansly, P. A.; Sparks Jr., A. N.; Knowles, T. C.; Chu, C. C. Application of Mycotrol and Naturalis-l (*Beauveria bassiana*) for Management of *Bemisia argentifolii* (Homoptera: Aleyrodidae) on Vegetables, Cotton and Ornamentals in Southern United States. *Subtrop. Plant Sci.* **1999,** *51*, 41–48.

Loria, R.; Galaini, S.; Roberts, D. W. Survival of Inoculum of the Entomopathogenic Fungus *Beauveria bassiana* as Influence by Fungicides. *Environ. Entomol.* **1983,** *12*, 1724–1726.

Lu, Z. X.; Laroche, A.; Huang H. C. Isolation and Characterization of Chitinases from *Verticillium lecanii*. *Can. J. Microbiol.* **2005,** *51*, 1045–1055. http://dx.doi.org/10.1139/w05-088

Lucero, H. A.; Ravizzini, R. A.; Vallejos, R. H. Inhibition of Spinach Chloroplasts Photophasphophrylation by the Antibiotics Leucinostatin and Efrapeptin. *FEBS Lett.* **1976,** *68*(1), 141–144.

Mathew, S. O.; Sandhu, S. S.; Rajak, R. C. "Bioactivity of *Nomuraea rileyi* against *Spilosoma obliqua*: Effect of Dosage, Temperature and Relative Humidity". *J. Ind. Bot. Soc.* **1998,** *77*, 23–25.

McCoy, C. W. Entomogenous Fungi as Microbial Pesticides. In *New Directions in Biological Control, Alternatives for Suppressing Agricultural Pests and Diseases, New Series,* UCLA Symposia on Molecular and Cellular Biology, UCLA Colloquium: Frisco, CO, USA, 1990, Vol. 112, pp 139–159.

McCoy, C. W.; Samson, R. A.; Boucias, D. G. Entomogenous Fungi. In *Handbook of Natural Pesticides; Microbial Insecticides, A: Entomogenous Protozoa and Fungi*; Ignoffo, C. M., Bhushan Mandava, N., Eds.; CRC Press: Boca Raton, FL, 1988, Vol. 5, pp 151–236.

Mcleod, D. M. "Investigations on the Genera *Beauveria* Vuill. and *Tritirachium* Limber". *Can. J. Bot.* **1954**, *32*, 818–890.

Meyling, N.; Eilenberg, J. Ecology of the Entomopathogenic Fungi *Beauveria bassiana* and *Metarhizium anisopliae* in Temperate Agroecosystems: Potential for Conservation Biological Control. *Biol. Control,* **2007**, *43*, 145–155.

Miętkiewski, R.; Zurek, M.; Tkaczuk, C.; Bałazy, S. Occurrence of Entomopathogenic Fungi in Arable Soil, Forest Soil and Litter. *Roczniki Nauk Rolniczych.* Ser. E. **1991**, *21*(1/2), 61–68, ISSN 0080–3693 (in Polish with English summary).

Murakoshi, S.; Ichinoe, M.; Suzuki, A.; Kanoka, M.; Isogai, A.; Tamura, S. Presence of Toxic Substance in Fungus Bodies of The Entomopathogenic Fungi, *Beauveria bassiana* and *Verticillum lecanii. Appl. Entomol. Zool.* **1978**, *13*(2), 97–102.

Mustafa, U.; Kaur, G. Extracellular Enzyme Production in *Metarhizium anisopliae* Isolates. *Folia Microbiol.* **2009**, *54*, 499–504. http://dx.doi.org/10.1007/s12223-009-0071-0

Nadeau, M. P.; Dunphy, G. B.; Boisvert, J. L. Development of *Erynia conica* (Zygomycetes: Entomophthorales) on the Cuticle of the Adult Black Flies *Simulium rostratum* and *Simulium decorum* (Diptera: Simuliidae). *J. Invertebr. Pathol.* **1996**, *68*, 50–58.

Neville, C. "Cuticle: Organisation". In *Biology of the Integument;* Bereiter-Hahn, J., Matolts, A. G., Richards, K. S., Eds.; Springer: Berlin, Germany, 1984; pp 611–625.

Noma, T.; Strickler, K. Effects of *Beauveria bassiana* on *Lygus hesperus* (Hemiptera: Miridae) Feeding and Oviposition. *Environ. Entomol.* **2000**, *29*(2), 394–402.

Nunez, E.; Iannacone, J.; Omez, H. G. "Effect of Two Entomopathogenic Fungi in Controlling *Aleurodicus cocois* (Curtis, 1846) (Hemiptera: Aleyrodidae)". *Chil. J. Agric. Res.* **2008**, *68*(1), 21–30.

Odier, F.; Vago, P.; Quiot, J.M.; Devauchelle, G.; Bureau, J.P. Etude Cytometrique Des Effect Da La Destruxine E Surdes Cellules Leucemiques Da Souris. *C. R. Acad. Sci., Ser. III.* **1987**, *305*, 575–578.

Osborne, L. S.; Hoemler, K.; Gerling, D. Prospects for Biological Control of *Bemisia tabaci. Bull. IOBC WPRS.* **1990a**, *13*, 153–160.

Ovichinnikov, Y. A.; Ivanov, V. T.; Mikhaleva, I. I. The Synthesis and Some Properties of Beauvericin. *Tethahedron Lett.* **1971**, *2*, 159–162.

Peeters, H.; Sozher, R.; Madry, N.; Oelrichs, P. B.; Kleinkauf, H. Kraepelin, G. Cell Free Synthesis of the Depsipeptide Beauvericin. *J. Antibiot.* **1983**, *26*(12), 1762–1766.

Pereira, R. M.; Stimac, J. L.; Alves, S. B. Soil Antagonism Affecting the Dose Response of Workers of the Red Imported Fire Ant, *Solenopsis invicta*, to *Beauveria bassiana* Conidia. *J. Invertebr. Pathol.* **1993**, *61*, 156–161.

Petch, T. Notes on Entomogenous Fungi. *Trans. Brit. Mycol. Soc.* **1938**, *21*, 34–67.

Phadke, C. H.; Rao, V. G.; Pawar, S. K. Natural Outbreak of the Mycordine Fungus, *Nomurea rileyi* (Farlow) Samson on Leaf Cating Caterpillar, *Sopdoptera litura* Fab. in Maharashtra. *Curr. Sci.* **1978**, *47*, 476–478.

Poprawski, T. J. Control of *Bemisia argentifoli* on Collards Using Different Formulations and Rates of *Beauveria bassiana*, 1997. *Arthropod Manage. Tests.* **1999**, *24*, 122–123.

Poprawski, T. J.; Greenberg, S. M.; Ciomperlik, M. A. Effect of Host Plant on *Beauveria bassiana* and *Paecilomyces fumosoroseus-Induced* Mortality of *Trialeurodes vaporariorum* (Homoptera: Aleyrodidae). *Environ. Entomol.* **2000**, *29*, 1048–1053.

Prince, R. C.; Crofts, A. R.; Steinrauf, L. K. A Comparison of Beauvericin, Enniatin and Valinomycin as Calcium Transporting Agents in Liposomes and Chromatophores. *Biophys. Res. Commun.* **1974**, *59*(2), 697–703.

Puzari, K. C.; Sarmah, D. K.; Hazarika, L. K. Medium for Mass Production of *Beauveria bassiana* (Balsamo) Vuillemin. *J. Biol. Control.* **1997**, *11*, 97–110.

Quesada-Moraga, E.; Navas-Cortés, J. A.; Maranhao, E. A. A.; Ortiz-Urquiza, A.; Santi-agoÁlvarez, C. Factors Affecting the Occurrence and Distribution of Entomopatho-genic Fungi in Natural and Cultivated Soils. *Mycol. Res.* **2007**, *111*(8), 947–966, ISSN 1469–8102.

Rajak, R. C.; Sandhu, S. S.; Mukherjee, S.; Kekre, S.; Gupta, A. "Natural Outbreak of *Nomuraea rileyi* on *Junonia orithyia*". *J. Biol. Control.* **1991**, *5*(2), 123– 124.

Rangel, D. E. N.; Braga, G. U. L.; Anderson, A. J.; Roberts, D. W. Variability in Conidial Thermo Tolerance of *Metarhizium anisopliae* Isolates from Different Geographic Origins. *J. Invertebr. Pathol.* **2005**, *88*, 116–125.

Rao, V. G.; Phadke, C. H. A Mustarding Disease of Tobacco Leaf Eating Caterpillar. *Curr. Sci.* **1977**, *46*, 648–649.

Rath, A. C.; Carr, C. J.; Graham, B. R. Characterization of *Metarhizium anisopliae* Strains by Carbohydrate Utilization (AP150CH). *J. Invertebr. Pathol.* **1995**, *65*, 152–161.

Ravensberg, W. J.; Malais, M.; van der Schaaf, D. A. Application of *Verticillium lecanii* in Tomatoes and Cucumber to Control Whitefly and Thrips. *Bull. IOBC WPRS.* **1990**, *13*(5), 173–178.

Roberts, D. W. Toxins from the Entomogenous Fungus *Metarhizium anisopliae* I. Produc-tion in Submerged and Surface Cultures, and in Inorganic and Organic Nitrogen Media. *J. Invertebr. Pathol.* **1966**, *8*, 212–221.

Roberts, D. W. Toxins from the Entomogenous Fungus *Metarhizium anisopliae*: Isolation of Destruxins from Submerged Cultures. *J. Invertebr. Pathol.* **1969**, *14*, 82–88.

Roberts, D. W.; Campbell, A. S. Stability of Entomopathogenic Fungi. *Misc. Publ. Entomol. Soc. Am.* **1977**, *10*, 19–76.

Roberts, D. W.; Campbell, A. S. Stability of Entomopathogenic Fungi. *Misc. Publ. Entomol. Soc. Am.* **1977**, *10*, 19–76.

Roberts, D. W.; Hajek, A. E. Entomopathogenic Fungi as Bioinsecticides. In *Frontiers in Industrial Mycology;* Springer: USA, 1992; pp 144–159.

Roberts, D. W.; Humber, R. A. Entomopathogenic Fungi. In *Infection Processes of Fungi;* Roberts, D. W., Aist, R., Eds.; Rockerfeller Fdn. New York,1984; pp 1–12.

Rossi, C; Benciari, Z.; Casinovi, C. G.; Tuttobello, L. Two Phytotoxic, Antibiotic Peptides Produced by Submerged Cultures of *Paecilomyces marquandii* (Masse) Hughes. *Phyto-pathol. Mediterr.* **1983**, *22*, 209–211.

Roy H. E.; Steinkraus D. C.; Eilenberg J.; Hajek A. E.; Pell J. K. Bizarre Interactions and Endgames: Entomopathogenic Fungi and Their Arthropod Hosts, *Annul. Rev. Entomol.* **2006**, *51*, 331–357. http://dx.doi.org/10.1146/annurev.ento.51.110104.150941 PMid: 16332215.

Rudgers, J. A.; Holah, J.; Orr. S. P.; Clay, K. Forest Succession Suppressed by an Intro-duced Plant-Fungal Symbiosis. *Ecology.* **2007**, *88*, 18–25.

Saikkonen, K.; Lehtonen, P.; Helander, M.; Koricheva, J.; Faeth, S. H. Model Systems in Ecology: Dissecting the Endophyte-Grass Literature. *Trends Plant Sci.* **2006**, *11*, 428–433.

Saito, T. In *Control of Two Whitefly Species, Bemisia tabaci and Trialeurodes vaporariorum, by A Preparation of Verticillium lecanii*, Proceedings of the Kanto Tosan Plant Protection Society, 1993; Vol. 40, pp 221–222.

Samson, R. A.; Evans, H. C.; Latge, J. P. *Atlas of Entomopathogenic Fungi*; Springer-Verlag GmbH & Co. KG: Berlin, New York, 1988.

Samuels, R. I.; Charnley, A. K.; Reynolds, S. E. Application of Reversedphase HPLC in Seperation and Detection of the Cyclodepsipeptide Toxins Produced by the Entomopathogenic Fungus *Metarhizium anisopliae*. *J. Chromatogr. Sci.* **1988b**, *26*, 15–19.

Samuels, R. I.; Charnley, A. K.; Reynolds, S. E. The Role of Destruxins in the Pathogenicity of Three Strains of *Metarhizium anisopliae* for the Tobacco Hornworm *Manduca sexta*. *Mycopathologia*. **1988a**, *104*, 51–58.

Samuels, R. I.; Reynolds, S. E.; Charnley, A. K. Calcium Channel Activation of the Insect Muscle by Destruxins, Insecticidal Compounds Produced by the Entomopathogenic Fungus, *Metarhizium anisopliae*. *Comp. Biochem. Physiol.* **1988c**, *90c*(2), 403–412.

Sandhu, S. S. "Effect of Physical Factors on Germination of Entomopathogenic Fungus *Beauveria bassiana* Conidia". *Nat. Acad. Sci. Lett.* **1995**, *18*(1–2), 1–5.

Sandhu, S. S.; Mishra, M. "In *Larvicidal Activity of Fungal Isolates Beauveria bassiana, Metarhizium anisopliae and Aspergillus flavus against Mosquito Sp. Culex Pipiens*", Proceedings of the National Symposium on Advances in Biological Control of Insect Pests, Muzaffarnagar, India, 1994; pp 145–150.

Sandhu, S. S.; Rajak, R. C.; Agarwal, G. P. "Studies on Prolonged Storage of *Beauveria bassiana* Conidia: Effects of Temperature and Relative Humidity on Conidial Viability and Virulence against Chickpea Borer *Helicoverpa armigera*". *Biocontrol Sci. Technol.* **1993**, 3, 47–53.

Sandhu, S. S.; Rajak, R. C.; Hasija, S. K. "Potential of Entomopathogens for the Biological Management of Medically Important Pest: Progress and Prospect". *Glimpses Plant Sci.* **2000**, *2000*, 110–117.

Sandhu, S. S.; Sharma, A. K.; Beniwal, V.; Goel, G.; Batra, P.; Kumar, A.; Jaglan, S.; Sharma, A. K.; Malhotra, S. "Myco-Biocontrol of Insect Pests: Factors Involved, Mechanism, and Regulation". *J. Pathog.* 2012, *2012*, Article ID 126819, 10. doi:10.1155/2012/126819

Sandhu, S. S.; Vikrant, P. "Evaluation of Mosquito Larvicidal Toxins in the Extra Cellular Metabolites of Two Fungal Genera *Beauveria* and *Trichoderma*". In *Emerging Trends in Mycology, Plant Pathology and Microbial Biotechnology*; Bagyanarayana, G., Bhadraiah, B., Kunwar, I. K., Eds.; BS: Hyderabad, India, 2006.

Scholte, E. J.; Knols, B. G.; Samson, R. A.; Takken, W.; Scholte, E. J.; Knols, B. G.; Takken, W. Entomopathogenic Fungi for Mosquito Control: A Review. *J. Insect Sci.* **2004**, *4*, 19.

Seidl, V. Chitinases of Filamentous Fungi: A Large Group of Diverse Proteins with Multiple Physiological Functions. *Fungal Biol. Rev.* **2008**, *22*, 36–42. http://dx.doi.org/10.1016/j.fbr.2008.03.002

Shah, P. A.; Pell, J. K. Entomopathogenic Fungi as Biological Control Agents. *Appl. Microbiol. Biotechnol.* **2003**, *61*(5–6), 413–423.

Shahid, A. A.; Rao, Q. A.; Bakhsh, A.; Husnain, T. Entomopathogenic Fungi as Biological Controllers: New Insights into Their Virulence and Pathogenicity. *Arch. Biol. Sci.* **2012**, *64*(1), 21–42.

Sheng, J.; An, K.; Deng, C.; Li, W.; Bao, X.; Qiu, D. Cloning A Cuticle-Degrading Serine Protease Gene with Biologic Control Function from *Beauveria brongniartii* and Its

Expression In *Escherichia Coli. Curr. Microbial.* **2006**, *53*, 124–128. http://dx.doi. org/10.1007/s00284-005-5336-5 PMid: 16832726

Singh, B. K.; Millard, P.; Whiteley, A. S.; Murrell, J. C. Unravelling Rhizosphere – Microbial Interactions: Opportunities and Limitations. *Trends Microbiol.* **2004**, *12*, 386–393.

Singkaravanit, S.; Kinoshita, H.; Ihara, F.; Nihira, T. Cloning and Functional Analysis of the Second Geranyl Diphosphate Synthase Gene Influencing Helvolic Acid Biosynthesis in *Metarhizium anisopliae. Appl. Microbiol. Biotechnol.* **2010**, *87*(3), 1077–1088.

Soper, R. S.; Holbrook, F. R.; Gordon, C. C. Comparative Pesticide Effects on *Entomophthora* and the Phytopathogen *Alternaria solani. Environ. Entomol.* **1974**, *3*, 560–562.

Sosnowska, D.; Bałazy, S.; Prishchepa, L.; Mikulskaya, N. Biodiversity of Arthropod Pathogens in the Białowieza Forest. *J. Plant Prot. Res.* **2004**, *44*(4), 313–321, ISSN 1427–4345.

St. Leger, R. J. Studies on Adaptation of *Metarhizium anisopliae* to Life in the Soil. *J. Invertebr. Pathol.* **2008**, *98*, 271–276.

St. Leger, R. J.; Bidochka, M. J.; Roberts, D. W. Isoforms of the Cuticle-Degrading *Pr*1 Proteinase and Production of a Metalloproteinase by *Metarhizium anisopliae. Arch. Biochem. Biophys.* **1994**, *313*, 1–7. http://dx.doi.org/10.1006/abbi.1994.1350

St. Leger, R. J.; Cooper, R. M.; Charnley, A. K. Characterization of Chitinase and Chitobiose Produced by the Entomopathogenic Fungus *Metarhizium anisopliae. J. Invertebr. Pathol.* **1991**, *58*, 415–426. http://dx.doi.org/10.1016/0022-2011(91)90188-V

Stansly, P. A.; Jacinto, G.; Orellana, M. Field Manipulation of *Nomuraea rileyi* (Moniliales: Moniliaceae) Effects on Soybean Defoliator in Coastal Ecuador. *J. Econ. Entomol.* **1990**, *83*, 2193–2195.

Steinhaus, E. A. Microbial Control, the Emergence of an Idea. *Hilgardia.* **1956**, *26*, 107–157.

Steinhaus, E. A. Microbial Diseases of Insects. In *Biological Control of Insect Pests and Weeds;* De-Bach, P., Ed.; Chapman and Hall: London, 1964; pp 515–547.

Steinhaus, E. A.; Steinhaus, E. A. *Disease in a Minor Chord*; Ohio State University Press: CO, USA, 1975.

Strasser, H.; Abendstein, D.; Stuppner, H.; Butt, T. M. Monitoring the Distribution of Secondary Metabolites Produced by the Entomogenous Fungus *Beauveria brongniartii* with Particular Reference to Oosporein. *Mycol. Res.* **2000**, *104*, 1227–1233.

Sun, B. D.; Liu, X. Z. Occurrence and Diversity of Insect-Associated Fungi in Natural Soils in China. *Appl. Soil Ecol.* **2008**, *39*(1), 100–108, ISSN 1873–0272.

Supakdamrongkul, P.; Bhumiratana, A.; Wiwat, C. Characterization of an Extracellular Lipase from the Biocontrol Fungus, *Nomuraea rileyi* MJ, and its Toxicity toward *Spodoptera litura. J. Invertebr. Pathol.* **2010**, *105*, 228–235.

Suzuki, A.; Kanaoka, M.; Isogai, A.; Murakoshi, S.; Ichinoe, M.; Tamara, S. Bassianolide, a New Insecticidal Cyclodepsipeptide from *Beauveria bassiana* and *Verticillium lacanii. Tetrahedron Lett.* **1977**, *25*, 2167–2170.

Tanabe, Y.; O'Donnell, K.; Saikawa, M.; Sugiyama, J. Molecular Phylogeny of Parasitic *Zygomycota* (Dimargaritales, Zoopagales) Based on Nuclear Small Subunit Ribosomal DNA Sequences. *Mol. Phylogenet. Evol.* **2000**, *16*(2), 253–262.

Tang, L.; Hou, R. F. Potential Application of Entomopathogenic Fungus, *Nomuraea rileyi* for Control of Corn Earworm, *Helicoverpa armigera. Entomol. Exper. ET Appl.* **1998**, *88*, 25–30.

Tellez, J. A.; Cruz, M. G.; Mercado, A.; Asaff, A. A. Mecanismos De Accion Y Respuesta En La Relacion De Hongos Entomopatogenos E Insectos. *Revista Mexicana de Micol.* **2009**, *30*, 73–80.

Thakur, R.; Jain, N.; Pathak, R.; Sandhu, S. S. "Practices in Wound Healing Studies of Plants". *Evid. Based Complement. Alternat. Med.* **2011**, *2011*, Article ID 438056, 17.

Thakur, R.; Sandhu, S. S. "Distribution, Occurrence and Natural Invertebrate Hosts of Indigenous Entomopathogenic Fungi of Central India". *Ind. J. Microbiol.* **2010**, *50*(1), 89–96.

Thaxter, R. The Entomophthoreae of the United States. *Mem. Boston Soc. Nat. Hist.* **1888**, *4*, 133–201.

Vandenberg, J. D.; Shelton, A. M.; Wraight, S. P. Application and Evaluation of Entomopathogens in Crucifers and Cucurbits. In *Field Manual of Techniques in Invertebrates Pathology*; Lacey, L. A., Kaya, H. K., Eds.; Kluwer Academic Publishers: Dordrecht, the Netherlands, 2000; pp 389–403.

Vänninen, I. Distribution and Occurrence of Four Entomopathogenic Fungi in Finland: Effect of Geographical Location, Habitat Type and Soil Type. *Mycol. Res.* **1995**, *100*(1), 93–101, ISSN 1469–8102.

Vänninen, I.; Husberg, G. B.; Hokkanen, H. M. T. Occurrence of Entomopathogenic Fungi and Entomopathogenic Nematodes in Cultivated Soils in Finland. *Acta Entomologica Fennica.* **1989**, *53*, 65–71, ISSN 0001–561X.

Veen, K. H.; Ferron, P. A Selective Medium for Isolation of *Beauveria tenella* and of *Metarhizium anisopliae*. *J. Invertebr. Pathol.* **1966**, *8*, 268–269.

Vega, F. E.; Goettel, M. S.; Blackwell, M.; Chandler, D.; Jackson, M. A.; Keller, S.; Koike, M.; Maniania, N. K.; Monzon, A.; Ownley, B. H.; Pell, J. K.; Rangel, D. E. N.; Roy, H. E. Fungal Entomopathogens: New Insights on Their Ecology. *Fungal Ecol.* **2009**, *2*, 149–159. http://dx.doi.org/10.1016/j.funeco.2009.05.001

Vega, F. E.; Posada, F.; Aime, M. C.; Pava-Ripoll, M.; Infante, F.; Rehner, S. A. Entomopathogenic Fungal Endophytes. *Biol. Control.* **2008**, *46*, 72–82.

Vestergaard, S.; Cherry, A.; Keller, S.; Goettel, M. Safety of Hyphomycete Fungi as Microbial Control Agents. In *Environmental Impacts of Microbial Insecticides;* Springer: Netherlands, 2003; pp 35–62.

Vey, A.; Quiot, J. M.; Vago, C. Mise En Evidence Et Etude De I' Action D'une Mycotoxine, La Beauvericine, Sur Des Cellules D'insectes Cultivees *in Vitro*. *Comp. Rendus De L'Academic Des Sci. De Paris*. **1973**, *276*(serie D), 2489–2492.

Vidal, C.; Osborne, L. S.; Lacey, L. A.; Fargues, J., Effect of Host Plant on the Potential of *Paecilomyces fumosoroseus* (Deuteromycotina: Hyphomycetes) for Controlling the Silverleaf Whitefly, *Bemisia argentifolii* (Homoptera: Aleyrodidae) in Greenhouses. *Biol. Control.* **1998b**, *12*, 191–199.

Vimala Devi, P. S. Conidia Production of the Entomopathogenic Fungus *N. rileyi* and its Evaluation for Control of *S. litura* on *Ricinus Communis*. *J. Invertebr. Pathol.* **1994**, *63*, 145–150.

Vincent, C.; Goettel, M. S.; Lazarovits, G. *Biological Control: A Global Perspective;* Vincent, C., Goettel, M. S., Lazarovits, G., Eds.; CAB International/AAFC: Wallingford, UK, 2007.

Wang, C.; Typas, M. A.; Butt, T. M. Detection and Characterization of *Pr*1 Virulent Gene Deficiencies in the Insect Pathogenic Fungus *Metarhizium anisopliae*. *FEMS Microbiol. Lett.* **2002**, *213*, 251–255. http://dx.doi.org/10.1111/j.1574-6968.2002.tb11314.x

Whipps, J. M. Microbial Interactions and Biocontrol in the Rhizosphere. *J. Exper. Bot.* **2001,** *52,* 487–511.

Wise, J. G.; Latchney, L. R.; Ferguson, A. M.; Senior. A. E. Defective Proton Atpase of Unca Mutants of *Escherichia Coli.* 5: Adenyl Imidodiphosphate Binding and ATP Hydrolysis. *Biochemistry.* **1984,** *23,* 1426–1432.

Wraight S. P.; Jackson M. A.; de Kock S. L. Production, Stabilization and Formulation of Fungal Biological Agents. In *Fungi as Biocontrol Agents*; Butt, T. M., Jackson, C., Magan, N. Eds.; CABI: Wallingford, 2001; pp 253–287. http://dx.doi.org/10.1079/9780851993560.0253

Wraight, S. P.; Carruthers, R. I. Production, Delivery, and Use of Mycoinsecticides for Control of Insect Pests of Field Crops. In *Methods in Biotechnology Biopesticides: Use and Delivery*; Hall, F. R., Menn, J. J., Eds.; Humana Press: Totowa, NJ, USA, 1999; Vol. 5, pp 233–269.

Wraight, S. P.; Carruthers, R. I.; Jaronski, S. T.; Bradley, C. A.; Garza, C. J.; Galaini-Wraight, S. "Evaluation of the Entomopathogenic Fungi *Beauveria bassiana* and *Paecilomyces fumosoroseus* for Microbial Control of the Silverleaf Whitefly, *Bemisia argentifolii*". *Biol. Control.* **2000,** *17*(3), 203–217.

Wraight, S. P.; Carruthers, R. I.; Bradley, C. A. In *Development of Entomopathogenic Fungi for Microbial Control of Whiteflies of the Bemisia tabaci Complex.* Fifth Simposio de Controle Biologico, Anais: Conferencias e Palestras, Embrapa-CNPSo, Foz do Igua-cu, Brazil,June9–14, 1996; Empresa Brasileira de Pesquisa Agropecuaria Centro National de Pesquisa de Soja: Londrina, PR, 1996; pp 28–34.

Yang, J.; Tian, B.; Liang, L.; Zhang, K. Q. Extracellular Enzymes and the Pathogenesis of Nematophagous Fungi. *Appl. Microbiol. Biotechnol.* **2007,** *75,* 21–31. http://dx.doi.org/10.1007/s00253-007-0881-4

Zattau, W. C.; McInnis, Jr, T. Life Cycle and Mode of Infection of *Leptolegnia chapmanii* Oomycetes) Parasitizing *Aedes aegypti. J. Invertebr. Pathol.* **1987,** *50*(2), 134–145.

Zhang, S.; Wideman, E.; Bernard, G.; Lesot, A.; Pinot, E.; Pedrini, N.; Keyhani, N. O. CYP52X1, Representing New Cytochrome P450 Subfamily, Displays Fatty Acid Hydroxylase Activity and Contributes to Virulence and Growth on Insect Cuticular Substrates in Entomopathogenic Fungus *Beauveria bassania. J. Biol. Chem.* **2012,** *287,* 13477–13486.

Zimmermann, G. Review on Safety of the Entomopathogenic Fungus *Beauveria bassiana* and *Beauveria brongniartii. Biocontrol Sci. Technol.* **2007,** *17,* 553. 596http://dx.doi.org/10.1080/09583150701309006

Zimmermann, G. The *Galleria* Bait Method for Detection of Entomopathogenic Fungi in Soil. *J. Appl. Entomol.* **1986,** *102,* 213–215.

Zocher, R.; Keller, U.; Kleinkauf, H. Enniatin Synthases, a Novel Type of Multifunctional Enzyme Catalyzing Depsipeptide Synthesis in *Fusarium oxysporum. Biochemistry.* **1982,** *21,* 43–48.

CHAPTER 6

ENTOMOPATHOGENS AND THEIR MASS PRODUCTION AND APPLICATION

KALMESH MANAGANVI and RAMANUJ VISHWAKARMA*

Department of Entomology, Bihar Agricultural University, Sabour, Bhagalpur 813210, India

Corresponding author. E-mail: entoramanuj@gmail.com

CONTENTS

ABSTRACT

Pest problems are an almost inevitable part of agriculture. Entomopathogens form an interesting area of research in pest management as being one of the most important tools of environmental friendly, applicable, and effective pest management strategies worldwide. In this respect, use of entomopathogenic organisms has turned out to be an important substitute in pest management. Entomopathogens are increasingly used in IPM in recent times and the increasing numbers of products are making their way to the marketplace. Entomopathogens have been suggested as controlling agents of insect pests for over a century, and biopesticides based on bacteria, viruses, entomopathogenic fungi, and nematodes are often considerable scope as plant protection agents against several insect pests and mites. Naturally occurring entomopathogens are important regulatory factors in insect populations. The knowledge of nutritional requirements is the main need in the cultivation of microorganisms using any cultural technique. The carbohydrates, proteins, lipids, and nucleic acids are made up of microelements like carbon, hydrogen, nitrogen, sulfur, and phosphorus, and these are involved in mechanisms like host pathogen interaction and self-defense mechanisms. Production of good quality inoculums in an adequate quantity is an essential component of the biocontrol program.

6.1 INTRODUCTION

Chemical pesticides may cause secondary pest outbreaks, accelerate the development of resistance, destroy natural enemies, create hazards for human's poisonings and are more troublesome because of their potentially bad effects on the environment and public health (Heckel, 2012). As a result, regulatory agencies in different countries have imposed restrictions on chemical pesticide usage. Moreover, chemical insecticides, many farmers and growers are now familiar with the use of predators and parasitoids for biological control of arthropod (insect and mite) pests, but it is also possible to use specific micro-organisms that kill arthropods.

The entomopathogens are a unique group of pesticides. Entomopathogens are very specific and some are less; they act as infectious organisms; some of them act through their toxins; they cover the entire range of microbial forms, from viruses to bacteria to fungi to protozoans, even some rickettsial diseases of insects are known (Engler & Rogoff, 1976).

Entomopathogens contribute to the natural regulation of many popula-tions of arthropods. Much of the research in this area concerns the causal agents of insect diseases and their exploitation for biological pest control. Many entomopathogens can be mass produced, formulated, and applied to pest populations in a manner analogous to chemical pesticides that is as non-persistent remedial treatments that are released inundatively. Ento-mopathogens have also been used as classical biological control agents of alien insect pests, and natural pest control by entomopathogens has been enhanced by habitat manipulation.

6.2   ENTOMOPATHOGENIC NEMATODES (EPNS)

EPNs are soil-inhabiting, lethal insect parasitoids that belong to the phylum Nematoda. Although many other parasitic nematodes cause diseases in plants, livestock and humans, EPNs as their name imply only infect insects. The most commonly studied EPNs include two families which are useful in the biological control of insect pests, the Steinerne-matidae and Heterorhabditidae (Gaugler & Han, 2002). These two fami-lies are not closely related phylogenetically but share similar life histories (Poinar, 1993). These are symbiotically associated with bacteria of the genera *Xenorhabdus* and *Photorhabdus*, respectively, are safe antagonists as a commercial bio-insecticide for many economically important insect pests in ornamentals, vegetables, fruits, and turf (Ehlers & Peters, 1995; Ehlers & Hokkanen, 1996).

Advances in mass production technology and the discovery of numerous isolates together with the desirability of reducing pesticides usage have resulted in a surge of commercial interest in EPNs of the genera *Steinernema* and *Heterorhabditis* (Bedding, 1981). Commercial-ization of EPNs has experienced highs and lows (Shapiro-Ilan & Gaugler, 2002). EPNs are pathogenic to over 200 insect hosts, yet nematodes have only been successfully marketed for a small fraction of these insects. In many cases the commercial success of EPNs was not achieved despite acceptable efficacy. Factors such as cost, shelf life, handling, mixing, covering, competition, compatibility, and profit margins to manufacturers and distributors contributed to the failure of nematodes to penetrate many markets. The current market is limited to certain insects such as citrus, turf and ornamental nurseries. EPNs have several advantages that qualify them as commercially valuable biocontrol agents. They recycle inside the host insect thus causing long-term, sustainable effects on the pest populations. The use of EPN is safe for both the user and the environment (Kaya & Gaugler, 1993).

6.3   METHODS OF MASS PRODUCTION OF EPNS

EPNs kill insects with the aid of mutualistic bacteria; the nematode–bacteria complex is mass produced for use as biopesticides using *in vivo* or *in vitro* methods (Shapiro-Ilan & Gaugler, 2002).

6.3.1   *IN VITRO METHODS OF MASS PRODUCTION OF EPNS*

6.3.1.1   *LIQUID CULTURE METHOD*

Liquid cultures of EPNs are particularly vulnerable to contamination. The presence of any non-symbiotic micro-organism will reduce nematode yields and prevent the subsequent scale-up. As a nematode process can last up to three weeks, maintenance of sterile conditions is a challenge for process engineers (Prabhu et al., 2007). The symbiotic bacteria can easily be isolated from nematode-infected insect larvae. Stock cultures are mixed with glycerol at 15% (v/v) and aliquots are frozen at –80 °C. Surface-sterilized dauer juvenile (DJ) should not be used, because this procedure cannot exclude the presence of contaminants. The preparation of nematode inoculums is preferably done with nematode eggs obtained from gravid female stages. Detailed descriptions about the production of monoxenic nematode inoculum are provided by Lunau et al. (1993) and Han and Ehlers (2000). Monoxenic cultures can be stored on shakers at 20rpm and 4 °C for several months until they are inoculated in to the bioreactor. Strain collections of nematodes can be kept in liquid nitrogen (Popiel & Vasquez, 1991). Owing to the potential of *Xenorhabdus* and *Photorhabdus* spp. to metabolize almost every kind of protein-rich medium, the selection of appropriate culture media for EPN production can largely follow economic aspects.

A standard medium to start with should contain a carbon source (e.g. glucose or glycerol), a variety of proteins of animal and plant origin, yeast extract and lipids of animal or plant origin (Pace et al., 1986; Han et al., 1993). When composing the concentration of different compounds and minerals, the osmotic strength of the medium must not surpass 600 milliosmol/kg. For commercial production, maximum yields are less important than mean yields and process stability. Therefore the influence of media compounds on DJ recovery and nematode population development takes priority. Essential amino acid requirements have only been defined for *Steinernema glaseri* (Jackson, 1973). Nematodes have nutritional demands for sterols, but they can metabolize the necessary sterols from a variety of steroid sources (Ritter, 1988), which are provided through the addition of lipids of animal or plant origin. In general, *S. carpocapsae* requires proteins of animal origin (Yang et al., 1997) and it is unable to reproduce without the addition of lipid sources to the medium, whereas *Heterorhabditis bacteriophora* produces offspring in a liquid medium without the addition of lipids (Han & Ehlers, 2001). However, it is not known whether variable fatty acid composition influences the nematode field performance. Equipments used in biotechnology, such as conventional bioreactors, stirred with flat-blade impellers, bubble columns, airlift and internal loop bioreactors, have been successfully tested (Pace et al., 1986; Ehlers et al., 1998). Cultures are always pre-incubated for 24–36 h with the specific symbiotic bacterium before DJ is inoculated.

The inoculum density for the symbiotic bacterium is between 0.5 and 1% of the culture volume. However, the adult density is defined by the percentage of DJ bound to recover. Usually, the nematode inoculum is between 5 and 10% of the culture volume. The optimum growth temperature for the symbiotic of *H. indica* was investigated under continuous culture conditions (Ehlers, 2001) and optimum growth was recorded between 35 and 37 °C. The culture medium should be between pH 5.5 and 7.0 when the culture is started. Until nematode harvest, the pH is constantly rising; and attempts to control the pH at 7.0 always had a negative influence on nematode yields. Oxygen supply must be maintained at approximately 30% saturation, also to prevent the bacteria from shifting to the secondary phase. An important parameter is the aeration rate; Ehlers et al. (2000) compared the yields of *H. megidis* in 10 l bioreactor cultures aerated at 0.3 and 0.7 vvm and they obtained a significantly higher number of adult nematodes eight days after DJ inoculation and a higher DJ final yield in the cultures aerated at 0.7 vvm. Increasing the aeration rate often increases foaming. The addition of silicon oil usually prevents foaming. However, it should be used carefully, because higher concentrations can be detrimental to the nematodes. Maximum yields of >500,000 DJ/ml were recorded by Ehlers et al. (2000) for *H. indica*. Yields showed a negative correlation with the body length of the DJ, which is first of all genetically defined and, although quite stable within a species, differs according to strain and culture conditions.

6.3.1.2 SOLID CULTURE METHOD

EPNs were first grown *in vitro* on a solid medium axenically. Thereafter, it was realized that growth increased with the presence of bacteria. Kaya and Gaugler (1993) given the importance of the natural symbiont which was recognized and monoxenic culture has been the basis for *in vitro* culture since to create monoxenic cultures surface sterilized nematodes were added to a lawn of bacterial symbionts. Lunau et al. (1993) suggested that surface sterilization of infective juveniles is insufficient to establish monoxenicity because contaminating bacteria survive beneath the nematode's cuticle. Therefore, an improved method has been developed where nematode eggs (which are axenic), obtained by rupturing gravid females in an alkaline solution, are placed on a pure culture of the symbiont. Wouts developed an improved medium that included yeast extract, nutrient broth, vegetable oil and soya flour. *In vitro* solid culture advanced considerably with the invention of a three dimensional rearing system involving nematode culture on crumbled polyether polyurethane foam.

Bacteria are inoculated first followed by the nematodes three days later. Nematodes can be harvested within 2–5 weeks by placing the foam onto sieves, which are immersed in water. Infective juveniles migrate out of the foam, settle downward and are pumped to a collection tank. The method developed by Bedding was first accomplished in Erlenmeyer flasks and then expanded to autoclavable bags with filtered air pumped in through a makeshift port. The bacteria were inoculated first followed by the nematodes several days after. Later it was realized that the two organisms could be added simultaneously, if a large concentration of bacteria is used (Gaugler & Han, 2002). The potential for large scale production was further advanced through several measures including using bags with a gas permeable Tyvac1 strip for ventilation (rather than forced air), automated mixing and autoclaving and harvest through centrifugal sifters. Additionally, limited expertise is required, and the logistics of production is flexible (Hazir et al., 2003).

6.3.2   IN VIVO METHOD OF MASS PRODUCTION OF EPNS

In vivo culture is a two-dimensional system that relies on production in trays and shelves. Production methods for culturing EPNs in insect hosts have been reported by various authors. A system based on the White trap, which takes advantage of the infective juvenile's natural migration away from the host cadaver upon emergence. The methods described consist of inoculation, harvest, concentration and decontamination. Insects are inoculated with nematodes on a dish or tray lined with absorbent paper or another substrate conducive to nematode infection such as soil or plaster of Paris. After 2–5 days, infected insects are transferred to the White traps; if infections are allowed to progress too long before transfer, harm to nematode reproductive stages may occur and the cadavers will be more likely to rupture. White traps consist of a dish on which the cadavers rest surrounded by water, which is contained by a larger dish or tray. The central dish provides a moist substrate for the nematodes to move upon, for example, an inverted Petri dish lid lined with filter paper or filled with plaster of Paris. The progeny infective juveniles that emerge migrate to the surrounding water where they are trapped and subsequently harvested.

In the White trap method, contamination is minimized because infective juveniles migrate away from the cadaver leaving most potential contaminants behind. However, some host material or microbial contamination is possible and can be reduced by repeatedly washing the harvested nematodes using the concentration methods described previously. Additionally, decontamination can be accomplished by the use of antimicrobial compounds such as streptomycin sulfate, hyamine1 (methyl benzehonium chloride), merthiolate, NaOCl, or $HgCl_2$ but the effects of these compounds on nematodes for commercial application have not been reported (Shapiro-Ilan & Gaugler, 2002).

6.3.2.1   LOTEK AN IMPROVED METHOD OF IN VIVO PRODUCTION OF EPNS

The LOTEK system (Gaugler & Brown, 2001) is one approach to increasing *in vivo* production efficiency and scalability. Technology for *in vitro* production of EPNs requires capital investment in sterilization equipment, as well as considerable technical expertise, and other disadvantages like expensive automated equipment, difficult to maintain in an aseptic state, difficulties in media preparation and nematode harvest, labor cost etc. offset the production of EPNs with this method. Now an improved method of *in vivo* production of EPNs has been described. LOTEK method in which harvest is achieved via a rinsing mist of water rather than active nematode migration to a water reservoir (Brown et al., 2004). Most procedures including harvest, separation and clean-up are automated to reduce labor costs.

LOTEK is a comprehensive rearing system of tools and procedures best described in five sequential stages.

i.   **Infection:** The first and most critical step is the infection or inoculation of insects. The goal of inoculation is to infect >95% of the insect hosts while minimizing labor inputs. Two methods of inoculation were developed depending on the insect nematode combination.

   a.   **Immersion**: Immersion requires dipping the tray in a concentrated nematode suspension. Each tray measured $30 \times 26 \times 4cm^3$ and accommodated 500 insects. *Tenebrio* and *Galleria* are inoculated with *S. coarpocapsae* by immersion in a suspension of 2.1×10^4 or 8×10^3/ml, respectively. Similarly *Galleria* was inoculated with *H. bacteriophora* by immersing insects in a nematode suspension of 1.2×10^4/ml.

   b.   **Pipette transfer:** For inoculation of *Tenebrio* with *H. bacteriophora* satisfactory levels of infection were only obtained using the pipette method. A total of 800 infective juvenile (IJ) were pipette on to an inoculation dishes to holding trays after infection is complete (3–5 days). Key disadvantages of the need for a substrate are substantially increased handling time and damage to the fragile cadavers.

ii.   **Conditioning:** Inoculated trays are immediately transferred to a humidity controlled chamber for conditioning. The components of the chamber include an air pump, a humidifier, water trap, and vented chamber. Inoculated trays are stacked tightly in the chamber to prevent insect escape. The goal of conditioning is to incubate the nematode killed hosts in a controlled environment and to synchronize their emergence. For *H. bacteriophora* infecting *Galleria* larvae, conditioning requires 14 days. One advantage of conditioning is major cost saving through a reduction in the amount of temperature controlled rearing space required the one capital expense that *in vivo* producers cannot escape. Without conditioning, seven times more space allotted for harvester units and the cost of constructing additional harvesters.

iii. **Harvest:** When conditioning and therefore nematode development is complete, the trays move from the incubator to the harvester. Two harvester designs were developed and tested.

    a. **Misting chamber:** In misting chamber each tray is suspended beneath two rigid plastic pipes that serve a dual function to support trays and deliver water via misting nozzles. Each pipe is equipped with a mist nozzle delivering water at a rate of 37 ml/min and 1.78 l of water/tray. A time controlled water supply periodic misting cycles (3 min duration at 6 h intervals) for each tray. The mist results in nematode emergence, and after it is rinsed through the holding tray perforations into the collection pan, it finally gets collected into a central storage tank. The process is mostly passive as the water flow removes the nematodes. Thus it is fundamental departure from the conventional White trap model which requires nematode migration to a water reservoir.

    b. **Drip irrigation:** The second harvester design employs a drip irrigation system suspended above sloping tray. The trays are constructed from a single piece of plastic coated wire mesh and lined with a cotton cloth substrate. The trays are held in place between two supports each bearing a row of 2.5 cm long pins. The trays simply hang passively between the pins at an angle of 20°. The water is delivered on to the upper edges of the trays via 30 cm lengths of soaker hose connected to the water source. The water (1 l/day/tray) drips on to the trays of cadavers and seeps through the cotton cloth, gently washing the emerged nematodes off the lower edge of the tray and into a storage tank. An average of 84.2 and 96.3% of emergent *H. bacteriphora* was harvested by each system after 24 and 48 h. The impressive emergence pattern is determined by the conditioning and the harvest system.

iv. **Separation:** Insufficient separation inevitably results in inferior stability due to microbial activity on non infective stages, media remains and other waste materials. The separator washes and concentrates harvested infective juveniles. Suspended nematodes from the holding tanks are pumped into a conical reservoir tank and gravity-fed to a distribution manifold that divides the flow in to a several small streams. The reservoir tank acts as a buffer ensuring a constant uninterrupted flow through the manifold. Streams are directed to a stainless steel screen oriented at a 38° angle. Waste water passes through the screen and is discarded, whereas most nematodes are deflected and collect on the screen surface. Eventually nematode slurry is generated that slowly flows by gravity off the screen to a collection gutter and in to a storage tank. The slurry can be further washed and concentrated by multiple passages through the separator.

Removal of bacteria, pigments, cadaver residues and other waste was measured by optical density of the harvest nematode suspension before and after one pass through separator. A single pass through the separator concentrated the nematodes eight fold and removed 87% of the water and associated waste with 4.5% loss of infective juveniles. After three passes, 97% of water had been removed and the nematodes concentrated 81 fold. Harvested nematodes are now ready for the formulation or aqueous storage.

v. **Clean-up:** The last step is to clean harvester. Each tray of spent cadavers is emptied and the tray placed back into the harvester. Detergents and disinfectants are then introduced in to the water line to clean the trays. Clean-up is the most laborious and tedious steps in conventional *in vivo* production. The capacity of the harvester to carry out washing and disinfection actions through the misting nozzles automates this step.

Advantages: *In vivo* production of EPNs offers several advantages and disadvantages relative to *in vitro* culture. *In vivo* production requires the least capital outlay and technical expertise. The quality of *in vivo* produced nematodes tends to be equal or greater than nematodes produced with other approaches. Cost of labor and insect tend to make *in vivo* culture the least cost efficient approach. Although *in vivo* production may not offer the same degree of economy of scale as *in vitro* approaches, some economy of scale can be obtained. Cost of space does not remain constant but decreases in relation to amount utilized. Labor cost can be reduced through the mechanization or streamlining process and insect cost may be reduced, if host are produced on-site and rearing process is mechanized.

The LOTEK system is one approach to increasing *in vivo* production efficiency and scalability (Gaugler & Brown, 2001).

6.4 ENTOMOPATHOGENIC PROTOZOA

The word "Protozoa" means first animal, primitive animal. The protozoa are small, unicellular micro-organisms. Protozoa are eukaryotes (organisms with a nucleus containing chromosomes encircled by a nuclear membrane), with numerous organelles whose functions are similar to the more complex, multi cellular organs of higher animals. The structure of a nucleus with a single set of chromosomes with one genome is called haploid; that with a double complement is called diploid; and that possessing several genomes, polyploidy. One cell can contain one or more nuclei. Protozoa reproduce both by sexual and asexual means, but some groups lack the ability for sexual or asexual reproduction. In some species, auto gamy (self fertilization) takes place.

Pasteur (1870) contributed to the establishment of insect pathology as a science through his classical study of the protozoan infection, pebrine, of the silkworm, *Bombyx mori*. The majority of entomogenous protozoa insect associations produce chronic, nonlethal infections of which a common feature is a reduction in the host reproductive output (Hurd, 1993). Most infections of protozoa exhibit nonspecific signs and symptoms of disease such as sluggishness, irregular growth, loss of appetite, malformed larvae, pupae, or adults, or adults with reduced vigor, fecundity, and longevity. Protozoa offer little potential as short-term, quick-acting microbial insecticides, but they are being considered as candidates for long-term application or introduction programs (Canning, 1982). Entomogenous protozoa are commonly transmitted on the surface of the egg (transovum) or within the egg (transovarial) of their host.

In the 1980 report of the Committee on Systematic and Evolution of the Society of Protozoologists, the protozoa are treated as a subkingdom, and seven phyla are recognized, five of which contain entomogenous species. Most members of protozoa found so far in heliothine larvae in Mississippi belong to the phylum microspora, class microsporea, order microsporida, and genus *Nosema*. Of some 14,000 described species of protozoa, about 500 are pathogens of insects. Many are chronic pathogens that may debilitate a host without producing obvious disease symptoms but some species are extremely virulent, causing stunted growth, slow development, and early death (Tanada & Kaya, 1993). Entry into the host is typically by ingestion, but some can invade through the cuticle. Some species may be transovarially transmitted from infected females to their offspring. Species that invade the cells of the host are usually found in the cell cytoplasm and are typically more pathogenic than extracellular species.

Some protozoans exhibit tissue tropism, infecting only certain tissues or organs, others are systemic. The neogregarines often exhibit tissue tropism, the tissue or tissues infected being a species specific characteristic. No toxins have been found to be associated with protozoa in insects. Death or debilitation of infected hosts may be, for example, the result of competition for metabolites, disruption of normal cell and tissue function, or blockage of the gut or other organs by extracellular species. Unlike the viruses and some bacteria, protozoans do not typically cause the rupture of the host's integument. Some insects such as mosquitoes become chlorotic or whitish in color when infected and some lepidopteran or coleopteran larvae may appear puffy. Many species of protozoan, however, do not typically cause outward signs of disease. The insect-pathogenic protozoa are currently recorded from four major groups of the protozoa: amoebas, gregarines, flagellates, ciliates.

6.4.1 AMOEBAS

Nearly all entomogenous amoeba species are in the family's Amoebidae (*Malameba, Malpighamoeba, Malpigiella*; strictly entomogenous genera) and Endamoebidae, found in roaches and various flies and fleas, as well as in other animals (Brooks, 1988).

Most insect-associated amoeba species are commensuals in digestive tracts of their hosts, but some are pathogens producing amoebiasis. Amoebas have a simple life cycle; there are no special stages with the exception of a cyst form in some species, and reproduction is by simple binary or multiple fission. The two most studied species, *Malpighamoeba mellificae* and *Malameba locustae*, both in the family Amoebidae, pack the Malpighian tubules, destroying the epithelia and producing chronic but debilitating effects (Tanada & Kaya, 1993).

M. mellificae, producing spring disease of honey bees, often occurs in mixed infections with *Nosema apis* a micro sporidium. Mixed infections are more severe than *M. mellificae* infections alone. *M. mellificae* infects only adult bees and is transmitted orally by ingestion of the cyst stage. The disease is usually chronic but Malpighian tubule epithelia can be destroyed and the lumens blocked by the pathogen, which may result in death of the bees. Infections are typically seen in the spring; far less occur in the summer. The cure is sanitation, cleaning and disinfecting hives, and a clean water supply (Bailey, 1963).

Malamoeba locustae (family Amoebidae) is a pathogen of approximately 50 species of grasshoppers and locusts. Prevalence in the field are usually low (Brooks, 1988), but this pathogen can devastate laboratory colonies. *M. locustae* primarily occurs in the lumens of Malpighian tubules, which can swell enormously. The pathogen is transmitted by ingestion of cysts on food and by cannibalism of infected hosts. Severe chronic effects result from infection including reduced adult longevity and fecundity, reduced activity, loss of appetite and death (Taylor & King, 1937).

Biological control: Amoebic diseases have been more of a problem in beneficial insects and insect colonies than useful as biological control agents. Some studies have been conducted and suggestions have been made regarding utilization of *M. locustae* as a control agent in range grasshoppers (Lange & Wittenstein, 1998).

6.4.2   GREGARINES AND COCCIDIA

The gregarines and coccidia have been placed in the phylum Apicomplexa and represent a group of protozoans that lack cilia, reproduce sexually, use micropores for feeding, move by body flexion or gliding, and produce oocysts containing sporozoites as the infective form (Tanada & Kaya, 1993).

The gregarines are divided into two groups, the eugregarines and neogregarines. The major difference between these groups is that eugregarines do not reproduce vegetatively in the host while the neogregarines do. The neogregarines are thought to be more primitive and are much more virulent since they can build up in huge numbers in the host tissues.

6.4.2.1   EUGREGARINES

About 1,400 species of eugregarines are known in two major groups-aseptate species that have one "body" compartment and septate species which have two. Both groups primarily consist of pathogens of annelids and arthropods (a few are commensuals living in the gut lumen). Typically, oocysts are ingested by the host and the spore wall dissolves in the host's gut juices, releasing sporozoites. If the eugregarine species is pathogenic, these forms enter the midgut epithelia and grow. At a certain stage, they exit the cell, killing it. Having reentered the gut lumen, the trophozoites mature into gamonts which pair in "syzygy" around which a cyst forms. One gamont produces microgametes and one produces macrogametes which fuse to form zygotes, the only diploid stage. Zygotes undergo division to form the eight new sporozoites per oocyst (Tanada & Kaya, 1993).

Eugregarines are often rather benign to the host because the infection level depends entirely on the number of oocysts ingested. Those species that inhabit the midgut are rarely seriously pathogenic; however, they have been reported to block the gut of the host (Harry, 1967). Some species inhabit the gastric cecum and these cause more serious pathologies than the midgut-inhabiting species (Tanada & Kaya, 1993). Ascogregarina (aseptate, in mosquitoes) and Gregarina (septate, in roaches) are the best known genera.

6.4.2.2   NEOGREGARINES

Because the neogregarines undergo multiple divisions after entering the host cells, their numbers are not necessarily related to the number of oocysts ingested by the host. The resulting "merozoites" spread the infection to other tissues in the host and undergo yet another division before undergoing sexual reproduction (Canning, 1964). Neogregarines are transmitted via contaminated food, or by cannibalism of infected hosts. Effects of neogregarine infection vary from a juvenile hormone effect on the host to host mortality. Some of the better known genera are *Mattesia* in beetles and moths, *Ophryocystis* in tenebrionid beetles, *Farinocystis* in *Tribolium*.

Biological control: It is generally agreed that eugregarines have little potential as biological control agents. They are typically commensuals and the lack of a merogony in the host is associated with lack of virulence and inability to efficiently produce sufficient quantities of infectious forms for use in biological control programs. However, some neogregarines, notably *Mattesia* sp. have been evaluated as biological control agents (Brooks, 1988). One species on dermestid pests of stored grain showed good potential in simulated warehouse conditions and another species may have some utility for control of sawtooth beetles (Lord, 2003).

6.4.2.3   COCCIDIA

Of the coccidia, 1% of species are restricted to insects, most are pathogens of vertebrates. They differ from the gregarines in that the mature gamonts are intracellular in host tissues. Like the gregarines, however, the life cycle is primarily haploid, with diploidy only occurring before meiosis.

Typically, the coccidia enter host midgut cells, migrate to the hemocoel, and then to the fat body tissues. The host is killed primarily due to multiple replication cycles in fat body and, in a few species, other tissues. Of the more common genera, *Adelina* is the best known. *Adelina* species infect Coleoptera, Lepidoptera, Orthoptera, Embioptera, Diptera, Collembola, and other invertebrates.

Biological control: Because coccidia are typically found in animals other than insects (including vertebrates), no field testing has been conducted on the few species for which there appears to be some potential for controlling host populations (Brooks, 1988). Recent studies on one species infecting crickets have reported physiological effects on the host (Dolgikh, 1998), but no recent studies have been conducted to utilize species in this pathogen group as biological control agents.

6.4.3   FLAGELLATES

Flagellates are characterized by adult morphology, body forms based on shape and the location from which the flagellum emerges from the kinetoplast a DNA rich organelle. Life cycles are varied and different stages usually have different body forms, including cysts. Sexual reproduction occurs but appears to be rare; most flagellates undergo simple binary fission (Tanada & Kaya, 1993).

Approximately 400 species of flagellates have been associated with insects, but only a few are pathogenic (Lipa, 1963). The majority is commensals or is vectored by the insects to hosts in other taxa. About 350 insect species have been found to host monoxenous, or one host, species of flagellates (Wallace, 1966). Higher trypanosomatids are heteroxenous, having more than one host. Most entomogenous flagellates are in the family trypanosomatidae. They have not been well studied, and there is limited knowledge about pathogenesis and transmission between hosts.

The disease, flagellatoses or flagelloses, is primarily localized in the digestive tract of the host, where some flagellates may form a carpet-like layer attached to the intestinal wall using their flagella to weave into the microvilli (Wallace, 1979). Other species may be free in the lumen. Some species, however, may enter the hemocoel, salivary glands or Malpighian tubules and these are typically more virulent, even fatal.

Most described species of trypanosomes occur in Hemiptera (35%) and Diptera (55%), and a few occur in Siphonaptera. In the laboratory, flagellates exhibit little host specificity. One species has been found infecting 26 species of flies in different families in the field (Wallace, 1966) although current molecular techniques could possibly differentiate some of these isolates. Transmission between hosts is usually by ingestion of feces, plant sap, cadavers, or by cannibalism. Infections in some species can persist through larval stages into the adult stage of the host.

The best known trypanosomatid flagellates are not insect pathogens but those vectored by insects to vertebrate animals. Some of these species do reproduce in the insect vector, and a few may also be pathogenic to the vector. A huge literature is available for study of insect vectored flagellates.

Biological control: Because of the weak pathology of flagellate diseases in insects, there is little interest in their use as biological control agents. Most interest in this group is focused on the role of hosts as vectors of vertebrate pathogens.

6.4.4   CILIATES

These protists use simple or compound cilia for locomotion and feeding, have a cytostome and have two types of nuclei, micro- and macro-nuclei. Sexual reproduction is by conjugation. Division is transverse rather than longitudinal (Tanada & Kaya, 1993).

Most ciliate species are commensuals in digestive tracts of insects (especially termites and roaches). Only a few are pathogenic (Corliss & Coats, 1976). Most of the pathogenic species are in the family tetrahymenidae. The hosts are typically mosquitoes and black flies, the larvae of which can be attacked in water. Pathogenic ciliates penetrate the host and multiply in the hemocoel, although other tissues may be invaded.

The mode of entry is unknown for most ciliate species; the mouth, cuticle, and wounds have been hypothesized to be entry points (Clark & Brandl, 1976; Washburn et al., 1988). The pathogens may utilize the host fat body tissues, in which case mortality rates are high. Ciliate disease is called ciliatoses.

Biological control: Because most ciliates are commensuals and because most pathogenic species do not cause fast, acute disease in their hosts, there has been little interest in using this group in biological control programs. However, some pathogenic species have been shown to be important natural enemies of some aquatic insect species, including mosquitoes.

6.4.5 MICROSPORA

Members of this group are commonly called microsporidia of the ento-mogenous protozoa, the members of microsporidia appear to have the best potential for use in insect control, although they generally induce mostly chronic infections and are difficult to mass produce (McLaughlin, 1971). However, current use and improvement of artificial diets, improvement in mass-rearing technologies, and an availability of trained personnel have made production of required numbers of *Heliothis virescens* or *Helicoverpa zea* larvae for production of *N. heliothidis* spores for large-scale experiments feasible. The microsporidia are known for their resistant spores, which survive well in the environment.

N. heliothidis was first described by Lutz and Splendore (1904) in Brazil from *H. zea* and was later found associated with both *H. zea* and *H. virescens* by Kramer (1959). Kramer also redescribed the life cycle of this protozoan. Lipa (1968) reported that *N. heliothidis* attacks the midgut epithelium, trachea and gonads of *H. zea*. Brooks (1968) presented data to show that *N. heliothidis* is transmitted transovarially in *H. zea* and that both the male and female can transmit the protozoan to the next generation. For example, under laboratory conditions, eggs produced by diseased *H. zea* (240 examined) were 100% infected with *N. heliothidis*. Lipa (1968) reported that under laboratory conditions a high percentage of diseased larvae failed to pupate or produced abnormal pupae.

In Mississippi, we have isolated *N. heliothidis* from both *H. zea* and *H. virescens* field collected larvae. In general, we found *N. heliothidis* more frequently in *H. zea* than in *H. virescens* larvae. Our studies showed that the average longevity of 20 pairs of healthy *H. zea* moths was 23.1 days (24.1 for male and 22.1 for female) and that for 20 pairs of diseased moths was 14.4 days (15.7 for male and 13.2 for female).

Thompson and Sikorowski (1979) reported that infected *H. zea* larvae accumulated fatty acids more rapidly than healthy larvae in the first 10 days, but then they decreased rapidly until death. Experimental infections have been obtained in *H. zea* and *H. virescens* with several other species of microsporidia (Brooks, 1988).

Signs and symptoms: Infected larvae of *H. zea* and *H. virescens* differ little, if any, from healthy larvae. Diseased larvae are not as active and lose their appetite. Infected larvae may fail to pupate, and some pupae and adults may be deformed. Thus, proper diagnosis requires microscopic examination of susceptible tissues such as fat bodies, midgut epithelium, etc.

Life cycle: The *Nosema* life cycle can be defined by two events: 1) mergony, the vegetative phase and 2) sporogony the production of spores. During merogony, the microsporidium multiplies rapidly by primary fission or multiple budding. From binucleate meronts, sporonts are formed. Nuclear division within the sporont gives rise to a tetranucleate sporont, which divides to produce two sporoblasts, each of which develops into a spore. Spores are ovoid, with one pole more pointed than the other. The fresh spores of *N. heliothids* measured in water were 3.3 to 6.0µm wide (Lipa, 1968).

Penetration: *N. heliothidis* can enter the host by three ways: 1) peros 2) ovarial and 3) cuticular. Entrance by the oral and cuticular means results in horizontal transmission. Oral entrances include feeding of the heliothine larvae on spore-contaminated food or feeding on dead or moribund insects. Entrance by the ovarial portal is accomplished either by contamination of the surface of the host eggs and infecting emerging larvae during eating of the contaminated egg chorion or by an ovarian transmission in which *N. heliothidis* meronts, sporonts and spores are integrated into the egg or embryo within the female's reproductive tract. Offspring from such females are also infected (Brooks, 1988). *H. zea* infected females may produce 100% infected offspring. Entrance by the cuticular portal is accomplished by inoculation of spores by the ovipositor of a *N. heliothidis* contaminated parasitoid (Brooks, 1973).

Infection: The infective stage of *N. heliothidis* is the spore. After ingestion by heliothine larvae, spores of *N. heliothidis* release a sporoplasm by injection through an everted polar tube. Only a short discussion of this topic is given in this publication. Microsporidian spores are among the smallest and most complicated of eukaryotic cells. The extrusion apparatus places the sporoplasm into a host cell. The sporoplasm is the infective unit of microsporidia. Once ingested, the spore is stimulated to evert its polar tube, which then serves as an injection needle through which the sporoplasm is injected. Under pressure, the rigid tube penetrates the peritrophic membrane and epithelial gut cells and deposits the sporoplasm directly into susceptible host cells. Afterward, the sporoplasm eventually undergoes merogony and sporogony to produce new spores (Brooks, 1988).

6.4.5.1   EFFECT OF N. HELIOTHIDIS ON HELICOVERA ZEA AS PREDATORS AND PARASITES

Nosema infected *H. zea* were fed to nymphs of *Nabis roseipennis, N. capsiformis, N. sordidus* and *N. alternatus* and *Chrysopa* sp. larvae. At maturity, the nymphs were crushed in drops of water on glass slides and then examined with a phase interference compound microscope to determine the presence or absence of *Nosema* spores. Brooks and Cranford (1972) documented the susceptibility of the braconid *Campoletis sonorensis* to *N. heliothidis.*

Infection: The insect fat bodies are the primary infection site. In advanced stages of disease, the fat body cells enlarge many times, giving the fat tissue a lobated appearance. The lobated fat tissues are filled with microsporidian spores. The abnormally large, white fat body is usually obvious through the integument. The degree of infection depends on spore dosage, temperature, and larval age. Feeding of the insect is usually normal at first, decreasing to very little for the last few days before death. *Vairimorpha* is transmitted to the next generation both on and in the eggs of infected adults.

Vairimorpha necatrix causes two different types of diseases resulting in mortality in its hosts, death that results from gut damage followed by bacterial septicemia, and death that results from microsporidiosis after ingestion of even a light spore dose (Maddox, 1966). A low dosage of *V. necatrix* results in a chronic infection of mainly the fat bodies and some muscular tissues, whereas high dosages produce an acute infection of mainly midgut tissues (Chu & Jaques, 1979). In the advanced stage of infection, the abnormally large, white fat body is usually conspicuous, and a dorsal swelling may appear on the last two or three abdominal segments (Pilley, 1976).

Spore production: *V. necatrix* develops only in living cells. Many species of noctuid larvae, in particular *H. zea* or *Trichoplusia ni*, are suitable hosts for spore production. Spore production per infected larva is about 2×10^{10} spores/g of host larva. Mass production and storage of *V. necatrix* spores were evaluated by Fuxa and Brooks (1979).

Production of microsporidian spores: Microsporidia are obligate pathogens and must be produced in laboratory-reared insects or in cell-tissue culture.

Laboratory insects: Generally, neonatal or early stage larvae are exposed to a low dose of spores placed on artificial diet or on a suitable food substrate. Several days later, depending on the rearing temperature, stage of insects, and diet, mature spores are harvested from insects by various methods as described by Brooks (1988). The habitual host(s) or permissive hosts, such as *H. zea* or *H. virescens* larvae that can support production of spores of various microsporidia are frequently used for spore production.

Tissue culture: Some species of microsporidia have been grown in tissue culture, but the high cost of culture media, the low yields in spores, and the imperfect production techniques for mass production still limit the usefulness of this method at this present time (Brooks, 1988). In Mississippi, we have been able to produce *N. heliothidis* and *V. necatrix* spores in *H. zea* larvae for large-scale field tests.

6.4.6   OTHER PROTOZOA

Class: Neogregarinida

The gregarines of the order neogregarindia are known as neogregarines. A number of entomopathogenic neogregarines produce lethal infections in important insect pests in the orders Diptera, Coleoptera, and Hemiptera.

Mattesia grandis: M. grandis, a neogregarine of the cotton boll weevil (*A. grandis*), was experimentally transmitted to *H. zea* and *H. virescens* by Ignoffo and Garcia (1965). The symptomology of this disease in the boll weevil is associated with a progressive destruction of the fat bodies and the reduction in the oviposition of eggs (McLaughlin, 1965).

Ophryocystis **sp.:** We observed the presence of this neogregarine in heliothine pupae that originated from an insectary reared population. Large amber-colored spores were occasionally found in the dark areas under translucent heliothine pupal cases. The spores can be seen on the scales of the moths under a dissecting microscope. Or, if the moths are allowed to emerge in clean plastic cups, the spores can be seen easily with a hand lens on the sides of the cups. McLaughlin and Myers (1970) described for the first time *Ophryocystis elektroscirrha* from naturally occurring populations of the monarch butterfly (*Danaus plexippus*) and the Florida queen butterfly *D. glippus berenice*. They reported that *O. elektroscirrha* infected the hypodermal tissue, that it remains in micronuclear schizogony until after pupation of the host, and then completes morphogenesis in the tissue that becomes the scales of the adult butterfly. The adult thus carries the spores externally.

6.5 RICKETTSIA

Rickettsia differ mainly from bacteria by,

- Gram negative.
- Obligate intracellular pathogens with typical bacterial cell walls and no flagella.
- Many are non pathogenic to insects.
- Occur as commensals or mutualists with in the tissues of insects.

Entomomogenous Rickettsia are in the

Phylum – Protophyta
Class   – Ricketsoidae
Order   – Rickettsioles
Family – Rickettsiaceae
Tribe   – wolbachieae, Rickettsiae
Genera – *Wolbachia* (seldom pathogenic)
        – *Rickettsia* (commonly pathogenic)

A number of Rickettsia species have been identified,

- *Rickettsia popilliae* from Japanese beetle
- *R. melolonthae* from *Melolontha melolontha*
- *R. tipulae* from family Tipulidae (Diptera)
- *R. grylli* from crickets, family Gryllidae (Orthoptera)
- *R. chironomi* from *Camtochironomus tentans*
- *R. blatae* from Cockroach
- *R. schistocercae* from *Schistocerca*
- *R. cetonidarum* and *R. armadillidii*

6.5.1 MODE OF INFECTION

Rickettsia species have not been cultivated in host cell free media. Some of them produce characteristic crystals. Infect mainly the fat and blood cells. They produce in general, a prolonged, chronic infection in an insect. *R. blatae* attacks other tissues and causes a systemic infection in the cockroach through its invasion of the hypodermis, trachea, malphigian tubule, muscle, nerve cell, blood cell, intestinal epithelium, etc.

6.5.2 DIFFERENCE FROM VERTEBRATE RICKETTSIA

Certain Rickettsia which are pathogens of insects and other invertebrates and are not transmitted to vertebrates. They are separable into distinct serological groups. The insect-pathogenic forms that produce crystals in one group; *R. grylli* on cricket in second group, and the Rickettsiae of scorpion in third group. These arthropod Rickettsiae are not related antigenically to Rickettsiae and chlamydia that infect vertebrates. On the other hand, *R. chironmi* appears morphologically and developmentally more closely related to chlamydia then to Rickettsia. The entomopathogenic Rickettsia posses a more complex developmental cycle than most bacteria and their forms change from typical elementary form (rod, coccoid, or kidney shaped) to spherical and giant forms. In addition, *R. chironmi* produces disc shaped bodies.

6.5.3 LIFE CYCLE

The life cycles are not well established and may vary with the Rickettsia species. Moreover the different terminology applied to Rickettsia developmental stages has caused difficulty in determining the relationship among various isolates. Huger and Kieg (1967) described the life cycles of *R. melolontha, R. tipulae, R. tenebrionis* and *R. blattae*. When the insect ingested the infectious elementary forms penetrate the insects midgut wall and replicate in tissues, especially the fat bodies, of the hemocoel and including gonads.

Small Rickettsia within the cytoplasmic vacuoles undergoes transformation to bacteria like forms that multiply by binary fission. Some cells are pleomorphic and form the characteristic protein crystals. The bacteria like forms enlarge to giant cells or to a rickettsiogenic stroma that reverts to small Rickettsia at the end of reproduction. Eventually all infected cells undergo lysis and masses of Rickettsia, crystals and Rickettsiae filled vacoules are released in the haemolymph.

6.5.4 SYMPTOMS

* In general, larvae at an advanced stage of infection may be blue grayish in colour.
* Those with sublethal infections develop into infected pupa and adults.
* Infected adults transmit the Rickettsia to their offspring through the egg.

R. popilliae: It infects the Japanese beetle, *Popillia japonica* and other beetles. An infected larva turns bluish and hence the name "Blue disease". The infected fat body contains crystals, when ingested by larvae, the Rickettsia cause death in about 45 days. When inoculated into hemocoel death occurs in 25 days. Large doses of *R. popilliae* are required to cause infection fed to a larva but less than six Rickettsia per larva is the LD_{50} by intra hemocoelic inoculation. The longevity of an infected larva is prolonged when the amount of available food is increased but this does not prevent its death.

R. melolontha: It infects the *M. melolontha* and other scarabid beetles. The disease in *Melolontha vulgaris* was discovered by Wille and Martignonoi (1952). It is called "Lorscher krankheit disease". It causes a chronic infection with apparent symptoms appearing after 2 to 3 months and death occurs 4 to 6 months. An infected larva develops white-bluish discoloration from the mass of Rickettsia in hemolymph. It is reduced in turgor and its reflexes diminish. An abnormal nervous reaction develops and an infected larva tends to rise to the surface of the soil when the temperature drops, where as an uninfected larva tends to burrow deeper in the soil. Such movement and death of infected larvae near soil surface may play an important role in epizootics through the transmission of the Rickettsia to young larvae that emerge from eggs laid in the upper region of the soil. Moreover, the wide host range including coleopteran families and Tipulidae (Diptera) may be important in the persistence of the Rickettsia in the host habitat. Rickettsial infection in chironomid was first reported in 1949 from larvae of camtochironomus tentans by Weiser.

6.6   CONCLUSION AND FUTURE PROSPECTS

Considerable progress has been made in recent years to evaluate the potential usefulness of EPNs and protozoa as microbial control agents. The present status of EPNs and protozoa as biological control agents can be summarized as follows: production, quality control, safety to vertebrates and beneficial insects, and large acreage field tests are all features that need further study before they are in general can be made more available for use as part of a new generation of microbial insecticides. However, the use of this species as a short-term microbial control agent on a large scale would not be practical.

KEYWORDS

- Biocontrol agents
- entomopathogenic nematodes
- arthropod
- protozoa
- insects
- rickettsia

REFERENCES

Bailey, L. *Infectious Diseases of the Honey-bee;* Land Books: London, 1963.

Bedding, R. A. Low Cost *in Vitro* Mass Production of *Heterorhabditis* Species for Field Control of Insect Pests. *Nematol. Mediterr.* **1981,** *27,* 109–114.

Brooks, W. M. Transovarian Transmission of *Nosema heliothidis* in Corn Earworm, *Heliothis Zea. J. Invertebr. Pathol.,* **1968,** *11,* 510–512.

Brooks, W. M. Protozoa: Most-Parasitic-Pathogen Interrelationships. *Misc. Publ. Entomol. Soc. Am.* **1973,** *9,* 105–111.

Brooks, W. M. *Entomogenous Protozoa.* In *Handbood of Natural Pesticides. Microbial Insecticides, Part A. Entomogenous Protozoa and Fungi;* Ignoffo, C. M., Ed.; CRC Press: Boca Raton, Florida, 1988; Vol. 5, p. 149.

Brooks, W. M.; Cranford, J. D. Microsporidoses of the Hymenopterous Parasites, *Campoletis sonorensis* and *Cardiochils nigriceps,* Larval Parasites of Heliothis Species. *J. Invertebr. Pathol.* **1972,** *20,* 77–94.

Brown, I.; Gaugler, R.; Shapiro-Ilan, D. LOTEK – An Improved Method for *in Vivo* Production of Entomopathogenic Nematodes. *Int. J. Nematol.* **2004,** *14,* 9–12.

Canning, E. U. Observations on the Life History of *Mattesia trogodermae* Sp. N., a Schizogregarine Parasite of the Fat Body of the Khapra Beetle, *Trogoderma graniarium* Everts. *J. Insect. Pathol.* **1964,** *6,* 305–317.

Canning, E. U. An Evaluation of Protozoal Characteristics in Relation to Biological Control of Pests. *Parasitology.* **1982,** *84,* 119–149.

Chu, W. H.; Jaques, R. P. Pathologie D'une Microsporidiose de L'arpenteuse du Chou, *Trichoplusia ni* (Lep.: Noctuidae), Par *Vairimorpha necatrix. Entomophaga.* **1979,** *24,* 229–235.

Clark, T. B.; Brandl, D. G. Observations on Tte Infection of *Aedes Sierrensis* by a Tetrahymenine Ciliate. *J. Invertebr. Pathol.* **1976,** *28,* 341–349.

Corliss, J. O.; Coats, D. W. A New Cuticular Cyst-Producing Tetrahymenid Ciliate, *Lambornella clarki* N. Sp. and the Current Status of Ciliatosis in Culicine Mosquitoes. *Trans. Am. Microsc. Soc.* **1976,** *95,* 725.

Dolgikh, V. V. The Effects of *Nosema grylli* and *Adelina grylli* on the Activities of Four Enzymes of Carbohydrate and Energy Metabolism in the Fat Body of the Crickets *Gryllus bimaculatus. Parazitologiya.* **1998,** *32*(5), 464–469.

Ehlers, R. Mass Production of Entomopathogenic Nematodes for Plant Protection. *App. Microbiol. Biotechnol.* **2001,** *56,* 623–633.

Ehlers, R. U.; Hokkanen, H. M. T. Insect Biocontrol with Non-Endemic Entomopathogenic Nematodes (*Steinernema* and *Heterorhabditis* Sp.): OECD and COST Workshop on Scientific and Regulation Policy Issue. *Biocontrol Sci. Technol.* **1996,** *16,* 295–302.

Ehlers, R. U.; Peters, A. Entomopathogenic Nematodes in Biological Control: Feasibility, Perspectives and Possible Risks. In *Biological Control: Benefits and Risks*; Hokkanen, H. M. T., Lynch, J. M., Eds.; Cambridge University Press: Cambridge, UK, 1995; pp 119–136.

Ehlers, R. U.; Lunau, S.; Krasomil-Osterfeld, K. C.; Osterfeld, K. H. Liquid Culture of the Entomopathogenic Nematode Bacterium Complex *Heterorhabditis megidis/ Photorhabdus luminescens. Biocontrol.* **1998,** *43,* 77–86.

Ehlers, R. U.; Niemann, I.; Hollmer, S.; Strauch, O.; Jende, D.; Shanmugasundaram, M.; Mehta, U. K; Easwaramoorthy, S. K.; Burnell, A. Mass Production Potential of the Bacto-Helminthic Biocontrol Complex *Heterorhabditis indica – Photorhabdus luminescens. Biocontrol Sci. Technol.* **2000,** *10,* 607–616.

Engler, R.; Rogoff, M. H. Entomopathogens: Ecological Manipulation of Natural Associations. *Environ. Health Perspect.* **1976,** *14,* 153–159.

Fuxa, J. R.; Brooks, W. M. Mass Production and Storage of *Vairimorpha necatrix* (Protozoa: Microsporida). *J. Invertebr. Pathol.* **1979,** *33,* 86–94.

Gaugler, R.; Brown, I. LOTEK – An Automated Apparatus for Production of Insecticidal Nematodes. U.S.S.N Patent 09/845,816, 2001.

Gaugler, R.; Han, R. C. Production Technology. In *Entomopathogenic Nematology*; Gaugler, R., Ed.; CABI: Wallingford, UK, 2002; p 289–310.

Han, R. C.; Ehlers, R. U. Pathogenicity, Development and Reproduction of *Heterorhabditis bacteriophora* and *Steinernema carpocapsae* Under Axenic in Vivo Conditions. *J. Invertebr. Pathol.* **2000,** *75,* 55–58.

Han, R. C.; Ehlers, R. U. Effect of *Photorhabdus luminescens* Phase Variants on the *in Vivo* and *in Vitro* Development and Reproduction of the Entomopathogenic Nematodes

Heterorhabditis bacteriophora and *Steinernema carpocapsae*. *Microb. Ecol.* **2001,** *35,* 239–247.

Han, R. C.; Cao, L.; Liu, X. Effects of Inoculum Size, Temperature and Time on *in Vitro* Production of *Steinernema carpocapsae* Agriotos. *Nematologica.* **1993,** *39,* 366–375.

Harry, O. G. The Effect of a Eugregarine *Gregarina polymorpha* (Hammerschmidt) on the Mealworm Larva of *Tenebrio molitor* (L.). *J. Eukaryot. Microbiol.* **1967,** *14,* 539–547.

Hazir, S.; Kaya, H. K.; Stock, P.; Keskin, N. Entomopathogenic Nematodes (Steinernematidae and Heterorhabditidae) for Biological Control of Soil Pests. *Turkish J. Biol.* **2003,** *27,* 181–202.

Heckel, D. G. Insecticide Resistance after Silent Spring. *Science.* **2012,** *337,* 1612–1614; 10.1126/science.1226994

Huger, A. M.; KRIEG, A. Ein Neuer Vermeh-Rungsmodus Von Rickettsien in 1nsekten. *Naturwissen Schaften.* **1967,** *54,* 475.

Hurd, H. Reproductive Disturbances Induced by Parasites and Pathogens of Insects. In *Parasites and Pathogens;* Beckage, N. E., Thompson, S. N., Federici, B. A., Eds.; Academic Press: New York, NY, 1993; pp 87–93.

Ignoffo, C. M.; Garcia. C. Infection of the Cabbage Looper, Bollworm, Tobacco Budworm, and Pink Bollworm with Spores of *Mattesia grandis* Mclaughlin Collected from Boll Weevils. *J. Invertebr. Pathol.* **1965,** *7,* 260–262.

Jackson, G. J. Neoaplectanaglaseri: Essential Amino Acids. *Exp. Parasitol.* **1973,** *34,* 111–114.

Kaya, H. K.; Gaugler, R. Entomopathogenic Nematodes. *Annu. Rev. Entomol.* **1993,** *38,* 181–206.

Kramer, J. P. Observations of the Seasonal Incidence of Microsporidosis in European Corn Borer Populations in Illinois. *Entomophaga.* **1959,** *4,* 37–42.

Lange, C. E.; Wittestein, Y. E. Susceptibilidad de la Langosta *Schistocerca cancellata* (Acrididae) a Diferentes Entomopatógenos. *Rev. Soc. Entomol. Arg.* **1998,** *57,* 19–22.

Lipa, J. J. Protozoan Infections other than *Sporozoan.* In *Insect Pathology: An Advanced Treatise*; Steinhaus, E. A., Ed.; Academic Press: New York, NY, 1963; Vol. 2.

Lipa, J. J. Some Observations on *Nosema heliothidis* Luts et Splendore, a Microsporidian Parasite of *Heliothis zea* (Boddie) (Lepidoptera, Noctuidae). *Acta Protozool.* **1968,** *4,* 237–278.

Lord, J. C. *Mattesia oryzaephili* (Neogregarinorida: Lipotrophidae), a Pathogen of Stored-Grain Insects: Virulence, Host Range and Comparison with *Mattesia dispora.* *Biocontrol Sci. Technol.* **2003,** *13,* 589–598.

Lunau, S.; Stoessel, S.; Schmidt-Peisker, A. J.; Ehlers, R. U. Establishment of Monoxenicinocula for Scaling up in Vitro Cultures of the Entomopathogenic Nematodes *Steinernema* Spp. and *Heterorhabditis* Spp. *Nematologica.* **1993,** *39,* 385–399.

Lutz, A.; Splendore, A. Uber Perbrine and Verwandte Mikorsporidien. Nachtrag Zur Ersten Mitteilung. *Zentr. Bakt. (I) Orig.* **1904,** *36,* 645–650.

Maddox, J. V., Studies on a Microsporidosis of the Armyworm, *Pseudaletia unipuncta* (Haworth). Ph.D. Thesis, University of Illinois, Urbana, IL, 1966.

McLaughlin, R. E. Some Relationships between the Boll Weevil, *Anthonomus grandis* Boheman, and *Mattesia grandis* Mclaughlin (Protozoa: Neogregarinida). *J. Invertebr. Pathol.* **1965,** *7,* 464–473.

McLaughlin, R. E. Use of Protozoans for Microbial Control of Insects. In *Microbial Control of Insects and Mites;* Burges, H. D., Hussey, N. W., Eds.; Academic Press: London, UK, 1971; pp 151–172.

McLaughlin, R. E.; Myers, J. *Ophryocystis elektroscirrha* Sp. N. a Neogregarine Pathogen of the Monarch Butterfly *Danaus plexippus* (L.) and the Florida Queen Butterfly *D. Glippus berenice* Cramer. *J. Protozool.* **1970,** *17,* 300–305.

Pace, G. W.; Grote, W.; Pitt, D. E.; Pitt, J. M. Liquid Culture of Nematodes. Int. Patent W.O 86/01074, 1986.

Pasteur, L. Etudes Sur La Maladie Des Vers a Soie, Tome I And II. Gauthier-Villars: Paris, France, 1870.

Pilley, B. M. A New Genus, *Vairimorpha* (Protozoa: Microsporida) for *Nosema necatrix*. Kramer 1965: Pathogenicity and Life Cycle in *Spodoptera exempta* (Lepidoptera: Noctuidae). *J. Invertebr. Pathol.* **1976,** *28,* 177–183.

Poinar, G. O. Jr. Origins and Phylogenetic Relationships of the Entomophilic Rhabditids, *Heterorhabditis* and *Steinernema. Fundam. Appl. Nematol.* **1993,** *16*(4), 333–338.

Popiel, I.; Vasquez, E. M. Cryo Preservation of *Steinernema carpocapsae* and *Heterorhabditis bacteriophora. J. Nematol.* **1991,** *23,* 432–437.

Prabhu, S.; Rajendran, G.; Subramanian, S. In Vitro Mass Production Technology for the Entomopathogenic Nematode, *Steinernema glaseri. Indian J. Nematol.* **2007,** *36,* 142–144.

Ritter, K. S. *Steinernema feltiae* (*Neoaplectana carpocapsae*): Effect of Sterols and Hypolipidemic Agents on Development. *Exp. Parasitol.* **1988,** *67,* 257–267.

Shapiro-Ilan, D. I.; Gaugler, R. Production Technology for Entomopathogenic Nematodes and their Bacterial Symbionts. *J. Ind. Microbiol. Biotechnol.* **2002,** *28,* 137–146.

Tanada, Y.; Kaya, H. K. *Insect Pathology;* Academic Press: New York, NY, 1993.

Taylor, A. B.; King, R. L. Further Studies on the Parasitic Amoebae Found in Grasshoppers. *Trans. Am. Microsc. Soc.* **1937,** *56,* 172–176.

Thompson, A. C.; Sikorowski, P. P. Effects of *Nosema heliothidis* on Fatty and Amino Acids in Larvae and Pupae of the Bollworm *Heliothis zea. Comp. Biochem. Physiol.* **1979,** *63A,* 325–328.

Wallace, F. G. The Trypanomatic Parasites of Insects and Arachnids. *Exp. Parasitol.* **1966,** *18,* 124–193.

Wallace, F. G. Biology of Kinetoplastida of Arthropods. In *Biology of the Kinetoplastida*; Lumsden, W. H. R.; Evans, D. A., Eds.; Academic Press: New York, 1979; Vol. 2, pp 213–240.

Washburn, J. O.; Egerter, D. E.; Anderson, J. R.; Saunders, G. A. Density Reduction in Larval Mosquito (Diptera: *Culicidae*) Populations by Interactions between a Parasitic Ciliate (Cilophora: Tetrahymenidae) and an Opportunistic Fungal (Oomycetes: *Pythiaceae*) Parasite. *J. Med. Entomol.* **1988,** *25,* 307–314.

Wille, R.; Martignoni, M. E.; Schweiz. Z. Allgem. *Pathol. U. Bakteriol.* **1952,** *15,* 470–474.

Yang, H.; Jian, H.; Zhang, S.; Zhang, G. Quality of the Entomopathogenic Nematode *Steinernema carpocapsae* Produced on Different Media. *Biol. Control,* **1997,** *10,* 193–198.

CHAPTER 7

ENTOMOPATHOGENIC NEMATODES: AN EMERGING BIOCONTROL AGENT FOR INSECT PESTS MANAGEMENT

RASHID PERVEZ* and SANTHOSH J. EAPEN

Division of Crop Protection, ICAR-Indian Institute of Spices Research, Kozhikode 673012, Kerala, India

Corresponding author. E-mail: rashid_pervez@rediffmail.com

CONTENTS

ABSTRACT

Entomopathogenic nematodes have been reported to occur in tropical, subtropical, and temperate regions except Antarctica, where they play an important role in soil and are also used as biological control agents against many insect pests. With the passing of time and realizing the harmful effects of insecticides, biocontrol agents are being deployed with varying degree of success. Entomopathogenic nematodes (EPNs) have got little attention by the researchers though they have a great potential in reducing pest population and with little manipulation their role can be enhanced. During the last decade testing of EPN based biopesticide have been found promising against variety of insect pests because of their desirable characters like high infectivity, prolonged storage (shelf life), heat and desiccation tolerant, and self-perpetuating in favorable conditions. Hence, an immediate need was felt to highlight the potentials of EPN in a comprehensive manner with facts and to make aware the farming community, extension workers, entomologists, and nematologists and at the same time acquaint them with the work done on this important aspect of biocontrol.

7.1 INTRODUCTION

Nematodes associated with insects are commonly regarded as entomopathogenic nematodes (EPNs). These when get entry into the insect body, parasitize, cause disease, and kill. Research work on EPNs progressed since 1932 with the discovery of *Steinernema glaseri* (Glaser, 1932) infecting Japanese beetle. Intense interest in EPNs for insect pest control has been fuelled mainly because of their potential efficacy and other impressive attributes. They have also stimulated strong commercial interest in development as biopesticides. These include their wide spectrum of insecticidal activity, ability to kill most insects within short periods, availability of efficient mass culturing techniques, safety to vertebrates, plants and non-target organisms, amenability for application using standard application equipment, and compatibility with many chemical pesticides. Besides, there has been growing concern on the need for ecological sustainable methods of pest management.

The interest in using EPN as biopesticide for pest control has increased exponentially over the past three decades. Globally, thousands of researchers are exploring the possibilities of the potential of EPN to manage harmful insects, mollusc, plant parasitic nematodes, and even soil-borne plant pathogens. The EPNs, specially, *Steinernema* and *Heterorhabditis* are considered successful and are commercially mass produced to manage insect pests problem in agricultural and horticultural crops. The ease of their production and exemption from registration requirements are the two big reasons for commercial developments of EPN. Therefore, demonstration of practical use specially in Europe and North America and later on in Japan, China, and Australia spurred development across the globe that have laid to the availability of EPNs against insect pests that were once thought impossible to control. This chapter brings together published information on diversity, management of insect pests, mass production, formulation, and application technology.

7.2   OCCURRENCE OF EPNS

EPNs have been reported to occur in tropical, subtropical, and temperate countries (Pervez et al., 2014a; Herrera et al., 2007; Griffin et al., 1991; Beavers et al., 1983) except Antarctica (Griffin, 1990), where they play an important role in soil (Kaya, 1990; Pervez et al., 2014b, 2015) and are also used as biological control agents against many insect pests (Pervez & Ali, 2013; Pervez et al., 2015; Kaya & Gaugler, 1993). The study of population dynamics of EPNs is fundamental to understanding their persistence, distribution, and effect on insect populations and for the development of predictive models for control programs (Hominick & Reid, 1990; Banu, 2003).

Initial research with EPNs in India was conducted primarily with exotic species viz., *S. carpocapsae*, *S. glaseri*, *S. feltiae*, and *H. bacteriophora* imported by researchers. In many cases, these EPNs yielded inconsistent results in field trials, probably due to their poor adaptability to the local agro-climatic conditions. India, as is the case with many other parts of the world, has a rich biodiversity resource because of its varied geographic, climatic, and weather conditions. It is divided into 15 agro-climatic and agro-ecological zones, which for the most part consist of tropical and subtropical areas. Therefore, a search for indigenous species/strains resulted in a number of nematode isolates from different parts of India (Ganguly, 2003; Ali et al., 2005b; Pervez et al., 2014b).

In addition, many surveys have revealed natural occurrence of several species/strains of EPNs in Andaman and Nicobar islands (Prasad et al.,

2001), Gujarat (Vyas, 2003), Kerala (Banu et al., 2004, 2005), New Delhi (Ganguly & Singh, 2000), Tamil Nadu (Josephrajkumar & Sivakumar, 1997; Bhaskaran et al., 1994), Meghalaya (Lal Ramliana & Yadav, 2010), and Uttar Pradesh (Ali et al., 2005b; Pervez & Ali, 2007).

7.2.1 DIVERSITY OF EPNS

Now, there are two genera *Steinernema* and *Neosteinernema* listed under Steinernematidae (Chitwood & Chitwood, 1937), only one genus *Heterorhabditis* under Heterorhabditidae (Poinar, 1976), and one genus *Oscheius* under Rhabditidae (Sudhaus, 1976). The genus *Steinernema* (Steiner, 1923 (Travassos, 1927) and *Heterorhabditis* (Poinar, 1976) are the most intensively studied group of nematodes associated with insects, although 79 species of *Steinernema*, one species of *Neosteinernema* (Nguyen & Smart, 1994), 19 species of *Heterorhabditis,* and 7 species of *Oscheius* (Weimin et al., 2010; Pervez et al., 2013) have been described.

Among the indigenous EPN isolates, one species of *Heterorhabditis*, five species of *Steinernema,* and two species of *Oscheius* have been described as new species from various parts of India (Ali et al., 2005a, 2005b; Pervez et al., 2013).

Other species identified as indigenous isolates include *S. carpocapsae* (Hussaini et al., 2001; Ali et al., 2005b; Pervez et al., 2014d), *S. abbasi* (Ghode et al., 1988; Hussaini et al., 2003), *S. tami* (Hussaini et al., 2001), *S. bicornutum* (Hussaini et al., 2001), *S. riobrave* (Ganguly, 2003), *S. feltiae* (Singh et al., 1992; Hussaini et al., 2003), *S. siamkayai* (Banu et al., 2005), *H. bacteriophora* (Sivakumar et al., 1989), and *H. indica* (Banu et al., 2005; Pervez et al., 2015).

Hominick et al. (1997) reported that *S. feltiae* and *H. bacteriophora* are widely distributed throughout the world, whereas *S. carpocapsae* and *H. indica* in India (Hussaini et al., 2003; Pervez et al., 2015), but Steinernematids are generally recovered more often than *Heterorhabditis* during non-targeted surveys (Hominick et al., 1997; Pervez et al., 2014b).

7.3 MANAGEMENT OF INSECT PESTS

Various options for insect pest management in agricultural and horticultural crops are available. However, in case of severe infestation, these measures fail miserably and farmers are left with no choice. Chemicals do act fast and effectively but due to faulty planning, use of wrong chemicals, and incorrect dosage and time of application, these become curse rather than boon. Very soon, they become ineffective due to development of resistance. The chemicals often do not give desired degree of control due to feeding habit of pests, weather conditions like rains, heavy wind, spurious products, and so on.

One of the ideal methods of pest management is biological control, where the various living organisms are deployed to check the damage caused by other living organisms. Most of the biocontrol agents such as fungi, viruses, and bacteria are conservative in nature. Fungi found effective and utilized effectively are *Beauveria bassiana* (Balsamo), *Nomuraea rileyi* (Farlow) Samson and *Metarhizium anisopliae* (Hetsch). These fungi have been utilized with a varying degree of success (Saxena & Ahmad, 1997), but suffer from some inherent deficiency like persistence, requirement of high humidity, formulations, and shelf life.

Among viruses, nuclear polyhedrosis virus of *Helicoverpa armigera* (*Ha*NPV) has been widely used with moderate success (Ahmad et al., 1999; Ali et al., 2005b). This virus is highly host specific and is comparatively easy to handle. But, it has limitations of multiplication in its hosts only, which in turn is a very cumbersome process. *Bacillus thuringiensis* (Bt) is a very good biocontrol agent but its production has to be done under specialized laboratories and thus its cost is highly prohibitive and can be used in case of high return cropping system.

Recently, EPNs are another important group of organism, which are emerging as potent biocontrol agent against insect pests because of desirable characters like high infectivity, prolonged storage (shelf life), heat and desiccation tolerance, and self-perpetuation in favorable conditions. It is hoped that in due course of time, EPN will be utilized in integrated pest management (IPM) program.

7.3.1   INFECTIVITY OF ENTOMOPATHOGENIC NEMATODES

During the last decade, testing of different EPNs against several insect pests of agricultural and horticultural crops has gained momentum and generated wide interest among plant protection workers. In India, several studies have been conducted on the pathogenicity of different EPN species against various insect pests.

Hussaini et al. (2000) reported that, six indigenous *Steinernema* isolates and two *Heterorhabditis* isolates against *Agrotis* spp. *S. bicornutum* (PDBC EN 3.2) and *S. carpocapsae* (PDBC EN 6.61) were found to be most effective against *A. ipsilon* and *A. segetum*, respectively, while both *Heterorhabditis indica* (PDBC EN 13.3 and PDBC EN 6.71) brought about 100% mortality of both *Agrotis* spp. In sand column assay, *S. bicornutum* (PDBC EN 3.1) and *H. indica* (PDBC EN 13.3 and PDBC EN 6.71) were promising against *A. ipsilon* larvae while *S. carpocapsae* (PDBC EN 6.61) and *H. indica* (PDBC EN 13.3 and PDBC EN 6.71) were virulent against *A. ipsilon* larvae.

S. mushtaqi was found more pathogenic to *Amsacta moorei* as it brought about 100% mortality within 48 h, followed by *Spodoptera litura, Lampides boeticus, Nezara viriduala,* and *Earias vittella* (within 72 h). Whereas, this level of mortality of *Clavigralla gibbosa, Centrococcus somatics,* and *Maruca vitrata* were recorded within 144 h (Pervez et al., 2008; Pervez & Ali, 2011). Rajkumar et al. (2002a) utilized strains of *Steinernema* and *Heterorhabditis* against *S. litura* and found caterpillar mortality ranged from 16.6 to 88.8%. However, 50% mortality was achieved within 72 h of exposure at an inoculum level of 100 infective juveniles (IJs)/100 g soil/ caterpillar. The LD_{50} values after 72, 96, and 120 h of exposure were 75, 56, and 39%, respectively. Bioefficacy studies of *Steinernema* (STUDP-1) and *Heterorhabditis* (HUDP-1) strains against insect pests viz. *Achaea janata, S. litura,* and *Chilo partellus,* have revealed that, the mortality was 92.5 and 97%, respectively, after 120 h of inoculation.

Abbas and Saleh (1998), reported pathogenicity of *S. riobrave* and *S. abbasi* against fully grown larvae of *S. littoralis*. The LD_{50} values were 148 and 101 IJs/larvae for *S. abbasi* and *S. riobrave* within 48–72 h post treatment, respectively. However, *Heterorhabditis* sp. (ELG), *H. indica,* and *Heterorhabditis* sp. (ELB) against *S. littoralis* recorded 100% mortality within 24 h (Abdel & Abd-Elgawad, 2007). *S. masoodi* and *S. mushtaqi* were found more pathogenic to *S. litura* and *H. armigera* as it brought about 100% mortality, followed by *S. seemae* and *S. carpocapsae,* where they gave 89% mortality after 72 h post exposure (Ali et al., 2009b; Pervez & Ali, 2009). *S. feltiae* found pathogenic to pre-pupa, pupa, and adult of *S. litura* (Narayanan & Gopalakrishnan, 1987) and *H. armigera* (Pervez & Ali, 2010).

Ali et al. (2009a) reported that *S. masoodi*-based liquid formulation against *Myllocerus* sp. infesting pigeon pea in pot, after 96 h, showed 88% mortality when 1.2×10^5 IJs was sprayed.

Ahmad et al. (2009) observed only 12.4% pod damage in chickpea in case of plots treated with *S. masoodi*-based liquid formulation compared to 34.8% pod damage in untreated control (resulting is an avoidable loss of 22.4%). Chickpea yield was 1475 kg/ha in treated, while 875 kg/ha in untreated. Thus, giving rise to 68.6% increase in yield.

Application of EPN alone and with the adjuvant significantly reduced *H. armigera* on pigeon pea compared to initial population. Both these treatments reduced pod damage with 66.9 and 97.3% increase in yield over control (Vyas et al., 2002a). Vyas et al. (2002b) studied 16 EPN isolates isolated from central and north Gujarat and tested for the management of *H. armigera*. Amongst the isolates, EPN-3 (Anand) and EPN-16 (Dehgam) were found highly virulent to *H. armigera* larvae, while remaining isolates induced low mortality of larvae. Bio-efficacy tests of promising strains EPN-3 and EPN-16 were carried out against *H. armigera* on pigeon pea and chickpea (Table 7.1). These induced maximum 96.8 and 70.9% larval mortality, respectively at a dose of 2000 IJs/five plants/pot treatment (Vyas et al., 2002a).

Field efficacy test results of *Steinernema thermophilum* against diamondback moth (DBM) (*Plutella xylostella*) infesting cabbage showed that *S. thermophilum* at 3000 IJs/ml caused the highest mortality of 46%, whereas at 2000 IJs/mL and the insecticide treatment caused 40.5 and 40% mortality in DBM larvae, respectively (Somvanshi et al., 2006).

S. carpocapsae and *S. kraussei* were used to control the larvae of the large pine weevil, *Hylobius abietis,* under field condition. Both EPNs infect and kill larvae reducing emergence of adult *Hericium abietis* (Torr et al., 2007).

S. carpocapsae was used to control larvae of the DBM, *P. xylostella*, infesting cabbage foliage. EPN-based formulation containing 0.3% of the surfactant Rimulgan and 0.3% of the polymer xanthan was applied. *S. carpocapsae* caused mortality decrease when DBM larvae were added 9 h after dauer juvenile (DJ) application, as 98% of the IJs were still alive after 9 h.

S. carpocapsae was used against shoot and fruit borer, *Leucinodes orbonalis* guenee, infesting brinjal pest under field conditions. Three rates of the EPN (1, 1.5, and 2 billion/ha) were evaluated by spraying infective juvenile stages 10–12 times, at 10-day intervals, during the brinjal growth cycle, starting at the 5–10% flowering stage. *S. carpocapsae* caused significant reduction in fruit borer damage and increased yield in the first 2 years.

TABLE 7.1   Field Application of EPNs against Various Insect Pests in India.

EPN species	Insect pests	Reference
1. *H. bacteriophora*	*Amsacta albistriga*	Bhaskaran et al. (1994)
2. *H. indica*	*H. armigera*	Anonymous (2001)
3. *S. carpocapsae*	*Scirpophaga incertulas*	Rao and Manjunath (1966)
	Scirpophaga incertulas, Mythimna separate	Israel et al. (1969a, 1969b)
	S. litura	Sitaramaiah et al. (2003)
	H. armigera	Hussaini et al. (2003), Ahmad et al. (2009)
	P. xylostella	Schroer and Ehlers (2005)
	albistriga	Bhaskaran et al. (1994)
	L. orbonalis	Ganga et al. (2009)
4. *S. glaseri*	*S. litura, Holotrichia consanguinea*	Vyas and Yadav (1993)
	Myllocerus discolor	Prabhuraj et al. (2000)
5. *S. masoodi*	*H. armigera, Myllocerus* sp.	Ali et al. (2009a), Ahmad et al. (2009)
6. *S. seemae*	*H. armigera*	Ahmad et al. (2009a)
7. *S. riobrave*	*H. armigera*	Vyas et al. (2002a, 2002b), Vyas et al. (2001b)
8. *S. thermophilum*	*P. xylostella*	Somvanshi et al. (2006)
9. *S. carpocapsae*	*H. abietis*	Torr et al. (2007)
10. *S. kraussei*	*H. abietis*	Torr et al. (2007)

7.3.2   INFECTIVITY OF EPNS AGAINST LEPIDOPTERAN INSECT PESTS

Karunakar et al. (1999a) reported that nine lepidopteran insects were susceptible to *S. glaseri, S. feltiae,* and *H. indicus.* EPNs, *S. masoodi, S. seemae,* and *S. carpocapsae* were found pathogenic to *H. armigera* bringing about mortality within 36, 38, and 48 h, respectively, while, *S. glaseri* and *S. thermophilum* killed the larvae of *H. armigera* within 56 h (Ali et al., 2008). First and second instar larvae of *H. armigera* were highly susceptible to *H. indica, S. masoodi,* and *S. mushtaqi* (Banu, 2001; Pervez,

2010). Among third and fourth stages, the third instar was highly susceptible to nematode infection with LC_{50} value of 2.43 and 3.69 IJ/larvae for *S. glaseri* and *H. indica*, respectively. The third instar larvae were highly susceptible followed by fourth and fifth instar larvae of *H. armigera* to *S. glaseri*, *H. indica*, and *H. bacteriophora* (Jothi & Mehta, 2003). Tahir et al. (1995) also reported that final instar larvae of the *H. armigera* were susceptible to *S. riobrave* than to *S. carpocapsae* and *Heterorhabditis* sp.

Pervez et al. (2014c) reported that, four native entomopathogenic nematodesEPNs, *Heterorhabditis* sp. (IISR-EPN 01), *Steinernema* sp. (IISR-EPN 02), *Oscheius* sp. (IISR-EPN 08), and *O. gingeri* found pathogenic against shoot borer larva (SBL), *Conogethes punctiferalis* Guen.), infesting ginger (*Zingiber officinale* Rosc.). LD_{50} of *Oscheius* sp. (IISR-EPN 08) was 48 IJs/larva, whereas LT_{50} of *Steinernema* sp. (IISR-EPN 02) was 29 h for desired mortality of SBL.

Pathogenicity of eight native EPNs was evaluated against larvae of hairy caterpillar, *Euproctis* sp., and larvae and pupae of the shoot borer, *C. punctiferalis*, infesting ginger. Of the tested EPNs, all isolates, except *Oscheius* sp. (IISR-EPN 08) caused 100% mortality to larvae of hairy caterpillar. *Heterorhabditis* sp. (IISR-EPN 01), *Steinernema* sp. (IISR-EPN 02), *Oscheius* sp. (IISR-EPN 08), and *O. gingeri* caused 100% mortality also to SBL. *O. gingeri* was the most virulent against the shoot borer pupae, causing 100% mortality, followed by *Steinernema* sp. (IISR-EPN 02) and *Oscheius* sp. (IISR-EPN 05) which killed 67% of the pupae (Pervez et al., 2012).

7.3.3 INFECTIVITY OF EPNS AGAINST STORED GRAIN PESTS

For the management of stored grain pests, preventive and curative measures *viz.*, fumigation and use of natural plant materials and biopesticides of plant origin have been common practices recommended earlier. This could not give desirable results. Therefore, efforts are now being made in this direction so that a package of EPN-based formulations, which are capable of suppressing insect's population to a safer limit, could be developed.

EPN infective against stored-product insects such as red flour beetle, *Tribolium castaneum* (Herbst), Indian meal moth, *Plodia interpunctella* (Hubner), Mediterranean flour moth, *Ephestia kuehniella* (Zeller), saw-toothed grain beetle, *Oryzaephilus surinamensis,* yellow meal worm,

Tenebrio molitor (L.), and the warehouse beetle, *Trogoderma variabile* (Ballion) (Athanassiou et al., 2008; Morris, 1985; Ramos-Rodriguez et al., 2006). Trdan et al. (2006) reported susceptibility of the young adult of granary weevil, *Sitophilus granarius* (L.), and the saw-toothed grain beetle, *O. surinamensis* (L.), to four EPN species.

Field trails stimulating empty grain bin treatments were conducted using red flour beetle and the Indian meal moth, *P. interpunctella*. Pathogenicity of EPNs in this trial ranged from 49% (*T. castaneum* pupae) to 99% (*P. interpunctella* larvae). *S. riobrave* had pathogenicity against many stored product insect species. In laboratory bioassays, this species was found to be effective against the red flour beetle, *T. castaneum*, at a range of different temperature and relative humidities (RH) associated with grain handling and processing facilities.

Ahmad et al. (2007) reported that *S. masoodi* brought about mortality in *Rhyzopertha dominica*, *Corcyra cephalonica*, and *Callosobruchus chinensis* within 48 h of the inoculation. *T. castaneum* being a pest of variety of grains and their processed product, showed comparative resistance against *S. masoodi*. Even after 150 h of treatment, only 30% mortality was recorded. *Neoaplectana carpocapsae* brought about 100% mortality of *Acanthoscelides obtectus* imago stage treated with 60 IJs/insect within 96 h, while, in case of *T. molitor* (L_2, L_3, and L_4), *R. dominica* adult, and *E. kuehniella* late caterpillar stage treated with 10–50 IJs/insect within 48 h brought mortality. Fayaz and Javed (2009) studied seven Pakistani strains of EPN belonging to genera *Steinernema* and *Heterorhabditis* against last instar and adult stages of the pulse beetle, *C. chinensis*, and found that *H. bacteriophora*, *S. siamkayai*, and *S. pakistanense* showed the highest virulence to pulse beetle larvae and adults. The last larval stage of the pulse beetle seems to be more susceptible than that of the adult. LC_{50} value in petri dish and concrete containers were 14–340 IJs/larvae and 41–441 IJs/larvae, respectively, and 59–1376 IJs/adult and 170–684 IJs/adult, respectively.

7.4   MASS PRODUCTION OF EPNS

Mass production on large scale and the ready availability of the organism in required quantity and at competitive cost make them acceptable among entrepreneurs and farmers. The EPNs are multiplied either on a suitable host (*in vivo*) or on a semi synthetic diet (*in vitro*). Both techniques of mass production of EPN have their own advantages and limitations.

7.4.1 IN VITRO PRODUCTION OF EPNS

Many studies have been conducted on the mass production of different EPN species in various media (Wouts, 1981; House et al., 1965) and modified media (Hussaini et al., 2002a, 2002b). Banu (2001) reported that different media *viz.*, Wouts medium, Wouts medium supplemented with Bengal gram flour or green gram flour instead of soy flour and coconut oil or groundnut oil instead of corn oil have been tried under laboratory condition at 25 °C. Maximum multiplication was recorded in Wouts media supplemented with Bengal gram flour and coconut oil.

Narkhedkar et al. (2001) studied on multiplication of 16 EPN isolates from cotton-growing zones of southern, central, and northern India. Three mass production media *viz.*, dog food medium, peptone medium, and Wouts medium were found to be favorable for nematode mass culturing. Vyas et al. (2001a) attempted *in vitro* mass production of native *Steinernema* sp. using 21 animal- and plant-protein-based media. Maximum production of nematodes was recorded in hen-egg yolk medium which was economically better than universally used dog food biscuit agar. Production of the IJs was poor in plant-protein-based medium compared to animal-protein-based medium.

Hussaini et al. (2002b) carried out mass multiplication of *Steinernema* sp. (SSL2) PDBC EN 13.21 in four combinations of dog biscuit medium (House et al., 1965) in comparison with Wouts medium (Wouts, 1981). The cost of production was also evaluated. After a culture time of 30 days and an initial inoculum of 500 IJs per 250 mL flask, a maximum yield of 30.58×10^5 IJs/flask was recorded from Wouts medium followed by dog biscuit + peptone + beef extract (24.5×10^5 IJs/flask), dog biscuit + beef extract (18.40×10^5 IJs/flask), dog biscuit + peptone (12.20×10^5 IJs/flask), and dog biscuit + bacterial culture (10.14×10^5 IJs/flask). The cost of production for 10 lakh IJs was highest for the dog biscuit + bacterial culture followed by dog biscuit + peptone + beef extract, dog biscuit + peptone, dog biscuit + beef extract, and Wouts media. Wouts medium was found to be the best both in terms of yield/gram of medium and cost of production per 10 lakh IJs.

Six artificial media *viz.*, Wouts medium, wheat flour medium, dog biscuit medium, egg yolk medium, nutrient agar medium, and agar–agar medium were evaluated *in vitro* for production of two species of EPNs, *Steinernema masoodi* and *S. seemae*. Maximum yield of *S. masoodi* and *S. seemae* were observed on egg yolk medium (3.9×10^6 and 4.9×10^6 IJs/ flask, respectively), followed by Wouts medium (2.8×10^6 and 3.6×10^6 IJs/flask, respectively). Both EPNs were unable to multiply on dog biscuit medium, wheat flour medium, and nutrient agar. Poor multiplication of *S. masoodi* (0.034×10^6) and *S. seemae* (0.039×10^6) were observed on agar–agar medium (Pervez & Ali, 2010).

7.4.2 IN VIVO PRODUCTION OF EPNS

Karunakar et al. (1992, 1999b) studied the production of *S. glaseri* and *H. indicus* on sugarcane internode borer, *Chilo sacchariphagus indicus* (Kapur). They recorded on an average production of 37,335.8 IJs/larva of *S. glaseri* and 210,283.3 IJs/larva of *H. indicus* when both were inoculated @ 25 IJs/larva. Hussaini and Singh (1998) reported the nutritional requirements of IJs of native *Steinernema* sp. from insect hosts *Galleria mellonella*, *A. ipsilon*, *S. litura*, *H. armigera*, and *C. cephalonica*. *G. mellonella* was found to be the most suitable host for *in vivo* production with yields of 514,383/g larva followed by *C. cephalonica*, *H. armigera*, *A. ipsilon*, and *S. litura*.

The LOTEK system of tools and procedures provides process technology for low cost, high efficiency mass production. The harvester collects 97% of *H. bacteriophora* that emerged from *G. mellonella* cadavers in 48 h. The separator removes 97.5% of the wastewater in three passes, while nematode concentration increased 81 fold. Gaugler et al. (2002) reported one component of nematode production in the United States in a cottage industry of low-volume producers using *in vivo* technology, based on a White trap method (White, 1927). They reported the first scalable system for *in vivo* nematode mass production. Unlike the White trap, there is no requirement for nematode migration to a water reservoir.

Elawad et al. (2001) investigated the progeny production of *S. abbasi* in various lepidopteran pre-pupae and found that mostly DJs developed in the greater wax moth, *G. mellonella* (2.34 lakh) followed by 2.20 lakh in boll worms *Helicoverpa virescens* and 1.66 lakh in *Spodoptera exigua*. Zaki et al. (2000) studied on *in vivo* production of *H. bacteriophora* and

S. carpocapsae on third instar larvae of silk worm *Bombyx mori* (L.). The average number of nematodes that emerged out from the third instar larva of silkworm was 2750. Progeny production of different *Steinernema* spp. and *Heterorhabditis* spp. were evaluated in different insect pests. Maximum yield of EPN was observed in final instar larvae of *G. mellonella*, 1–3.5 × 10^5 IJs/cadaver (Rajkumar et al., 2002b; Singh & Yadava, 2002; Ali et al., 2008), while, for *H. armigera* and *C. cephalonica* larvae, the yield of IJs was 1–3 × 10^5 IJs/cadaver and 0.5–1.0 × 10^5 IJs/cadaver, respectively (Ali et al., 2008).

EPN species, *S. masoodi, S. seemae, S. carpocapsae, S. feltiae*, and *Oscheius* sp. on mustard saw fly, *Athalia proxima*, were multiplied. Result shows production of IJs of *Oscheius* sp. was 0.65 × 10^5 IJs/cadaver, followed by *S. feltiae* (0.62 × 10^5 IJs/cadaver), *S. carpocapsae* (0.57 × 10^5 IJs/cadaver), *S. seemae* (0.53 × 10^5 IJs/cadaver), and *S. masoodi* (0.50 × 10^5 IJs/cadaver) (Pervez et al., 2007). Vyas and Yadav (1993) tested the suitability of mulberry silkworm, *B. mori*, and root grub, *Holotrichia serrata*, for *in vivo* mass production of *S. glaseri* and *Heterorhabditis* sp. They studied the economics in terms of IJs obtained on expenditure of rupees. *B. mori* was the suitable host for the production of *S. glaseri* and *Heterorhabditis* sp. and yielded 1,191,032.6 and 8,807,110.9 IJs, respectively, for one rupee spent. *G. mellonella* was the next best alternative, which produced 983,611.5 and 5,080,930.8 IJs of *S. glaseri* and *Heterorhabditis* sp., respectively, for one rupee spent. *Hydrangea serrata* was the least suitable host, which yielded 896,038.7 and 128,399.8 IJs of *Heterorhabditis* sp. and *S. glaseri*, respectively, for one rupee spent.

7.5   EPN-BASED FORMULATIONS AND STORAGE

EPN-based pesticides are nowadays largely marketed in the Western countries in comparison to other organisms, apart from bacteria *B. thuringiensis*.

Formulation refers to the preparation of a product from an active ingredient by the addition of certain active (functional) and non-active (inert) substances. Formulation is intended to improve activity, absorption, uniform application, delivery, and ease of use or storage stabilities of an active ingredient (EPN). High oxygen and moisture requirements, sensitivity to temperature exposures, and behavior of IJs limit the choice of the formulation and its ingredients in case of EPN.

EPN can be formulated either with active nematodes in various substrates or with reduced mobility. In case of latter, the EPNs are kept in partially anhydrobiotic condition until these are used for application. When these come in contact with the medium of application, that is, water or inert material, EPN become active (Womersley, 1990).

7.5.1 FORMULATION WITH ACTIVELY MOVING EPNS

Soil has been found a good substrate as short-term storage of EPN. Concentrated EPNs are mixed with soil or put in earthen pot and covered with muslin cloth. As an alternative, the earthen pot can be buried in soil so that the brim of the pot is six inches below the ground level (Ali, 2007). This way, EPNs will be saved to a great extent from the extremities of fluctuating weather conditions. EPN mixed with the soil can also be stored in sealed polythene bags at normal condition for considerably long time. This type of storage requires less space and bags can be stacked over one another.

The nematode may be stored in distilled water with a drop of Triton X-100 (a wetting agent that prevents nematodes from sticking to the side of the container) or 0.1% formalin (if frequent contamination seems to be a problem). If stored without aeration, the nematodes should be concentrated to no more than 10,000 IJs/mL, and the water depth should be 1 cm or less. Tissue culture flasks are ideal for this type of storage since there is large surface-to-volume ratio. Higher nematode concentration (100,000 IJs/mL) will not be detrimental in aeration suspension. An aquarium pump or any forced air supply can be attached to the nematode suspension. Steinernematids can be stored at 4–10 °C for 6–12 months without much loss of activity, whereas *Heterorhabditis* can be stored for 2–4 months at 4–10 °C.

Bedding (1984, 1988) proposed that a well-rinsed semisolid paste of Steinernematids IJs may be stored on moist autoclaved polyether polyurethane foam which is kept in a sterilized container for up to one year. The IJs can be extracted later by squeezing the foam. Storage in activated charcoal or in an inactive dried state may be the other alternatives. Polyether polyurethane sponge-based formulations are also widely used to store and to ship small quantities of EPN.

Vermiculite formulation is a significant improvement over sponge. Its advantages include more concentrated EPN product, longer storage stabilities, and more convenient application. Normally, an aqueous EPN suspension is mixed homogenously with vermiculite and the mixture is placed in the thin polyethylene bags. The vermiculite–nematode mixture is added to the spray tank directly, mixed in water, and applied as either spray or drench.

7.5.2 FORMULATIONS WITH REDUCED MOBILITY

The formulations containing partially anhydrobiotic nematodes include gel, powders, granules, cadavers, capsules, and baits (Gaugler & Kaya, 1990).

7.5.2.1 GEL FORMULATION

EPN slurry is mixed in anhydrous polyacrylamide. The EPNs are partially desiccated, but survival at room temperature is low.

7.5.2.2 POWDER FORMULATION

EPNs are mixed in clay to remove excess surface moisture and to cause partial desiccation. The formulation, termed as "sandwich" consists of a layer of nematode between two layers of clay. Many wettable powder formulations have been developed. EPN species can be stored up to three months at 22 °C without loss of viability (Grewal, 2002).

7.5.2.3 GRANULAR FORMULATION

EPNs are partially encapsulated in Lucerne, sodium alginate, and wheat flour. EPNs are distributed throughout a wheat gluten matrix. The "Pesta" formulation includes a filter and a humectant to enhance nematode survival (Connick et al., 1993). The process involves drying of granules to low moisture to prevent EPN migration and reduce risk of contamination.

7.5.2.4   DESICCATED CADAVERS

EPN applied in the form of infected *Galleria* larvae was found to be as effective as aqueous suspension against soil pests. Use of EPN-infected cadavers can be superior to aqueous suspensions. A formulation based on desiccated cadaver coated with clay has been developed that allow storage without rupturing cadavers or adhering together (Shapiro & Glazer, 1996). Such desiccated cadavers are put in water before spraying and EPN comes out of it and are used as sprays, mixed with irrigation water or dispersed in soil. There is advantage of ease of handling, storage, transport, and so on in this type of EPN storage techniques. However, there is gradual depletion of moisture from the cadaver and they are good for storage for a period of 2–3 months only.

Two formulations enabled cadavers to be partially desiccated without affecting reproduction; other formulations and non-formulated cadavers exhibited reduced reproduction upon desiccation. Four-day-old cadavers were more amenable to desiccation than eight-day-old cadavers. Formulated cadavers were more resistant to rupturing and sticking together during agitation than non-formulated cadavers (Shapiro et al., 2001).

7.5.2.5   CAPSULES

Macrogels containing encapsulated EPN have been suggested as delivery systems for the control of soil and foliar pests. Encapsulation of EPN in calcium alginate gel beads was advocated for their use as baits, or for soil applications. When the alginate capsules were placed in soil with adequate moisture, most IJs migrated out of the granules within a week. Navon et al. (1999) evaluated calcium alginate gel for *S. riobrave* and reported high nematode survival for 48 h in the gel at 61% RH.

7.5.2.6   BAITS

Baits containing IJs of EPN, an inert carrier (corn cob grits, groundnut hull, or wheat bran) and a feeding stimulant, for example, glucose, malt extract, molasses or sucrose, or a sex pheromone have been developed by Georgis (1990). *S. carpocapsae* and *S. scapterisci* have shown promising results in these baits. EPN due to "sit-and-wait" foraging strategy, do not escape the formulation and are more tolerant to desiccation than other species. Yet only moderate control of cutworms, grasshoppers, and mole cricket was achieved. When trap stations were used, these ensured EPN contact with the target pest and also protected them from sunlight and desiccation.

7.6　APPLICATION TECHNOLOGY

Application technology is one of the most important aspects in pest management, which is often ignored resulting in reduced degree of pest control. EPN being live material has to be handled very carefully and with caution so that the deleterious effect of surrounding is reduced to the minimum and they work efficiently. This is one of the important limiting factors for its success.

A variety of application methods have been used, depending upon the target host and environment. Syringes (Lindegren et al., 1981, 1987; Deseo, 1982), handguns (Deseo & Miller, 1985), cotton swabs dipped into a suspension of nematodes (Foschi & Deseo, 1983), and oil cans and backpack sprayers (Lindegren & Barnett, 1982) have been employed to inoculate individual insect galleries. EPNs have been applied for the control of soil insects by injecting them with handgun sprayers by sending IJs through drip irrigation systems by drenching the soil with a sprinkling can and by applying with liquid formulation (Reed et al., 1986). EPN containing baits have been successfully employed with insects such as mole cricket. Other foliage applications can be accomplished with many types of commercial ground and aerial sprayers such as mist blower, fan sprayer, pressurized sprayer, and electrostatic sprayers.

EPN can be applied using most conventional liquid application system designed to deliver pesticides, fertilizers, and irrigation. Various considerations involved in the selection of an application system is volume, agitation system, pressure and recycling time, environmental conditions, and spray distribution pattern. EPN can be applied through low- or high-volume sprayer, but the pressure in the spray tank should not be too high, otherwise, EPN will shred into pieces. In general, EPN should not be subjected to a pressure exceeding 300 psi (2070 kPa) (Grewal, 2002).

7.6.1 SOIL APPLICATION

Spraying the soil surface is the most common method of EPN application. Both soil type and moisture influence their survival and movement. In general, EPN mobility and survival is lower in clay than in sandy loam soil. Soil temperature between 12 and 30 °C is considered favorable for application of most EPN species. If soil temperature is above 30 °C, a pre-application irrigation is usually recommended to reduce soil temperature before EPN application. However, with the discovery of heat-tolerant species of EPN, there is a hope to overcome this problem. Often it is recommended that after applying EPN in the soil, a light irrigation should be given so that these go down deeper into the soil where they act efficiently and are away from the adverse environmental extremities.

7.6.2 FOLIAR APPLICATION

Droplet size and spray distribution are important considerations when applying EPN in the foliar environment. Conventional hydraulic nozzles are known to produce a wide range of droplet size many being too small to carry IJs. Type of nozzle used has been found to affect the EPN deposit on plant surface. Solid cone nozzle or flat fan nozzles deposit greater number of EPN on to leaves and give higher mortality to target insect (Lello et al., 1996; Ali et al., 2009c). Increased flow rate through the nozzles results in greater number of EPN deposit on plant surface (Akhtar et al., 2012). The deposition of EPNs on foliage is generally increased by addition of adjuvant to the spray solution. Often the results obtained from foliar application of EPN are not up to the desired level due to intolerance of EPN to extremes of desiccation, temperature, and UV radiation. Effective protection against these external factors can be achieved by addition of florescent brightener (Tinopal, Ujala, Ranipal, etc.), application during dusk has been advocated to overcome these problems to some extent. However, if a large area is to be treated, this may not be practical due to prevailing darkness just in one or one-and-half hour.

7.7   CONCLUSION AND FUTURE PROSPECTS

The use of EPN is more meaningful in the present context of World Trade Organisation (WTO) and emphasis on organic farming where in no chemicals are sprayed on plants and additives are permitted to food grains and the products are cost effective and competitive as well. In order to make the EPN as a component of IPM program, concerted efforts are needed with respect to EPN species, insect, and its environments. Some of the points which need immediate attention are:

- Efforts on isolating indigenous EPN to adapted to high temperatures and environmental conditions.
- Identification of EPN species based on morphological and molecular characterization.
- Working out proper dose, method, time, and frequency of applications for effective control of insect pests.
- Small-scale cottage industry for EPN production at village level may compete with the conventional pesticide.
- Search for better and efficient adjuvants, surfactants, anti-caking substances, anti-desiccant, and UV retardant is important for better and efficient formulation.
- Application technology for aerial and soil insect pests as well as stored grain pests so as to incorporate this important bioagent with the existing IPM modules for various crops.
- The quality of commercial EPN product needs standardization in various types of formulations.

KEYWORDS

- entomopathogenic nematodes
- biopesticide
- insect pests
- pest control
- infectivity

REFERENCES

Abbas, M. S. T.; Saleh, M. M. E. Comparative Pathogenicity of *Steinernema abbasi* and *S. riobravae* to *Spodoptera littoralis* (Lepidoptera: Noctuidae). *Int. J. Nematol.* **1998,** *8* (1), 43–45.

Abdel-Razek, A. S.; Abd-Elgawad, M. M. Investigations on the Efficacy of Entomopathogenic Nematodes against *Spodoptera littoralis* (Biosd.) and *Galleria mellonella* (L.). *Arch. Phytopathol. Plant Protect.* **2007,** *40* (6), 414–422.

Ahmad, R.; Ali, S. S.; Pervez, R. Field Efficacy of *Steinernema masoodi* Based Biopesticide against *Helicoverpa armigera* (Hubner) infesting Chickpea. *Trends Biosci.* **2009,** *2* (1), 23–24.

Ahmad, R.; Ali, S. S.; Pervez, R. Management of Bruchids Infesting Pulses in Storage – Feasibility of Exploitation of Entomopathogenic Nematodes (Abstract), Biopesticide International Conference (BIOCICON), Palayamkottai, Tamil Nadu, 2007; pp 113–114.

Ahmad, R.; Yadava, C. P.; Lal S. S. Efficacy of Nuclear Polyhedrosis Virus for the Management of *Helicoverpa armigera* (Hubner) Infesting Chickpea. *Ind. J. Pulses Res.* **1999,** *12* (2), 92–96.

Akhtar, M. H.; Asif, M.; Ali, S. S.; Pervez, R. Performance of Different Sprayer Nozzles on Deposition of Infective Juveniles of Three Entomopathogenic Nematodes Species on Pigeonpea. *World J. Appl. Sci. Res.* **2012,** *2* (1), 28–31.

Ali, S. S.; Ahmad, R.; Pervez, R. Entomopathogenic Nematodes: An Emerging Option for Pest Management. In *Milestones in Food Legumes Research;* Ali, M., Kumar, S., Eds.; Indian Institute of Pulses Research: Kanpur, 2009b; pp 290–309.

Ali, S. S.; Ahmad, R.; Akhtar, M. H.; Asif, M.; Pervez, R. EPN Based Biopesticide, Performance of Different Sprayer Nozzles of Deposition of Infective Juveniles on Pigeonpea Leaf (Abstract), Fifth International Conference on Biopesticides: Stakeholders Perspectives, New Delhi, India, 2009c; p 78.

Ali, S. S.; Pervez, R.; Ahmad, R. Efficacy of *Steinernema masoodi* (Ali et al., 2005) against Grey Weevil *Myllocerus* sp. Infesting Early Pigeonpea. *Trends Biosci.* **2009a,** *2* (1), 40–41.

Ali, S. S.; Pervez, R.; Ahmad, R. Longevity of *Steinermema seemae* (Nematoda: Rhabditida) and its Infectivity against *Galleria mellonella* (Abstract), National Symposium on Legume Research: Recent Trends and Future Prospects in Post Genomic Era; S. V. B. P. University of Agriculture and Technology, Meerut, India, 2007; p 34.

Ali, S. S.; Pervez, R.; Hussain, M. A.; Ahmad, R. Susceptibility of Three Lepidopteran Pest to Five Entomopathogenic Nematodes and *In Vivo* Mass Production of These Nematodes. *Arch. Phytopathol. Plant Protect.* **2008,** *41* (4), 300–304.

Ali, S. S.; Ahmad, R.; Hussain, M. A.; Pervez, R. Pest Management through Entomopathogenic Nematodes. Indian Institute of Pulses Research: Kanpur, 2005b; p 58.

Ali, S. S.; Shaheen, A.; Pervez, R.; Hussain, M. A. *Steinernema masoodi* sp. n. and *Steinernema seemae* sp. n. (Rhabditida: Sternernematidae) from Uttar Pradesh, India. *Int. J. Nematol.* **2005a,** *15* (1), 89–99.

Anonymous. *Survey, Identification and Utilization of Entomopathogenic Nematodes against Some Lepidopterous and Coleopterous Insect Pests;* Annual Report, Project Directorate of Biological Control: Bangalore, 2001; p 209.

Athanassiou, C. G.; Palyvos, N. E.; Kokouli-Durate, T. Insecticidal Effect of *Steinernema feltiae* (Filipjev) (Nematoda: Steinernematidae) against *Tribolium confusum* du val (Coleptera: Tenebrionidae) and *Ephestia kuehniella* (Zeller) (Lepidoptera: Pyralidae) in Stored Wheat. *J. Stored Prod. Res.* **2008,** *44,* 52–57.

Banu, G. J.; Subahasan, K.; Iyer, R. Occurrence and Distribution of Entomopathogenic Nematodes in White Grub Endemic Areas of Kerala. *J. Plant Crop.* **2004,** *32,* 333–334.

Banu, J. G. Studies on Entomopathogenic Nematodes from Kerala. Dissertation Submitted for Ph. D. (Nematology) to Tamil Nadu Agricultural University, Coimbatore, Tamil Nadu, India, 2001.

Banu, J. S. Scope of Entomopathogenic Nematodes for the Management of Pests of Intercrops in Coconut Based Cropping System. In *Current Status of Research on Entomopathogenic nematodes in India;* Hussaini, S. S., Rabindra, R. J., Nagesh, M., Eds.; Project Directorate of Biological Control: Bangalore, 2003; pp 179–184.

Banu, J. G.; Nguyen, K. B.; Rajendran, G. Occurrence and Distribution of Entomopathogenic Nematodes in Kerala, India. *Int. J. Nematol.* **2005,** *15,* 9–16.

Beavers, J. B.; McCoy, C. W.; Kaplan, D. T. Natural Enemies of Subterranean *Diaprepes abbreviatus* (*Coleoptera*: Curculionidae) Larvae in Florida. *Environ. Entomol.* **1983,** *12* (3), 840–843.

Bedding, R. A. Large Scale Production, Storage and Transport of the Insect Parasitic Nematodes *Neoaplectana* spp. and *Heterorhabditis* spp. *Ann. Appl. Biol.* **1984,** *104,* 117–120.

Bedding, R. A. Storage of Insecticidal Nematodes. World Patent No. WO 88/08668, 1988.

Bhaskaran, R. K. M.; Sivakumar, C. V.; Venugopal, M. S. Biocontrol Potential of Entomopathogenic Nematode in Controlling Red Hairy Caterpillar, *Amsacta albistriga* (Lepidoptera: Arctiidae) of groundnut. *Ind. J. Agr. Sci.* **1994,** *64,* 655–657.

Chitwood, B. G.; Chitwood, B. G. An Introduction to Nematology. Monumental Printing Company: Baltimore, MD, 1937; p 213.

Connick, W. J. Jr.; Nickle, W. R.; Vinyard, B. T. "Pesta", New Granular Formulation for *Steinernema carpocapsae*. *J. Nematol.* **1993,** *25* (2), 198–203.

Deseo, K. V. Prove di lotta col nematode entomopatogeno (*Neoaplectana carpocapsae* Weiser) contro i rodilegno *Cossus cossus* L. e *Zeuzera pyrina* L. (Lepidoptera; Cossidae). *Atti Giornate Fitopatologiche.* **1982,** *3,* 3–10.

Deseo, K. V.; Miller, L. A. Efficacy of Entomogenous Nematodes, *Steinernema* spp. against Clearing Moth, *Synanthedon* spp. in North Italian Application in Orchards. *Nematologica.* **1985,** *31,* 100–108.

Elawad, S. A.; Gowen, S. R.; Hague, N. G. M. Progeny Production of *Steinernema abbasi* in Lepidopterous Larvae. *Int. J. Pest Manage.* **2001,** *47,* 17–21.

Fayaz, S.; Javed, S. Laboratory Evaluation of Seven Pakistani Strains of Entomopathogenic Nematodes against a Stored Grain Insect Pest, Pulse Beetle *Callosobruchus chienensis* (L.). *Nematology.* **2009,** *41,* 255–260

Foschi, S. Y.; Deseo, K. V. Risulatati di lotta con Nematode Entomopatogenei su *Zeuzera pyrina* L. (Lepidopt; Cossidae nel 1982). *La difesa delle piante.* **1983,** *3* (4), 153–156.

Ganga Visalakshy, P. N.; Krishnamoorthy, A.; Hussaini, S. S. Field Efficacy of the Entomopathogenic Nematode *Steinernema carpocapsae* (Weiser, 1955) against Brinjal Shoot and Fruit Borer, *Leucinodes orbonalis* guenee. *Nematol. Medit.* **2009,** *37,* 133–137.

Ganguly, S. Taxonomy of Entomopathogenic Nematodes and Work Done in India. In *Current Status of Research on Entomopathogenic Nematodes in India;* Hussaini, S. S.,

Rabindra, R. J., Nagesh, M., Eds.; Project Directorate of Biological Control: Bangalore, 2003; pp 69–108.

Ganguly, S.; Singh, L. K. Entomopathogenic Nematodes Distributed in Delhi and Adjoining Areas (Abstract), National Congress on Centenary of Nematology in India, Appraisal and Future Plans; IARI: New Delhi, 2000; p 118.

Gaugler, R.; Kaya, H. K. Entomopathogenic Nematodes in Biological Control. CRC Press: Boca Raton, FL, 1990; p 365.

Gaugler, R.; Brown, I.; Shapiro-Ilan, D.; Atwa, A. Automated Technology for *In Vivo* Mass Production of Entomopathogenic Nematodes. *Biol. Control.* **2002,** *24,* 199–206.

Georgis, R. Formulation and Application Technology. In *Entomopathogenic Nematodes in Biological Control;* Gaugler, R., Kaya, H. K., Eds.; CRC Press: Boca Raton, FL, 1990; pp 179–191.

Ghode, M. K.; Mishra, R. P.; Swain, P. K.; Nayak, U. K.; Pawar, A. D. Pathogenicity of *Steinernema feltiae* Weiser to *Heliothis armigera* Hubner. *Plant Protection Bulletin, India.* **1988,** *40* (2), 7–8.

Glaser, R. W. A Pathogenic Nematode of the Japanese Beetle. *J. Parasitol.* **1932,** *18,* 119.

Grewal, P. S. Formulation and Application Technology. In *Entomopathogenic Nematology;* Gaugler, R., Ed.; CABI Publishing: Wallingford, UK, 2002; pp 265–287.

Griffin, C. T.; Moore, J. F.; Downes, M. J. Occurrence of Insect-Parasitic Nematodes (Steinernematidae, Heterorhabditidae) in the Republic of Ireland. *Nematologica.* **1991,** *37* (1), 92–100.

Griffin, C. T.; Downes, M. J.; Block, W. Tests of Antarctic Soils for Insect Parasitic Nematodes. *Antar. Sci.* **1990,** *2,* 221–222.

Herrera, R. C.; Escuer, M.; Labrador, S.; Robertson, L.; Barrios, L.; Gutie´rrez, C. Distribution of the Entomopathogenic Nematodes from La Rioja (Northern Spain). *J. Invertebr. Pathol.* **2007,** *95,* 125–139.

Hominick, W. M.; Reid, A. P. Perspectives on Entomopathogenic Nematology. In *Entomopathogenic Nematodes in Biological Control;* Gaugler, R., Kaya, H. K., Eds.; CRC Press: Boca Raton, FL, 1990; pp 327–345.

Hominick, W. M.; Briscoe, B. R.; del Pino, F. G.; Heng, J.; Hunt, D. J.; Kozodoi, E.; Mracek, Z.; Nguyen, K. B.; Reid, A. P.; Spiridonov, S.; Stock, S. P.; Sturhan, D.; Waturu, C.; Yoshida, M. Biosystematics of Entomopathogenic Nematodes: Current Status, Protocols and Definitions. *J. Helminthol.* **1997,** *71,* 271–298.

House, H. L.; Welch, H. E.; Cleugh, T. R. A Food Medium of Prepared Dog Biscuit for the Mass Production of the Nematode DD-136 (Nematoda: Steinernematidae). *Nature.* **1965,** *206,* 847.

Hussaini, S. S.; Singh, S. P. Entomophilic Nematodes for Insect Control. In *Biological Suppression of Plant Diseases, Phytoparasitic Nematodes and Weeds;* Singh, S. P., Hussaini, S. S., Eds.; PDBC: Bangalore, 1998; pp 238–267.

Hussaini, S. S.; Singh, S. P.; Nagesh, M. *In Vitro* and Field Evaluation of Some Indigenous Isolates of *Steinernema* and *Heterorhabditis Indica* against Shoot and Fruit Borer, *Leucinodes orbonalis. Indian J. Nematol.* **2002a,** *32* (1), 63–65.

Hussaini, S. S.; Singh, S. P.; Parthasarathy, R.; Shakeela, V. Virulence of Native Entomopathogenic Nematodes against Black Cutworms, *Agrotis ipsilon* (Hufnagel) and *A. segetum* (Noctuidae : Lepidoptera). *Indian J. Nematol.* **2000,** *30* (1), 103–105.

Hussaini, S. S.; Singh, S. P.; Parthasarathy, R.; Shakeela, V. *In Vitro* Production of Ento-mopathogenic Nematodes in Different Artificial Media. *Indian J. Nematology.* **2002b,** *30* (1), 103–105.

Hussaini, S. S.; Ansari, M. A.; Ahmad, W.; Subbotin, S. A. Identification of Some Indian Populations of *Steinernema* species (Nematoda) by RFLP Analysis of the ITS Region of rDNA. *Int. J. Nematol.* **2001,** *11,* 73–76.

Hussaini, S. S.; Rabindra, R. J.; Nagesh, M. Current Status of Research on Entomopathogenic Nematodes in India. Project Directorate of Biological Control: Bangalore, 2003; p 205.

Israel, P.; Prakash Rao, P. S.; Rao, Y. R. V. J.; Verma, A. Effectiveness of DD 136, an Insect Parasitic Nematodes against Pests of Rice and Its Mass Multiplication in the Laboratory (Abstract). All Indian Nematology Symposium: New Delhi, 1969a; p 36.

Israel, P.; Rao, Y. R. V. J.; Prakash Rao, P. S.; Verma, A. Control of Paddy Cutworms by DD-136, a Parasitic Nematode. *Curr. Sci.* **1969b,** *16,* 390–391.

Josephrajkumar, A.; Sivakumar, C. V. A Survey for Entomopathogenic Nematodes in Kanyakumari District, Tamil Nadu, India. *Indian J. Entomol.* **1997,** *59* (1), 45–50.

Jothi, B. D.; Mehta, U. Efficacy of Three Species of Entomopathogenic Nematodes with Antidesiccants against *Helicoverpa armigera. Int. J. Nematol.* **2003,** *13* (2), 292–232.

Karunakar, G.; David, H.; Easwaramoorthy, S. Influence of Dosage of *Steinernema carpo-capsae, S. glaseri* and *Heterorhabditis indicus* on Mortality of the Host and Multiplica-tion of Infective Juveniles in Sugarcane Internode Borer, *Chilo sacchariphagus indicus. J. Biol. Control.* **1992,** *6,* 26–28.

Karunakar, G.; Easwaramoorthy, S.; David, H. Susceptibility of Nine Lepidopteran Insects to *Steinernema glaseri, S. feltiae* and *Heterorhabditis indicus* Infection. *Int. J. Nematol.* **1999a,** *9,* 68–71.

Karunakar, G.; Easwaramoorthy, S.; David, H. Influence of Temperature on Infectivity, Penetration and Multiplication of *Steinernema glaseri, S. feltiae* and *Heterorhabditis indicus. Int. J. Nematol.* **1999b,** *9* (2), 126–129.

Kaya, H. K.; Gaugler, R. Entomopathogenic Nematodes. *Annu. Rev. Entomol.* **1993,** *38,* 181–206.

Kaya, H. K. Soil Ecology. In *Entomopathogenic Nematodes in Biological Control;* Gaugler, R., Kaya, H. K., Eds.; CRC Press: Boca Raton, FL, 1990; pp 93–115.

Lal Ramliana; Yadav, A. K. Occurrence of Entomopathogenic Nematodes (Rhabditida: Steinernematidae and Heterorhabditidae) in Meghalaya, NE India. *Sci. Vis.* **2010,** *10* (3), 89–100.

Lello, E. R.; Patel, M. N.; Matthews, G. A.; Wright, D. J. Application Technology for Ento-mopathogenic Nematodes against Foliar Pests. *Crop Protect.* **1996,** *15* (6), 567–574.

Lindegren, J. E.; Barnett, W. W. Applying Parasitic Nematodes to Control Carpenter Worms. *Calif. Agric.* **1982,** *36* (11&12), 7–8.

Lindegren, J. E.; Silva, A. F.; Valero, K. A.; Curtis, C. E. Comparative Small Scale of Field Application of *Steinernema feltiae* for Navel Orange Worm control. *J. Nematol.* **1987,** *19,* 503–504.

Lindegren, J. E.; Yamashita, T. T.; Barnet, W. W. Parasitic Nematodes May Control Carpenter Worms in Fig Trees. *Calif. Agric.* **1981,** *35,* 25–26.

Morris, O. N. Susceptibility of 31 Species of Agricultural Pets to Entomogenous Nema-todes *Steinernema feltiae* and *Heterorhabditids bacteriophora. Can. Entomol.* **1985,** *122,* 309–320.

Narayanan, K.; Gopalakrishnan, C. Effect of Entomogenous Nematode *Steinernema feltiae* (Rhabditida: Steinernematidae) to the Pre-Pupa, Pupa and Adult *Spodoptera litura* (Noctuidae; *Lepidoptera*). *Indian J. Nematol.* **1987,** *17* (2), 273–276.

Narkhedkar G. Nandini; Lavhe, N. V.; Mayee, C. D. Entomopathogenic Nematodes-Potential Agent against American Bollworm of Cotton (*Helicoverpa armigera*). *SAIC Newsletter.* **2001,** *11*, 4.

Navon, A.; Keren, S.; Salame, L.; Glazer, I. An Edible-to-Insects Calcium Alginate Gel as a Carrier for Entomopathogenic Nematodes. *Biocontrol Sci. Techn.* **1999,** *8* (3), 429–437.

Nguyen, K. B.; Smart, G. C. Jr. *Neosteinernema longicurvicauda* n. gen., n. sp. (Rhabditida: Steinernematidae), a Parasite of the Termite *Reticulitermes flavipes* (Koller). *J. Nematol.* **1994,** *26*(2), 162–174.

Pervez, R. Biocontrol Potential of Entomopathogenic Nematodes against Different Instar Larvae of Gram Pod Borer, *Helicoverpa armigera* Infesting Chickpea. *Curr. Nematol.* **2010,** *21* (2), 17–21.

Pervez, R.; Ali, S. S. Natural Occurrence of Entomopathogenic Nematodes Associated with Chickpea Ecosystem. *Curr. Nematol.* **2007,** *18* (2), 19–22.

Pervez, R.; Ali, S. S. Infectivity of *Steinernema mushtaqi* (Rhabditida: Steinernematidae) against Insect Pests and Their Mass Production. *Arch. Phytopathol. Plant Protect.* **2011,** *44* (14), 1352–1355.

Pervez, R.; Ali, S. S. Entomopathogenic Nematodes as a Potent Biopesticides against Insect Pests. In *Newer Approaches to Biotechnology;* Behra, K. K. Ed.; Narendra Publishing House Publishers and Distributors: New Delhi, 2013; pp 213–235.

Pervez, R.; Ali, S. S. Infectivity of *Spodoptera litura* (F.) (Lepidoptera: Noctuidae) by Certain Native Entomopathogenic Nematodes and Their Penetration in Test Insect and *In Vivo* Production. *Trends Biosci.* **2009,** *2* (2), 70–73.

Pervez, R.; Ali, S. S. *In Vitro* Production of Entomopathogenic Nematodes in Different Artificial Media (Nematoda: Rhabditida). *Curr. Nematol.* **2010,** *21* (2), 9–12.

Pervez, R.; Eapen, S. J.; Devasahayam, S. Biodiversity of Entomopathogenic Nematodes. In *Advances Frontier of Biotechnology;* Kambasa Kumar Behra, Ed.; Jaya Publishing House: Delhi, 2014b; pp 125–143.

Pervez, R.; Eapen, S. J.; Devasahayam, S. Entomopathogenic Nematodes in Management of Insect Pests of Spice Crops. *Indian J. Arecanut Spices Med. Plants.* **2015,** *17* (2), 24–26.

Pervez, R.; Ali, S. S.; Ahmad, R. Efficacy of Some Entomopathogenic Nematodes against Mustard Saw Fly and *In Vivo* Production of These Nematodes. *Int. J. Nematol.* **2007,** *17* (1), 55–58.

Pervez, R.; Ali, S. S.; Ahmad, R. Efficacy of Entomopathogenic Nematodes against Green Bug, *Nezara viridula* (L.) and Their *In Vivo* Mass Production. *Trends Biosci.* **2008,** *1* (1–2), 49–51.

Pervez, R.; Devasahayam, S.; Eapen, S. J. Determination of LD_{50} and LT_{50} of Entomopathogenic Nematodes against Shoot Borer (*Conogethes punctiferalis* guen.) Infesting Ginger. *Ann. Plant Protect. Sci.* **2014c,** *22* (1), 169–173.

Pervez, R.; Eapen, S. J.; Devasahayam, S.; Jacob, T. K. Efficacy of Some Entomopathogenic Nematodes against Insect Pests of Ginger and Their Multiplication. *Nematol. Medit.* **2012,** *40* (1), 39–44.

Pervez, R.; Eapen, S. J.; Devasahayam, S.; Dinsha, M. Characterization of Entomopatho-genic Nematode *Steinernema carpocapsae* From Ginger (*Zingiber officinale* Rosc.) Rhizosphere in India. *J. Plant Protect. Sci.* **2014d**, *6* (1), 13–20.

Pervez, R.; Eapen, S. J.; Devasahayam, S.; Jacob, T. K. Natural Occurrence of Entomo-pathogenic Nematodes Associated with Ginger (*Zingiber officinale* Rosc.) Ecosystem in India. *Indian J. Nematol.* **2014a**, *42* (2), 238–245.

Pervez, R.; Eapen, S. J.; Devasahayam, S.; Jacob, T. K. A New Species of Entomopatho-genic Nematode *Oscheius gingeri* sp. n. from Ginger Rhizosphere. *Arch. Phytopathol. Plant Protect.* **2013**, *46* (5), 526–535.

Poinar, G. O. Jr. Description and Biology of a New Insect Parasitic Rhabditoid, *Heter-orhabditis bacteriophora* N. Gen, N. Sp. (Rhabditida: Heterorhabditidae N. Fam.). *Nematologica.* **1976**, *21,* 463–470.

Prabhuraj, A.; Viraktamath, C. A.; David, H. Field Evaluation of EPN against Brinjal Root Weevil, *Myllocerus discolor*, Boh. *Pest Manag. Hort. Ecosyst.* **2000**, *6,* 149–151.

Prasad, G. S.; Ranganath, H. R.; Singh, P. K. Occurrence of the Entomopathogenic Nema-tode in Parts of South Andamans. *Curr. Sci.* **2001**, *80* (4), 501–502.

Rajkumar, M.; Parihar, A.; Siddiqui, A. U. Effect of Entomopathogenic Nematode, *Heter-orhabditis* sp. against Tobacco Caterpillar, *Spodoptera litura* (F.) (Abstract), National Symposium on Biodiversity and Management of Nematodes in Cropping Systems for Sustainable Agriculture, ARS, Durgapura, Jaipur, 2002a; p 75.

Rajkumar, M.; Parihar, A.; Siddiqui, A. U. Mass Multiplication of Indigenous Entomo-pathogenic Nematode from Udaipur (Abstract), National Symposium on Biodiversity and Management of Nematodes in Cropping Systems for Sustainable Agriculture, ARS, Durgapura, Jaipur, 2002b; p 77.

Ramos-Rodriguez, O.; Campbell, J. F.; Ramaswamy, S. Pathogenicity of Three Species of Entomopathogenic Nematodes to Some Major Stored Product Insect Pest. *J. Stored Prod. Res.* **2006**, *42,* 241–252.

Rao, V. P.; Manjunath, T. M. DD-136 Nematode That Can Kill Many Insect Pests. *Indian J. Entomol.* **1966**, *33,* 215–217.

Reed, D. K.; Reed, G. L.; Creighton, C. S. Introduction of Entomogenous Nematodes into Trickle Irrigation Systems to Control Striped Cucumber Beetle (*Coleoptera*: Chrysome-lidae). *J. Econ. Entomol.* **1986**, *79* (5), 1330–1333.

Saxena, H.; Ahmad, R. Field Evaluation of *Beauveria bassiana* (Bals) Wuillemin against *Helicoverpa armigera* (Hubner) Infesting Chickpea. *J. Biol. Control.* **1997**, 11 (1&2), 93–96.

Schroer, S.; Ehlers, R. U. Foliar Application of the Entomopathogenic Nematode *Steiner-nema carpocapsae* for Biological Control of Diamondback Moth Larvae (*Plutella xylo-stella*). *Biol. Control.* **2005**, *33,* 81–86.

Shapiro, D. I.; Glazer, I. Comparison of Entomopathogenic Nematode Dispersal from Infected Hosts versus Aqueous Suspension. *Environ. Entomol.* **1996**, *25* (6), 1455–1461.

Shapiro-Ilan, D. I.; Lewis, E. E.; Behle, R. B.; McGuire, M. R. Formulation of Entomo-pathogenic Nematode Infected Cadavers. *J. Invertebr. Pathol.* **2001**, 78, 17–23.

Singh, M.; Sharma, S. B.; Ganga Rao, G. V. Occurrence of Entomopathogenic Nematodes at ICRISAT Center. *Int. Arach. Newslett.* **1992**, *12,* 15–16.

Singh, V.; Yadava, C. P. S. Mass Production Technique of Entomopathogenic Nematode, *Heterorhabditis bacteriophora* on Greater Wax Moth, *Galleria mellonella* for Insect

Control (Abstract), National Symposium on Biodiversity and Management of Nematodes in Cropping Systems for Sustainable Agriculture, ARS, Durgapura, Jaipur, 2002, p 80.

Sitaramaiah, S.; Gunneswara Rao, S.; Hussaini, S. S.; Venkateswarh, P.; Nageshwara Rao, S. Use of Entomopathogenic Nematodes *Steinernema carpocapsae* against *Spodoptera litura* Fab. in Tobacco Nursery. In *Botanical Control of Lepidopteran Pests;* Tandon, P. L., Ballal, C. R., Jalali, S. K., Rabindra, R. J., Eds.; Project Directorate of Biological Control: Bangalore, 2003; pp 211–214.

Sivakumar, C. V.; Jayaraj, S.; Subramanian, S. Observation of Indian Population of the Entomopathogenic Nematodes, *Heterorhabditis bacteriophora*. *J. Biol. Control.* **1989,** *2,* 112–113.

Somvanshi, V. S.; Ganguly S.; Paul A. V. N. Field Efficacy of the Entomopathogenic Nematode *Steinernema thermophilum* Ganguly and Singh (Rhabditida: Steinernematidae) against Diamondback Moth (*Plutella xylostella* L.) Infesting Cabbage. *Biol. Control.* **2006,** *37,* 9–15.

Steiner, G. *Aplectana kraussei* n. sp., eine in der Blattwespe Lyda sp. Parasitiernde Nematodenform, nebst Bemerkungen uber das Steitenorgan der parasitischen Nematoden. Zbl. Bakt. *Parasitenk. Infektionskrank, Abstract,* **1923,** *2* (59), 14–18.

Sudhaus, W. Verg Leichende Untersuchngen zur Phylogenie, Systematic, Okologie, Biologie und Ethologie der Rhabditidae (Nematoda). *Zoologica.* **1976,** *43,* 1–229.

Tahir, H. I.; Otto, A. A.; Hague, N. G. M. The Susceptibility of Helicoverpa (Heliothis) armigera and Erias vitella Larvae to Entomopathogenic Nematodes. *Afro-Asian J. Nematol.* **1995,** *5* (2), 161–165.

Torr, P.; Heritage, S.; Wilson, M. J. *Steinernema kraussei*, an Indigenous Nematode Found in Coniferous Forests: Efficacy and Field Persistence against *Hylobius abietis. Agr. Forest Entomol.* **2007,** *9,* 181–188.

Travassos, L. Sobre o genero Oxysomatium. *Bol. Biol. Sao Paolo.* **1927,** *5,* 20–21.

Trdan, S.; Vidric, M.; Valic, N. Activity of Four Entomopathogenic Nematodes against Young Adult of *Sitophilus granarius* (Coleoptera: Curculionidae) and *Oryzaephilus surinamensis* (Coleoptera: Silvanidae) Under Laboratory Condition. *Plant Dis. Protect.* **2006,** *113,* 168–173.

Vyas, R. V. Entomopathogenic Nematodes – A New Tool for Management of Insect Pests of Crops. In *Current Status of Research on Entomopathogenic Nematodes in India;* Hussaini, S. S., Rabindra, R. J., Nagesh, M. Eds.; Project Directorate of Biological Control: Bangalore, 2003; pp 69–108.

Vyas, R. V.; Yadav, D. N. *Steinernema glaseri* (Steiner) Travassos a Biological Control Agent of Root Grub *Holotrichia consanguinea* Blanchard. *Pak. J. Nematol.* **1993,** *11,* 41–44.

Vyas, R. V.; Pharindera, Y.; Ghelani, Y. H.; Chaudhary, R. K.; Patel, N. B.; Patel, D. J. *In Vitro* Mass Production of Native *Steinernema* sp. *Ann. Plant Protect. Sci.* **2001a,** *9,* 77–78.

Vyas, R. V.; Patel, N. B.; Parul, P.; Patel, D. J. Efficacy of Entomopathogenic Nematode against *Helicoverpa armigera* on Pigeonpea. *Int. Chickpea Pigeonpea Newslett.* **2002a,** *9,* 43–44.

Vyas, R. V.; Patel, N. B.; Patel, P. D.; Patel, D. J. Field Efficacy of Entomopathogenic Nematode (*Steinernema riobrave*) against Cotton Bollworms (Abstract), National Congress on Centenary of Nematology in India – Appraisal and Future Plans, IARI, New Delhi, 2001b; p 115.

Vyas, R. V.; Patel, N. B.; Yadav, P.; Ghelani, Y. H.; Patel, D. J. Performance of Entomo-pathogenic Nematode for Management of Gram Pod Borer *Helicoverpa armigera. Ann. Plant Protect. Sci.* **2002b,** *11* (1), 107–109.

Weimin, Y. E.; Barragan, A. T.; Cardoza, Y. Z. *Oscheius carolinensis* N. Sp. (Nematoda: Rhabditidae), a Potential Entomopathogenic Nematode from Vermicompost. *Nematology.* **2010,** *12* (1), 121–135.

White, G. F. A Method for Obtaining Infective Nematode Larvae from Cultures. *Science.* **1927,** *66,* 302–303.

Womersley, C. Z. Dehydration Survival and Anhydrobiotic Potential. In *Entomopathogenic Nematodes in Biological Control;* Gaugler, R., Kaya, H. K., Eds.; CRC Press: Boca Raton, FL, 1990; pp 117–137.

Wouts, M. W. Mass Production of the Entomopathogenic Nematodes *Heterorhabditis heliothidis* (Nematoda: Heterorhabditidae) on Artificial Media. *J. Nematol.* **1981,** *13* (4), 467–69.

Zaki, F. A.; Mantoo, M. A.; Gul, S. *In Vivo* Culturing of Entomopathogenic Nematodes, *Heterorhabditis bacteriophora* and *Steinernema carpocapsae* on Silk Worm (*Bombyx mori*) and Their Effect on Some Lepidopterous Insects. *Indian J. Nematol.* **2000,** *30,* 1–4.

CHAPTER 8

ROLE OF INSECT VIRUSES IN THE MANAGEMENT OF INSECT PESTS

CHANDRA SHEKHAR PRABHAKAR[1*], AMIT CHOUDHARY[2], JAIPAL SINGH CHOUDHARY[3], PANKAJ SOOD[4], and PAWAN KUMAR MEHTA[5]

[1]Department of Entomology, Veer Kunwar Singh College of Agriculture, Bihar Agricultural University, Dumroan, Buxar 802136, Bihar, India

[2]Department of Entomology, Punjab Agricultural University, Ludhiana 141004, Punjab, India

[3]Research Centre, ICAR-Research Complex for Eastern Region, Plandu, Ranchi 834010, Jharkhand, India

[4]Krishi Vigyan Kendra, CSK Himachal Pradesh Krishi Vishvavidyalaya, Sundernagar, Mandi 175019, Himachal Pradesh, India

[5]Department of Entomology, CSK Himachal Pradesh Krishi Vishvavidyalaya, Palampur, Kangra 176062, Himachal Pradesh, India

*Corresponding author. E-mail: csprabhakar.ento@gmail.com

CONTENTS

ABSTRACT

The grave negative impacts of insecticides usage on the environment and public health led to concerted efforts directed toward reduction in chemical control of insect pests, diseases, and weeds. Integrated pest management has been endorsed by essentially all the multilateral environmental agreements that have transformed the global policy framework of natural resource management, agriculture, and trade. IPM has remained dominant paradigm of pest control over the last 50 years. Now ecologically based pest management (EBPM) programs are taking shapes for the sustainable insect pest management. There are indeed needs of safer biological insecticides for the successful implementation of EBPM programs worldwide. Amongst various alternatives to the EBPM, insect pathogens in particular insect pathogenic viruses have great potential for the sustainable insect pest management. Although there is a great diversity of insect viruses, only a few are exploited commercially for pest management. Among them insect viruses belonging to family *Baculoviridae* are highly explored. Since the discovery of nuclear polyhedrosis viruses causing jaundice in silkworm a lot of research work has been carried out on various aspects of viruses like their taxonomic features, mode of action and application and development of suitable formulations for pest management. Hence in this chapter an attempt has been made to summarize the essential information on these subjects. Besides such an extensive research on various aspects, insect viruses are still not so common in EBPM strategies. Hence there is a dire need to have collaborative efforts among various stakeholders *viz.* corporate sector, scientist and ultimately farmers to make best use of available insect viruses to curtail pesticide load on the environment.

8.1 INTRODUCTION

In the past few decades, appreciation of the negative impacts of insecticide usage on the environment and public health led to concerted efforts directed toward a reduction in chemical control of insect pests, diseases, and weeds (Haase et al., 2015). Apart from these issues, the problems pertaining to pest resistance toward the regular use of pesticides or different pesticides having similar mode of action, loss of target and non-target insect species, and so forth have also been debated worldwide. Thus, the necessity to develop alternate management strategies led to the evolvement of the concept of ecologically based pest management (EBPM) with main objective to minimize the usage of chemical pesticides to a significant level. Amongst various alternatives, insect pathogens in particular have great potential as integral component of EBPM. The populations of many pests under natural conditions are regulated by epizootics caused by entomopathogens such as bacteria, fungi, and viruses. Entomopathogens have also been deployed successfully as classical biological control agents of alien insect pests and their efficacy can be enhanced through habitat manipulation (Kalha et al., 2014). Among insect pathogens, insect viruses have been found quite promising and can play a vital role in devising suitable EBPM. The study of entomopathogenic viruses and their usage for pest management is comparatively older than that of the new generation synthetic pesticides. Among various families of insect viruses, viruses belonging to the family Baculoviridae are highly explored for the development of commercial virus based biopesticides owing to high host specificity and vertebrate safety.

8.2   HISTORY OF INSECT VIRUSES

Insects, like other organisms, also suffer from many diseases caused by various microorganisms. In the early history of insect pathology the diseases of beneficial insects particularly of honey bees and silkworm drew attention of the scientific community. Aristotle way back in the year 330 BC highlighted the diseases of honey bees which were speculated to be due to the foraging of honey bees on sick plants. Earlier diseases were described on the basis of symptoms they produce, for example wilting, jaundice, and so forth. The causal organism behind such symptoms could only be described with the invention of microscope in the late seventeenth century. However, the diseases caused by viruses were still defined on the basis of symptoms it produced because these are sub-microscopic entities. During 1527, Vida Marco Girolamo published the poem "De Bombyce" which probably described an infection of the silk moth by a nuclear polyhedrosis virus (NPV). The infection caused larvae to swell, while skin produced shining appearance and infected larvae showed melting like symptoms through which a puss like fluid exudates. Similar symptoms were also observed by Merian (1679). It was, however Maestri (1856) who observed NPV causing jaundice in silkworm which at that time was observed as crystal-like bodies under microscope. The infective nature of these particles was later studied by Bolle (1894). Similar symptoms, that is, wilting was also observed in gypsy moth by Glaser and Chapman (1913) and was reported to be caused by particles, capable of passing through filters. Later on White (1917) described the first virus disease in honey bees. Paillot (1933) published *L'Infection chez les Insectes* describing granulosis virus (GV). The first cytoplasmic polyhedrosis virus (CPV) was, however, described by Ishimori (1934). Later on with the advent of more sophisticated microscope, that is electron microscope, Bergold (1947) published the micrograph of NPV. Steinhaus (1949) published a landmark book "Principles of Insect Pathology" which described the classification of viruses. Next described insect virus was entomopoxvirus (EPV) by Vago (1963), while Tanada described GV infecting codling moth in the same year. Huger (1966) found a beetle-specific virus from Malaysia named *Rhabdionvirus oryctes* now referred to as *Oryctes* virus. Since various viruses are capable of causing disease in some of the agriculturally important insect pests, scientists started to explore their utilization for pest management. Consequently, the first viral formulation was recommended against *Helicoverpa/Heliothis zea* (Boddie) was a NPV registered under trade name Viron/H during 1973 and later Elcar by International Minerals and Chemical Corporation under the supervision of Ignoffo. Later in 1976, TM BioContro-1 a NPV based product was registered against Douglas fir tussock moth, while Gypcheck (a NPV based product) was registered against gypsy moth during 1978. The main disadvantage of these pesticides is the fast degradation upon exposure to direct sunlight (ultraviolet (UV) radiations). Therefore, studies were conducted to explore many adjuvants for sustaining insecticidal activity for longer time. Simultaneously, efforts were also started for genetic improvement of insect viruses. In this direction Smith, Fraser, and Summers manipulated the NPV genes in 1983. First field trial of a transgenic viral insecticide took place in the United Kingdom during 1993. The whole genome of NPV, however was first sequenced by Ayres et al. (1994), which caused disease in *Autographa californica* Speyer. Understanding and detection of insect viruses have been increased many fold with the development of molecular biology and biotechnological tools. Novel insect picorna-like virus has been identified in the brains of aggressive worker honey bees by Fujiyuki et al. (2004). Recent discoveries revealing the mutualistic relationship of one insect virus with host insects that lead to the resistance against second insect virus used as microbial insecticides in pest management system added new vistas in insect–virus interactions. *Helicoverpa armigera* densovirus-1 (HaDNV-1) is widespread in wild populations of *H. armigera* (Hubner) adults (Xu et al., 2014) having a negative interaction with *H. armigera* nuclear polyhedrosis virus (HaNPV) resulting in enhanced resistance to HaNPV.

8.3   TAXONOMY, CLASSIFICATION, AND DIVERSITY OF INSECT VIRUSES

The viruses are classified using Linnaean (binomial) taxonomy that was initially based on morphological characteristics and then stretched to take into account genomic phylogeny. Viruses do not encode their own protein synthesis machinery; therefore, all viruses must produce messenger ribonucleic acid (mRNA) in their life cycle for translation by host machinery in order to replicate. How the viruses do this is the basis for another useful categorization of viruses, the Baltimore classification (Sparks, 2010). This classification separates viruses into seven classes based upon the type of nucleic acid present in the capsid that functions as the viral genome and how mRNA is produced for protein production in the cell (Baltimore, 1971). This classification provides insight into class-specific strategies, as well as overall mechanisms shared by all viruses. The Baltimore classes are as follows: (a) dsDNA viruses (herpesviruses, baculoviruses, and poxviruses), (b) ssDNA viruses (+) sense DNA (parvoviruses), (c) dsRNA viruses (reoviruses), (d) (+)ssRNA viruses (picornaviruses, togaviruses), (e) (−)ssRNA viruses (orthomyxoviruses, filoviruses), (f) ssRNA-RT viruses (positive sense RNA with a DNA intermediate such as retroviruses), and (g) dsDNA-RT (DNA viruses that utilize reverse transcription such as hepadnaviruses) (Sparks, 2010).

Out of 73 known virus families, insect viruses have been listed in 13 families as described by Murphy et al. (1995). Since 1966, the International Committee on Taxonomy of Viruses (ICTV) has developed a sound framework for the nomenclature and classification of viruses. In view of the general approval accorded by virologists to this uniform taxonomy of viruses, insect virus classification also follows the direction of the ICTV (van Regenmortel et al., 2000). As type of genetic material (single- or double-stranded DNA, single-or double-stranded RNA, and positive or negative strand), virion morphology and size (icosahedral, rod-shaped), presence of an envelope surrounding the virion, presence of an occlusion body engulfing the virions, and host and host range are the characteristics for the classification of any other viruses, the same criteria is being followed to classify the diversity of viral groups that attack insect species. Insect pathogenic viruses are named in acronyms, according to their host and the viral group to which it belongs. For example, the cabbage butter fly, *Pieris brassicae* granulosis virus is named PbGV. Therefore, all nucleopolyhedrosis viruses are named NPV, just as the granulosis viruses are named GV. Further, lepidopteran baculoviruses are divided in NPVs (group I and II) and GVs by genomic nucleic acids sequence data (van Oers & Vlak, 2007). Likewise, the entomopoxviruses are named EPV, the iridoviruses as IV, and the cytoplasmic polyhedrosis viruses (cypoviruses) as CPV. Insect viruses are highly diverse; however, viruses from few families are frequently used in insect pest management and have shown potential to be used as biological control agents such as viruses from the family Baculoviridae, Poxviridae, and Reoviridae (Table 8.1).

8.3.1 FAMILY: BACULOVIRIDAE

Baculoviruses make up a family of insect viruses named as Baculoviridae. Members of the family Baculoviridae contain circular dsDNA with a size fluctuating between 80 and 180 kbp, which is condensed within nucleocapsids and are predicted to encode from about 90 to 180 genes (Okano et al., 2006; van Oers & Vlak, 2007). Enveloped rod-shaped nucleocapsids are referred to as virions. These are occluded either within polyhedral or within granular OBs mainly consisting of polyhedrin or granulin protein. Baculoviruses are grouped into two main groups or genera: NPVs and GVs. NPVs are further divided into single nucleopolyhedroviruses (SNPVs) with only one virion per envelope; and multiple nucleopolyhedroviruses (MNPVs) with several virions per envelope. The silkworms are infected with SNPV type species whereas larvae of *A. californica* with MNPV. More than 600 isolates of NPVs have been reported from a variety of insect species, 90% of these have lepidopteran hosts (Faulkner, 1981; Adams & McClintock, 1991; Vlak, 1992). NPV virions replicate only in the nuclei of susceptible cells and their OBs fluctuate between 1 and15 μm

TABLE 8.1 Diversity and Characteristics of Insect Viruses.

Family	Type of genetic materials/ nucleic acids	Genome size	Generic name of virus	Virion shape and diameter	Virion envelop	Examples of insects species	Code number for viral disease (according to ICTV)
Poxviridae	dsDNA	Linear, 250 + kbp	Poxvirus, Entomopoxvirus	Ovoid ; 160–250 nm × 300–470 nm	Yes	Amsacta moorei	19
Baculoviridae	dsDNA	Circular, 80–180 kbp	Nucleopolyhedrosis virus, Granulovirus	Rod-shaped; 30–60 nm × 250–300 nm	Yes	Bombyx mori	12, 9
		Circular, 80–180 kbp	–	Rod shaped; 30–60 nm × 250–300 nm, nuclear virus	No	Autographa califormica	11, 13
Reoviridae	dsRNA	10 segments	Cytoplasmic polyhedrosis virus	Isomeric, 80 nm	No	Bombyx mori	5, 14
Iridoviridae	dsDNA	140–210 kbp	Iridovirus, Chloriridovirus	Icosahedral, 120–180 nm	No	Asiatic rice Borer, Chilo suppressalis	10
Parvoviridae	ssDNA	5–6 kbp	Densovirus	Icosahedral, 18–26 nm	No	Heliothis virescens	6
Picornaviridae	ssRNA	–	Picornaviruses, Cricket paralysis virus, Drosophila C virus Gonometa virus, Rhopalosiphum padi virus	Spherical	–	Drosophila	13, 15, 7
Ascoviridae	dsDNA	Circular 120–180 kbp	Ascovirus	Allantoid, reniform, 130 × 200–400 nm	Yes	Heliothis virescens	–

TABLE 8.1 *(Continued)*

Family	Type of genetic materials/ nucleic acids	Genome size	Generic name of virus	Virion shape and diameter	Virion envelop	Examples of insects species	Code number for viral disease (according to ICTV)
Polydnaviridae	dsDNA	150–250 kbp	*Polydnavirus*	Irregular/prolate ellipsoid; >200 nm in length	Yes	*Cotesia melano-cella, Campoletis sonorensis*	13
Rhabdoviridae	ssRNA –ve	11–15 kbp	*Sigmavirus*	Bullet-shaped virus; 45–100 nm × 100–430 nm	Yes	*Spodoptera frugiperda*	3
Nodaviridae	ssRNA +ve	2 segments 4.5 kbp	*Nodavirus, Nodamura virus, Black beetle virus*	Icosahedral; 32–33 nm	No	*Oncopera intricoides*	13
Birnaviridae	dsRNA +ve	2 segments	*Birnavirus, Drosophila X virus*	Bisegmented; 60 nm	No	*Drosophila*	13
Caliciviridae	ssRNA +ve	7.5–8.5 kbp	*Caliciviruses, Amyelois chronic stunt virus*	Icosahedral; 35–40 nm	No	*Amyelois transitella*	13
Iflaviridae	ssRNA +ve	8.5–9.5 kbp	*Iflavirus*	Icosahedral; 30 nm	No	*Chinese oak silk-moth, Antheraea perry; leafhopper, Graminella nigrifrons*	–
Dicistroviridae	ssRNA +ve	9–10 kbp	*Cripavirus*	Icosahedral; 30 nm	No	Glassywinged Sharpshooter, *Homalodisca vitripennis*	–
Tetraviridae	ssRNA +ve	1–2 Segments; 6.5–8 kbp	*Betatetravirus*	Icosahedral; 40 nm	No	Slug caterpillar, *Setora nitens*	–

TABLE 8.1 (*Continued*)

Family	Type of genetic materials/nucleic acids	Genome size	Generic name of virus	Virion shape and diameter	Virion envelop	Examples of insects species	Code number for viral disease (according to ICTV)
Other unclassified viruses							
Arkansas bee virus	ssRNA	–	–	Small spherical	No	*Apis mellifera*	12
Bee acute paralysis virus	ssRNA	–	–	Small spherical	No	*Apis mellifera*	1
Bee chronic paralysis virus	ssRNA	–	–	Small spherical	No	*Apis mellifera*	2, 13
Bee slow paralysis virus	ssRNA	–	–	Small spherical	No	*Apis mellifera*	15
Bee virus X	ssRNA	–	–	Small spherical	No	*Apis mellifera*	13
Bee virus Y	ssRNA	–	–	Small spherical	No	*Apis mellifera*	13
Black queen cell virus	ssRNA	–	–	Small spherical	No	*Apis mellifera*	13
Cloudy wing virus	ssRNA	–	–	Small spherical	No	*Apis mellifera*	13
Crystalline array virus	ssRNA	–	–	Small spherical	No	*Apis mellifera*	4
Drosophila A virus	ssRNA	–	–	small spherical	No	*Drosophila*	13
Drosophila P virus	ssRNA	–	Small spherical	Small spherical	No	*Drosophila*	13
Kashmir bee virus	ssRNA	–	–	Small spherical	No	*Apis mellifera*	13

TABLE 8.1 *(Continued)*

Family	Type of genetic materials/ nucleic acids	Genome size	Generic name of virus	Virion shape and diameter	Virion envelop	Examples of insects species	Code number for viral disease (according to ICTV)
Kelp fly virus	ssRNA	11 kb	–	Small spherical 29 nm	No	Kelp fly	13
Sacbrood virus	ssRNA	–	–	Small spherical	No	*Apis mellifera*	18
Bee chronic paralysis virus	ssRNA	–	–	Small ovoid	No	*Apis mellifera*	2, 21
Drosophila RS virus	ssRNA	–	–	Small ovoid	No	*Drosophila melanogaster*	13
Cricket macro-nuclesus virus	ssRNA	–	–	Rod-shaped enveloped	No	Australian field crickets, *Teleogryllus commodus*	13
Hypera virus	ssRNA	–	–	Rod-shaped enveloped	No	*Hypera brunneipennis*	14
Bee filamentous virus	ssRNA	–	–	Long flexuous rod enveloped	No	*Apis mellifera*	22

ds = double stranded, ss = single stranded.

in size (Tinsley & Kelly, 1985). Granulovirus virions are always singly embedded within the envelope, consequently the OBs are very small (0.2–0.5 μm) compared to NPV polyhedra. Virions replicate in the cytoplasm of susceptible cells. Additionally, there is another group not belonging to the family Baculoviridae, but placed under this family due to many similarities with baculoviruses and its use as a biological control agent. This is the nudivirus or non-occluded baculovirus (NOB) specific to the coconut rhinoceros beetle, *Oryctes rhinoceros* L. (Coleoptera: Scarabaeidae). The rod-shaped virions are very similar to those of baculoviruses, except that they are not occluded within an OB.

8.3.2   FAMILY: POXVIRIDAE

The Poxviridae family is separated into two sub-families: the *Entomopoxvirinae,* which comprise of insect poxviruses; and the *Chordopoxvirinae,* which comprise of vertebrate poxviruses (Goodwin et al., 1991). The well-known chickenpox and smallpox viruses belong to this family. They show allantoid- to brick-shaped virions, occluded within ovoid OBs called spheroids (Tinsley & Kelly, 1985; Adams, 1991). The virions measure up to 400 nm in length, 250 nm in width and contain dsDNA ranging in size from 270 to 320 kbp. Virions replicate in the cytoplasm of susceptible cells. EPVs have been isolated from 27 insect species of Orthoptera, Lepidoptera, Diptera, and Coleoptera orders. The first sub-family of EPV comprises three genera based on host insect and virion morphology named as *Entomopoxvirus* A, *Entomopoxvirus* B, and *Entomopoxvirus* C (Arif, 1984; Arif & Kurstak, 1991). *Entomopoxvirus* A infects only Coleopteran species and the type species infects *Melolontha melolontha*; *Entomopoxvirus* B infects Lepidopteran and Coleopteran species and the type species infects *Amsacta moorei;* and *Entomopoxvirus* C infects only Dipteran species and the type species infects *Chironomus luridus.* A fourth group (D), which attacks hymenopteran insects has been proposed to the ICTV for consideration.

8.3.3   FAMILY: REOVIRIDAE

Reoviridae is a family of segmented dsRNA viruses with 12 genera; some of them infect mammals (Evans & Entwistle, 1987). The viruses of this family isolated from insects are called CPVs. These viruses have linear dsRNA genomes divided into 10–12 segments of about 12–32 kb in total. Icosahedral virions show 12 lateral projections and are occluded within large isometrical polyhedra of up to 10 μm in size (Hukuhara & Bonami, 1991). Virions replicate in the cytoplasm of the midgut epithelial cells (Hukuhara, 1985). CPVs are mainly isolated from insects belonging to order Lepidoptera, and occasionally from Diptera or Hymenoptera, and rarely from Coleoptera or Neuroptera orders. The *Bombyx mori* cytoplasmic polyhedrosis virus (BmCPV) was the first report of an occluded cytoplasmic insect virus noticed first by Ishimori (1934). ICTV assemble all the insect-specific reoviruses within the genus *Cypovirus*, and recognizes 70 species all hosted by *lepidopteran* species.

8.4   MODE OF ACTION OF INSECT VIRUSES

Insect viruses, as they occur in nature, consist of double-stranded and single-stranded circular/straight genomes of viral DNA/RNA with/without a double membrane envelop. Among various entomopathogenic viruses, the mode of action of baculoviruses is well studied. Baculoviruses consist of double-stranded and circular viral DNA with a double membrane envelop. Baculoviruses typically have narrow host ranges, often limited to just one or a few related insect species, although the most intensely studied member of the family, *A. californica* multiple nucleopolyhedrovirus (AcMNPV), is able to infect as many as 30 species from several *lepidopteran* genera. These occlusion-derived viruses (ODV) contain one or multiple nucleocapsids with each capsid holding a single viral genome. One or more ODV, in turn, are "occluded" within a protein matrix, referred to as occlusion bodies, that protect the virus from the environment with the condition that the occlusion bodies eventually degrade under UV light (Cory & Myers, 2003; Harrison & Hoover 2012). Baculoviruses are transmitted orally in nature, and the target of primary infection is the larval midgut epithelium. After being consumed, occlusion bodies get dissolved in the midgut lumen, releasing the embedded ODV, which infect midgut epithelial cells (Clem & Passarelli, 2013). ODV bind and fuse with midgut cells, releasing the nucleocapsids into the cytoplasm. The nucleocapsids are transported to the nucleus, uncoat, and begin replication, producing budded virus particles (Miller, 1997; Elderd, 2013). Progeny BV bud from the basal surface of the epithelium, cross the basal lamina, and infect most of the remaining tissues of the larva. The process of ODV attachment and entry depends on several viral proteins found in the ODV envelope, referred to as *per os* infectivity factors (PIFs) (Clem & Passarelli, 2013). Recent findings indicated that some of the PIFs form a complex in the ODV envelope, likely interact with an unknown midgut receptor, and mediate fusion with the plasma membrane (Peng et al., 2012). Thus the PIFs constitute a novel virus attachment/fusion protein complex. Interestingly, PIFs appear to be ancient genes that are conserved among three related insect virus families, as PIF homologs are also found in the nudiviruses and the polydnaviruses, the latter of which form mutualistic relationships with parasitic wasps (Bezier et al., 2009; Burke et al., 2013). In most baculovirus infections, the budded virus then spreads throughout the larva via the tracheal system and hemocytes. Late in infection, the host tissues are filled with virions that are occluded in millions of occlusion bodies, which are released upon death when the host liquefies (Harrison & Hoover, 2012).

8.5　FIELD APPLICATION OF INSECT VIRUSES FOR PEST MANAGEMENT

Earlier record of utilization of viruses was found with the usage of polyhedrosis virus causing *Wipfelkrankheit* in nun moth by Ruzicka (1924) in Europe. Though control was highly dependent upon application methods, yet it opened the new vista in plant protection for the usage of viruses. This was followed by the success recorded in the management of European pine sawfly, *Neodiprion sertifer*, in North America in the early years of twentieth century with a NPV specific to this insect brought from Finland to Canada. The introduction of this entomopathogenic virus in 1938 reduced the pest infestation rapidly which just disappeared from previously infested 12,000 square miles area by 1945 (Balch & Bird, 1944). Similarly, Lecontvirus® was registered for use in Canada in 1983 and is providing effective control of *N. lecontei* (Bird, 1961).

Reiff (1911) reported periodic epozootics caused by *Lymantria dispar* multicapsid nuclear polyhedrosis virus (LdMNPV) in the population of gypsy moth (*Lymantria dispar*), an exotic insect pest of forest plants in the United States. However, field application of this virus for the control of pest was made by former Soviet Union as early as 1959 (Lipa, 1998).

Later on the United States developed a formulation based on this virus named Gypcheck® which substantially reduced infestation of the pest.

Helicoverpa comprising *armigera, virescens,* and *zea* form a pest complex are infesting a wide range of crops in a wider area across the continents. NPV infecting *H. zea* was identified by Glaser and Chapman (1915) who reported "wilt" symptoms occurring in larvae, however the field trials were conducted by Chamberlin and Dutky (1958) and Tanada and Reiner (1962). The popular formulation of this virus in different countries are Viron H, Elcar®, Biotrol-VHZ®, and Gemstar® in the United States whereas, NPV of *H. armigera* (HearSNPV) are Helicovex® in Europe and VIVUS MAX in Australia (Vega & Kaya, 2012) (Table 8.2).

A *Cydia pomonella* granulovirus (CpGV) was identified by Tanada (1964) capable of causing epizootics in *C. pomonella* populations infesting pome fruits and walnuts. Results of field tests proved this virus to be a potential candidate in effective alternative for the management of codling moth larvae in integrated and organic pome fruit and walnut production in several countries (Huber & Dickler, 1977). In Argentina, Carpovirus Plus® (Quintana & Alvarado, 2004) and Madex® (Haase et al., 2015) are providing highly satisfactory results when the virus was applied at doses of 10^{13} OBs/hectare at intervals of 8–10 days.

During 1970s, a nucleopolyhedrosis virus of important soybean pest velvet bean caterpillar, *Anticarsia gemmatalis* (AgMNPV) was isolated in different regions of Brazil (Allen & Knell, 1977; Carner & Turnipseed, 1977). This virus was developed and used as a potential biopesticide in soybean integrated pest management (IPM) programs on a pilot scale during 1980/1981 and 1981/1982 under the coordination of Dr. Flavio Moscardi (EMBRAPA) (Moscardi, 1999). These trials, conducted with virus produced by collecting dead larvae in the fields produced promising results. Presently, this virus is applied in about 200,000 hectares per year in Brazil (Levy et al., 2009) along with 1,00,000 hectares in Paraguay (Kokubu, 2015).

Phthorimaea opercullela, an important pest of potato worldwide is infected with a granulovirus (PhopGV). In Peru, a PhopGV isolate has been developed as a microbial pesticide through an initiative of the International Potato Center (CIP). This virus provided effective control of this pest when 20 virus-infected larvae were mixed with 1 kg of talc and used as a suspension in 1 L of water. A dry product applied at a dose of 5 kg per ton of stored potatoes too provided high levels of control (ca. 95% mortality) (Raman et al., 1987).

TABLE 8.2 List of Insect Virus Based Biopesticide Products Commercialized Worldwide (modified after Haase et al., 2015).

Virus name	Target insect	Crop	Product name	Producer company	Country
Spodoptera litura NPV	*Spodoptera litura*	Vegetables and tobacco	Spodo-Cide	M/s Biocontrol Research Laboratory (BCRL), Bangalore	India
Helicoverpa armigera NPV	*Helicoverpa armigera*	Vegetables, tobacco, and cotton	Heli-Cide	M/s Biocontrol Research Laboratory (BCRL), Bangalore	India
Spodoptera litura NPV	*Spodoptera litura*	Vegetables, tobacco	Spodopterin	M/s Multiplex, Bangalore	India
Helicoverpa armigera NPV	*Helicoverpa armigera*	Vegetables and tobacco, cotton	Heliokill	M/s Multiplex, Bangalore	India
Helicoverpa zea NPV	*Helicoverpa armigera Heliothis virescens, Helicoverpa zea*	Cotton, tobacco, and vegetables	Elcar	SANDOZ, Novartis	USA
Anticarsia gemmatalis MNPV	*Anticarsia gemmatalis*	Soybean	Baculo-soja, Baculovirus Nitra, Coopervirus SC, protégé, Multigen	Nova Era Biotecnologia Agricola, Nitral Urbana, COODETEC, Milenia, EMBRAPA	Brazil
Autographa californica MNPV +	*Autographa califórnica Trichoplusia ni Pseudoplusia includens*	Alfalfa and vegetable crops	VPN-ULTRA	Agricola El So	Guatemala
Spodoptera albula NPV	*Heliothis virescens Spodoptera exigua Estigmene acrea Plutella xylostella*				

TABLE 8.2 (*Continued*)

Virus name	Target insect	Crop	Product name	Producer company	Country
Spodoptera sunia NPV	*Spodoptera* spp.	Vegetables	VPN 82	Agricola El So	Guatemala
Cydia pomonella GV	*Cydia pomonella, C. pomonella, Grapholita molesta*	Apple, pear, walnut, apple, and peach	Carpovirus Plus, Madex, Carpovirusine, Madex Twin	NPP-Arysta Life Science, Andermatt Biocontrol	Argentina
Cydia pomonella GV	*Cydia pomonella, C. pomonella, Grapholita molesta*	Apple, pear, walnut, apple, and peach	Madex Twin	Andermatt Biocontrol	Uruguay
Cydia pomonella GV	*Cydia pomonella, C. pomonella, Grapholita molesta*	Apple, pear, walnut apple, and peach	Carpovirusine	NPP-Arysta Life Science	Chile
Erinnyis ello GV	*Erinnyis ello*	Cassava	Baculovirus erinnyis	Empresa de Pesquisa Agropecuária e Extensão Rural de Santa Catarina S.A. BioCaribe	Brazil
Erinnyis ello GV	*Erinnyis ello*	Rubber trees	Baculovirus erinnyis	BioCaribe SA CORPOICA	Colombia
Helicoverpa zea SNPV	*Heliothis* and *Helicoverpa* spp.	Maize, tomato, Cotton, and tobacco	Gemstar	Certis USA	Mexico

TABLE 8.2 (Continued)

Virus name	Target insect	Crop	Product name	Producer company	Country
Helicoverpa zea SNPV	Heliothis and Helicoverpa spp.	Maize, tomato, Cotton, and tobacco	HzNPV CCAB	AgBiTech Australia	Brazil
Helicoverpa armigera NPV	Heliothis and Helicoverpa spp.	Tomato, sweet pepper, maize, soybean, tobacco, and vegetable crops	Diplomata / Helicovex	Koppert / Andermatt Biocontrol	Brazil
Phthorimaea operculella GV	Phthorimaea operculella Tecia solanivora	Potato	Baculovirus Corpoica	CORPOICA	Colombia
Phthorimaea operculella GV	Phthorimae operculella Tecia solanivora	Potato	PTM baculovirus	SENASA Peru	Peru
Phthorimaea operculella GV	Phthorimae operculella Tecia solanivora	Potato	PTM baculovirus	INTA Costa Rica	Costa Rica
Phthorimaea operculella GV + Bacillus thuringiensis	Phthorimaea operculella Tecia solanivora Symmetrischema tangolias	Potato	Matapol Plus	PROINPA Foundation	Bolivia
Phthorimaea operculella GV + Bacillus thuringiensis	Phthorimaea operculella Tecia solanivora Symmetrischema tangolias	Potato	Bacu-Turin	INIAP	Ecuador

TABLE 8.2 (Continued)

Virus name	Target insect	Crop	Product name	Producer company	Country
Spodoptera exigua NPV	Spodoptera exigua	Tomato, chili, and eggplant	SPOD-X LC	Certis USA—SUMMIT AGRO	Mexico
Lymantria dispar NPV	Gypsy moth caterpillars	Forest	Gypchek	USDA	USA
Orygia pseudotsugata NPV	Tussock moth, Orygia pseudotsugata	Forest	TM Biocontrol-1	USDA	USA
Orygia pseudotsugata NPV	Tussock moth, Orygia pseudotsugata	Forest	Virtuss		Canada
European Sawfly NPV	European Sawfly	Forest	Neochek-S	US forest service	USA
European Sawfly NPV	European Sawfly	Forest	Virox		UK
Pine caterpillar CPV	Pine caterpillar	forest	Matsukemin		Japan
Codling moth GV	Codling moth	Vegetable crops	Madmex	Andermatt Biocontrol,	Switzerland
Codling moth GV	Codling moth	Vegetable crops	Granupom		Germany
Red-headed sawfly NPV	Red-headed sawfly	Forest	Lecontvirus		Canada
Spodoptera exigua NPV	Beet armyworm	Beet and vegetables	Spod-X	ThermoTrilogy	—
Helicoverpa zea NPV	Helicoverpa armigera Heliothis virescens, Helicoverpa zea	Cotton, tobacco, and vegetables	GemStar	Thermo Trilogy	—
Spodoptera litura NPV	Spodoptera litura	Vegetables	Spodoterin	NPP, France	France

The fall armyworm, *Spodoptera frugiperda* is a polyphagous insect that causes economic losses in several important crops, such as maize, sorghum, rice, cotton, and pastures. Its control was based on the use of broad-spectrum chemical insecticides or *Bt* transgenic crops, having variable environmental effects and control failure due to development of resistance in the target pest (Yu, 1992). Thus baculoviruses were evaluated in countries like Argentina (Yasem de Romero, 2009), Brazil (Valicente & da Costa, 1995), Mexico (Ríos-Velasco et al., 2012), and Peru (Vásquez, 2002) for their effectiveness. The efficacy in controlling the pest was found to be similar to that of chemical insecticides at 22 days post emergence of the pest (Gómez et al., 2013).

In India, GV of *Chilo infuscatellus* was used for control of sugarcane shoot borer in Tamil Nadu (Easwaramoorthy, 2002). In India, two companies namely M/s Biocontrol Research Laboratory (BCRL) and M/s Multiplex at Bangalore are involved in mass production of baculoviruses belonging to HaNPV and SlNPV. The HaNPV formulation @ 0.43% in cotton and tomato and @ 2.0% in pigeon pea, chickpea, and tomato were approved for use to manage *H. armigera* larvae. Similarly, SINPV formulation (0.5% AS) was registered for use against the larvae of *Spodoptera litura* on tobacco and other crops. The NPV based biopesticides are marketed under different trade names such as Spodo-Cide®, Spodopterin®, Heli-Cide®, Heliokill®, and so forth in the Indian market (Ramanujam et al., 2014).

Other biopesticides currently under development include *Hyblea puera* NPV for controlling teak defoliator (Biji et al., 2006), *Amsacta albistriga* NPV for controlling the pest on groundnuts. The following formulations of entomopathogenic viruses are registered in India *viz.* HaNPV as Helicide, Virin-H, Helocide, Biovirus-H, Helicop, and Heligard and *S. litura* nucleopolyhedrosis virus as Spodocide, Spodoterin, Spodi-cide, and Biovirus-S (Rabindra & Grzywacz, 2010).

Cabbage butterfly, *P. brassicae* (Linnaeus) (Lepidoptera: Pieridae), is a serious pest of cabbage, cauliflowers, and many crucifer crops of the world (Bhandari et al., 2009; Feltwell, 1978). A PbGV strain was isolated and characterized from Himachal Pradesh, India by Sood (2004) during 2004 which was found effective in managing *P. brassicae* larvae (Sood et al., 2011; Sood & Prabhakar, 2012).

The effective formulation technology is another challenge for the development of a stable insect virus formulation under field conditions.

Usually, active ingredients, that is, virus OBs/POBs are mixed with various adjuvant for improving its efficacy, stability, and handling. The most common process of formulating a virus formulation is drying the infected larvae (through dehydration, lyophilization or by a humid air flow technique) to generate a powder. Lactose is added to improve the stability and infectivity of the virus. A suitable carrier usually silica and clays depending upon compatibility with the virus particles is then added to achieve the uniform concentration of the formulation. Surfactants, adherents, thickeners, binders, phagostimulants, UV protectants or optical brighteners, boric acid as stress causative, and so forth are added to make a good formulation (Sood et al., 2013a, 2013b). Formulation in the form of microencapsulation with enhanced efficacy has also been developed recently (Arthurs et al., 2006).

8.6   TRANSMISSION AND PERSISTENCE OF INSECT VIRUSES

Transmission of insect viruses to new hosts is a fundamental process in virus ecology as it determines the long-term persistence of a virus within its host population. Virus transmission can be divided into two broad categories: vertical transmission, the transfer of a virus pathogen between parent and offspring and horizontal transmission, the passage of virus among individuals who are not parent and offspring (Cory, 2015). Transmission of virus is the way to evolve themselves to a new strain of virus and find new host. Development of molecular biology techniques helped in more detailed ecological studies of virus communities in natural populations for latent infection in vertically transmitted virus pathogen as well as persistence in the natural conditions (Sood et al., 2010; Virto et al., 2014).

8.6.1   HORIZONTAL TRANSMISSION

Horizontal transmission in insect viruses often occurs through the ingestion of contaminated food via the susceptible feeding stages of insects, although examples of cannibalism, sexual transmission, and even vectoring occur (Parker et al., 2010; Fuller et al., 2012). Horizontal transmission can be modulated by a variety of environmental factors, one of which is host plant (Cory, 2010). Baculoviruses, and other pathogens that must be ingested, have an intimate relationship with their host plant (Cory & Hoover, 2006). Host plants can impact transmission by interacting directly with the virus, altering the resistance of the host or even interacting with the gut biota (Stevenson et al., 2010; Shikano et al., 2010). However, a recent experimental and modeling study of gypsy moth–NPV interaction has implicated variation in induced defenses as one possible explanation for complex patterns seen in gypsy moth dynamics (Elderd et al., 2013). Thus host plant–virus interactions could have a more significant impact on host dynamics than had been previously envisaged (Cory, 2015).

8.6.2   VERTICAL TRANSMISSION

It is not clear how many viruses rely solely on vertical transmission for their persistence. At one extreme there are the polydnaviruses, a large group of viruses found in a mutualistic relationship with braconid and ichneumonid wasps (Roossinck, 2011), whereas the transmission of viruses is totally tied to the reproduction of their wasp host as a provirus is incorporated in the wasp genome (known as an endogenized viral element) and only replicates in the wasp. The wasp injects polydnavirus particles containing virulent genes into the lepidopteran host which are expressed and manipulate its immune system and alter growth, allowing the development of the larval wasp (Strand et al, 2013; Drezen et al., 2014; Cory, 2015). Virus and wasp are thus entirely dependent on each other for survival. Only a few viruses have been confirmed as being transmitted only vertically without integration. The best studied of these are the sigma viruses (Family Rhabdoviridae) in *Drosophila* species. Sigma viruses are transmitted vertically by both males and females, although female transmission is more efficient (Longdon & Jiggins, 2012).

8.7   ADVANTAGES AND DISADVANTAGES OF INSECT VIRUSES

8.7.1   ADVANTAGES

- Insect viruses are environmentally safe and the formulations of these viruses are easily degraded and do not leave any harmful residues.

- Most insect viruses are relatively specific therefore the risk to non-target/beneficial insects is very low.
- Many viruses occur naturally and may already be present in the environment and can be utilized for the development of bio-pesticides.
- Insect viruses are highly effective hence like other control tactics these can successfully be incorporated in IPM strategies.
- Viruses transmit either horizontally or vertically hence are self-perpetuating to various extent in nature limiting the repeated applications under field conditions.
- Insect viruses can safely be incorporated into organic farming protocols.
- Some of the baculoviruses can be mass-produced by simple cottage industries which do not need much investment hence are cost effective.
- Insect viruses do not show cross infectivity against vertebrates, hence these do not pose health hazards in the workers unlike chemical pesticides.
- Insect virus is formulated in such a way that these do not require any special apparatus for application.
- Insect viruses are compatible with other management practices hence can be effectively integrated in IPM strategies.

8.7.2   DISADVANTAGES

- Viruses do not have quick knockdown effect, that is, most of the insect viruses take several days to kill the host insect during which the pest continuously damage the crop.
- Most of the insect viruses are quite effective against younger larvae, hence as the larval stage progresses they become less susceptible to virus infection, necessitating the application at a particular stage. Therefore, farmers need to be educated before using these types of pesticides.
- Although viruses can persist in the environment for months or years, exposed virus particles, like those on the surface of plants, are quickly inactivated by direct sunlight or high temperatures, which can limit their persistence within a given season.
- Also, some agricultural practices can reduce persistence between seasons, such as tillage, which buries virus particles in the soil.

- Still there are very few formulations available in the market because the registration process for a biopesticide is similar to that of synthetic chemical pesticide. This process requires detailed data on various aspects that too require capital investment which the biopesticide producing companies can hardly pay.

8.8 SUGGESTIONS FOR APPLICATION OF INSECT VIRUSES

Viruses are usually not "stand alone" solutions to an insect pest problem, but are most effective in conjunction with other management strategies. Although there are not many commercial products available in the market right now, several are being developed and may become available in the near future. Biopesticides present themselves as the useful option in organic farming, yet all the commercially available insect viruses are not allowed for use in organic systems. Hence, before application one must discuss with the certifying agency to plan the usage of a particular product. Since the effectiveness of a viral biopesticide largely depend upon prevailing abiotic and biotic conditions, there are several things a farmer can do to increase the effectiveness like

A. Insect viruses are fairly specific, be sure that the target pest is correctly identified.
B. Scout the fields carefully before application and apply virus when the target pests are in susceptible stage, that is, young and actively feeding. Scouting can also help to discover natural viral or other disease outbreaks developing in your crop, which depending on their extent, could influence other control decisions.
C. Apply virus to maximize the longevity and effectiveness of virus particles:
    - Apply in the morning or evening or on cloudy days when degradation from sunlight is reduced.
    - Avoid applying on rainy days, as rain will wash virus particles off the leaf surfaces.
    - Use formulations with UV light blockers and sticking agents to increase longevity.
    - Mixed cropping and reduce soil disturbance after application may enhance the effectiveness and may lead to better control within and between growing seasons.

KEYWORDS

- biopesticides
- entomopathogens
- insect virus
- insect pest
- IPM

REFERENCES

Adams, J. R.; McClintock, J. T. Baculoviridae: Nuclear Polyhedrosis Viruses. In *Atlas of Invertebrate Viruses*; Adams, J. R., Bonami, J. R., Eds.; CRC Press: Boca Ratón, FL, 1991; pp 87–204.

Adams, J. R. Introduction and Classification of Viruses of in Vertebrates. In *Atlas of Invertebrate Viruses*; Adams, J. R., Bonami, J. R., Eds.; CRC Press: Boca Ratón, FL, 1991; pp 1–8.

Allen, G. E.; Knell, J. D. A Nuclear Polyhedrosis Virus of *Anticarsia gemmatalis*: Ultrastructure, Replication, and Pathogenicity. *Fla. Entomol.* **1977**, *60*, 233–240.

Arif, B. M. The Entomopoxviruses. *Adv. Virus. Res.* **1984**, *29*, 195–201.

Arif, B. M.; Kurstak, E. Entomopoxviruses. In *Viruses of Invertebrates*; Kurstak, E., Ed.; Marcel Dekker: New York, 1991; pp 179–195.

Arthurs, S. P.; Lacey, L. A.; Behle, R. W. Evaluation of Spray-Dried Lignin-Based Formulations and Adjuvants as Solar Protectants for the Granulovirus of the Codling Moth, *Cydia Pomonella* (L). *J. Invertebr. Pathol.* **2006**, *93*, 88–95.

Ayres, M. D.; Howard, S. C.; Kuzio, J.; Lopez-Ferber, M.; Possee, R. D. The Complete DNA Sequence of *Autographa californica* Nuclear Polyhedrosis Virus. *Virology.* **1994**, *202*, 586–605.

Balch, R. E.; Bird, F. T. A Disease of the European Saw Fly, *Gilpinia hercyniae* (HTG), and Its Place in Natural Control. *Sci. Agric.* **1944**, *65*, 65–80.

Baltimore, D. Expression of Animal Virus Genomes. *Bacteriol. Rev.* **1971**, *35*, 235–241.

Bergold, G. H. Die Isolierung des Polyeder-Virus und die Natur der Polyeder. *Z. Naturforsch.* **1947**, *2b*, 122–143.

Bezier, A.; Annaheim. M.; Herbiniere, J.; Wetterwal, C.; Gyapay, G.; et al. Polydnaviruses of Braconid Wasps Derive from an Ancestral Nudivirus. *Science.* **2009**, *323*, 926–930.

Bhandari, K.; Sood, P.; Mehta, P. K.; Choudhary, A.; Prabhakar, C. S. Effect of Botanical Extracts on the Biological Activity of Granulosis Virus Against *Pieris brassicae*. *Phytoparasitica.* **2009**, *37*, 317–322.

Biji, C. P.; Suheendrakumar, W.; Sajeev, T. V. Quantitative Estimation of Production of *Hyblea puera* NPV in Three Stages of Teak Defoliator. *J. Virol. Methods.* **2006**, *136*, 78–82.

Bird, F. T. Transmission of Some Inset Viruses with Particular Reference to Ovarial Transmission and Its Importance in the Development of Epizootics. *J. Insect Pathol.* **1961,** *3,* 352–380.

Bolle, G. Giallume od il Mal del Grasso del Baco da Seta. Communicazione Preliminare. Atti e Mem, dell' I. *R. Soc. Agr. Gorizia.* **1894,** *34,* 133–136.

Burke, G. R.; Thomas, S. A.; Eum, J. H.; Strand, M. R. Mutualistic Polydnaviruses Share Essential Replication Gene Functions with Pathogenic Ancestors. *PLoS Pathog.* **2013,** *9,* e1003348. doi:10.1371/journal.ppat.1003348.

Carner, G. R.; Turnipseed, S. G. Potential of a Nuclearpolyhedrosis Virus for the Control of the Velvetbean Caterpillar in Soybean. *J. Econ. Entomol.* **1977,** *70,* 608–610.

Chamberlin, F. S.; Dutky, S. R. Tests of Pathogens for the Control of Tobacco Insects. *J. Econ. Entomol.* **1958,** *51,* 560.

Clem, R. J.; Passarelli, A. L. Baculoviruses: Sophisticated Pathogens of Insects. *PLoS Pathog.* **2013,** *9*(11), e1003729. doi:10.1371/journal.ppat.1003729

Cory, J. S. Insect Virus Transmission: Different Routes to Persistence. *Curr. Opin. Ins. Sci.* **2015,** *8,* 130–135.

Cory, J. S. The Ecology of Baculoviruses. In *Insect Virology*; Asgari, S., Johnson, K. N., Eds.; Academic Press: New York, 2010; pp 405–421.

Cory, J. S.; Hoover, K. Plant-Mediated Effects in Insect–Pathogen Interactions. *Trends Ecol. Evol.* **2006,** *21,* 278–286.

Cory, J.; Myers, J. The Ecology and Evolution of Insect Baculoviruses. *Annu. Rev. Ecol. Evol. Syst.* **2003,** *34,* 239–272.

Drezen, J. M.; Chevignon, G.; Louis, F.; Huguet, E. Origin and Evolution of Symbiotic Viruses Associated with Parasitoid Wasps. *Curr. Opin. Ins. Sci.* **2014,** *6,* 35–43.

Easwaramoorthy, S. Granulovirus Formulation in Pest Management in India. *Proceedings of the ICAR-CABI Workshop on Biopesticide Formulations and Applications*; Project Directorate of Biological Control: Bangalore, India, 2002; pp 41–48.

Elderd, B. D. Developing Models of Disease Transmission: Insights from Ecological Studies of Insects and Their Baculoviruses. *PLoS Pathog.* **2013,** *9*(6), e1003372.

Elderd, B. D.; Rehill, B. J.; Haynes, K. J.; Dwyer, G. Induced Plant Defences, Host–Pathogen Interactions, and Forest Insect Outbreaks. *Proc. Natl. Acad. Sci. USA.* **2013,** *110,* 14978–14983.

Evans, H. F.; Entwistle, P. Viral Diseases. In *Epizootiology of Insect Diseases*; Fuxa, J. R., Tanada, Y., Eds.; John Wiley & Sons: New York, 1987; pp 257–322.

Faulkner, P. Baculovirus. In *Pathogenesis of Invertebrate Microbial Diseases;* Davidson, E. W., Ed.; Allanheld Osmun CO Publishers, INC: Totowa, NJ, 1981; pp 3–37.

Feltwell, J. The Depredations of the Large White Butterfly (*Pieris brassicae*) (Pieridae). *J. Res. Lepidoptera.* **1978,** *17,* 218–225.

Fujiyuki, T.; Takeuchi, H.; Ono, M.; Ohka, S.; Sasaki, T.; Nomoto, A.; Kubo, T. Novel Insect Picorna-Like Virus Identified in the Brains of Aggressive Worker Honeybees. *J. Virol.* **2004,** *78*(3), 1093–1100.

Fuller, E.; Elderd, B. D.; Dwyer, G. Pathogen Persistence in the Environment and Insect-Baculovirus Interactions: Disease-Density Thresholds, Epidemic Burn Out and Insect Outbreaks. *Am. Nat.* **2012,** *179,* E70–E96.

Glaser, R. W.; Chapman, J. W. The Wilt Disease of Gypsy Moth Caterpillars. *J. Econ. Entomol.* **1913,** *6,* 479–488.

Glaser, R. W.; Chapman, J. W. A Preliminary List of Insects Which have Wilt with a Comparative Study of Their Polyhedra. *J. Econ. Entomol.* **1915**, *8*, 140–150.

Gómez, J.; Guevara, J.; Cuartas, P.; Espinel, C.; Villamizar, L. Microencapsulated *Spodoptera frugiperda* Nucleopolyhedrovirus: Insecticidal Activity and Effect on Arthropod Populations in Maize. *Biocontrol. Sci. Technol.* **2013**, *23*, 829–847.

Goodwin, R. H.; Milner, R. J.; Beaton, C. D. Entomopoxvirinae. In *Atlas of Invertebrate Viruses*; Adams, J. R., Bonami, J. R., Eds.; CRC Press: Boca Ratón, FL, 1991; pp 259–285.

Haase, S.; Sciocco-Cap, A.; Romanowski, V. Baculovirus Insecticides in Latin America: Historical Overview, Current Status and Future Perspectives. *Viruses.* **2015**, *7*, 2230–2267.

Harrison, R.; Hoover, K. Baculoviruses and Other Occluded Insect Viruses. In *Insect Pathology*; Vega, F., Kaya, H., Eds.; Academic Press: London, 2012.

Huber, J.; Dickler, E. Codling Moth Granulosis Virus: Its Efficacy in the Field in Comparison with Organophosphorus Insecticides. *J. Econ. Entomol.* **1977**, *70*, 557–61.

Huger, A. M. A Virus Disease of the Indian Rhinoceros Beetle, *Oryctes Rhinoceros* (Linnaeus), Caused by a New Type of Insect Virus, *Rhabdinovirus Oryctes* gen. n., sp. n. *J. Invertebr. Pathol.* **1966**, *8*, 38–51.

Hukuhara, T. Pathology Associated with Cytoplasmic Polyhedrosis Viruses. In *Viral insecticides for biological control;* Maramorosh, K., Sherman, K. E. Eds.; Academic Press: Orlando, FL, 1985; pp 121–162.

Hukuhara, T.; Bonami, J. R. Reoviridae. In *Atlas of Invertebrate viruses*; Adams, J. R., Bonami, J. R., Eds.; CRC Press: Boca Raton, FL, 1991; pp 393–434.

Ishimori, N. Contribution a l'etude de la Grasserie du ver a Soie. C. R. *Seances Soc. Biol. Fil.* **1934**, *116*, 1169–1174.

Kalha, C. S.; Singh, P. P.; Kang, S. S.; Hunjan, M. S.; Gupta, V.; Sharma, R. Entomopathogenic Viruses and Bacteria for Insect-Pest Control. In *Integrated Pest Management: Current Concepts and Ecological Perspective*; Abrol, D. P., Ed.; Academic Press: San Diego, CA, 2014; pp 225–2244.

Kokubu, H. Brief Overview of Microorganisms Used Against Agricultural Insect Pests. 2015. Available online: http://e-cucba.cucba.udg.mx/index.php/e-ucba/article/view/10/pdf_brief

Levy, S. M.; Moscardi, F.; Falleiros, A. M.; Silva, R. J.; Gregorio, E. A. A Morphometric Study of the Midgut in Resistant and Non-Resistant *Anticarsia gemmatalis* (Hubner) (Lepidopetra: Noctuidae) Larvae to Its Nucleopolyhedrosis (AgMNPV). *J. Invertebr. Pathol.* **2009**, *101*, 17–22.

Lipa, J. J. Eastern Europe and the Former Soviet Union. In *Insect Viruses and Pest Management*; Hunter-Fujita, F. E., Entwhistle, P. F., Evans, H. F., Crook, N. E., Eds.; John Wiley & Sons: Chichester, UK, 1998; pp 216–231.

Longdon, B., Jiggins, F. M. Vertically Viral Endosymbionts of Insects: Do Sigma Viruses Walk Alone? *Proc. R. Soc. B.* **2012**, *279*, 3889–3898.

Maestri, A. Del Giallume. In *Frammenti Anatomici Fisiologici e Patologici sul baco da seta*; Fusi: Pavia, 1856; pp 117–120.

Merian, M. S. *Del' Raupen Wunderbare Verwandelung, und Sonderbare Blumennahrung*; J. A. Graff: Niirnberg, 1679; p 155.

Miller, L. K. *The Baculoviruses*; Plenum Press: New York, 1997.

Moscardi, F. Assessment of the Applications of Baculoviruses for Control of Lepidoptera. *Annu. Rev. Entomol.* **1999**, *74*, 480–485.

Murphy, F. A.; Faquet, C. M.; Bishop, D. H. L.; Gabrial, S. A.; Jarvis, A. W.; Martelli, G. P. *Classification and Nomenclature of Viruses;* Sixth Report of the International Committee on Taxonomy of Viruse. Springer-Verlag: Berlin, 1995.

Okano, K.; Vanarsdall, A. L.; Mikhailov, V. S.; Rohrmann, G. F. Conserved Molecular Systems of the Baculoviridae. *Virology.* **2006**, *344*, 77–87.

Paillot, A. *L'infection Chez Les Insectes*; G. Patissier: 'l'revoux, 1933; p 535.

Parker, B. J.; Elderd, B. D.; Dwyer, G. Host Behaviour and Exposure Risk in an Insect-Pathogen Interaction. *J. Anim. Ecol.* **2010**, *79*, 863–870.

Peng, K.; van Lent, J. W.; Boeren, S.; Fang, M.; Theilmann, D. Characterization of Novel Components of the Baculovirus Per os Infectivity Factor Complex. *J. Virol.* **2012**, *86*, 4981–4988.

Quintana, G.; Alvarado, L. Carpovirus Plus: Primer Insecticida Biológico Para el Control de Cydia Pomonella en Montes Comerciales de Pera, Manzana y Nogal. In *AgroInnova—La Innovación Tecnológica para Mejorar la Competitividad*; SECyT-INTA: Rosario, Argentina, 2004; pp 15–17.

Rabindra, R. J.; Grzywacz, D. India. In *The Use and Regulation of Microbial Pesticides in Representative Jurisdictions Worldwide*; Todd, J. K., Svircev, A. M., Goettel, M. S., Woo, S. G., Eds.; IOBC Global, 2010; pp 12–17. http://www.iobc-global.org/download/Microbial_Regulation_Book_Kabaluk_et_al_2010.pdf

Raman, K. V.; Booth, R. H.; Palacios, M. Control of Potato Tuber Moth *Phthorimaea operculella* (Zeller) in Rustic Potato Stores. *Trop. Sci.* **1987**, *27*, 175–194.

Ramanujam, B.; Rangeshwaran, R.; Sivakmar, G.; Mohan, M.; Yandigeri, M. S. Management of Insect Pests by Microorganisms. *Proc. Indian Natn. Sci. Acad.* **2014**, *80*(2), 455–471.

Reiff, W. *The Wilt Disease or Flacherie of the Gypsy Moth*; Wright & Potter Printing Company: Boston, 1911.

Ríos-Velasco, C.; Gallegos-Morales, G.; Berlanga-Reyes, D.; Cambero-Campos, J.; Romo-Chacón, A. Mortality and Production of Occlusion Bodies in *Spodoptera frugiperda* Larvae (Lepidoptera: Noctuidae) Treated with Nucleopolyhedrovirus. *Florida Entomol.* **2012**, *95*, 752–757.

Roossinck, M. J. The Good Viruses: Viral Mutualistic Symbioses. *Nat. Rev.* **2011**, *9*, 99–108.

Ruzicka, J. Die Neusten Erfahrungen Uber Die Nonne in Bohmen. *Cent. Ges. Forsta.* **1924**, *50*, 33–68.

Shikano, I.; Ericsson, J. D.; Cory, J. S.; Myers, J. H. Indirect Plant Mediated Effects on Insect Immunity and Disease Resistance in a Tritrophic System. *Basic Appl. Ecol.* **2010**, *11*, 15–22.

Sood, P. New Record of Granulovirus on Cabbage White Butterfly, *Pieris Brassicae* Linn. from Dry Temperate Regions of Himachal Pradesh. *Himachal J. Agri. Res.* **2004**, *30*, 146–148.

Sood, P.; Prabhakar, C. S.; Mehta, P. K. Field Evaluation of an Indigenous Granulosis Virus Isolate for *Pieris brassicae* (Linnaeus) Management under North Western Himalayan Conditions. *J. Biol. Con.* **2011**, *25*(3), 217–222.

Sood, P.; Choudhary, A.; Prabhakar, C. S.; Mehta, P. K. Effect of Feeding Stimulants on the Insecticidal Properties of *Pieris brassicae* Granulovirus (PbGV) Against *Pieris brassicae. Phytoparasitica.* **2013a,** *41*(4), 483–490.

Sood, P.; Mehta, P. K.; Bhandari, K.; Prabhakar, C. S. Transmission and Effect of Sublethal Infection of Granulosis Virus (PbGV) on *Pieris brassicae* Linn. (Pieridae: Lepidoptera). *J. App. Entomol.* **2010,** *134* (9–10), 774–780.

Sood, P.; Mehta, P. K.; Prabhakar, C. S. Effect of UV Protectants on the Efficacy of *Pieris brassicae* Granulovirus. *Biol. Agri. Horti.* **2013b,** *29,* 69–81.

Sood, P.; Prabhakar, C. S. Molecular Characterization of *Pieris brassicae* Granulosis Virus (PbGV) from India. *J. Biol. Cont.* **2012,** *26*(2), 131–137.

Sparks, W. O. Interaction of the Baculovirus Occlusion-Derived Virus Envelope Proteins ODV-E56 and ODV-E66 with the Midgut Brush Border Microvilli of the Tobacco Budworm, *Heliothis virescens* (Fabricius). Ph.D. Dissertation, Iowa State University, USA, 2010.

Steinhaus, E. A. *Principles of Insect Pathology;* McGraw-Hill Book Co. Inc: New York, 1949; p 757.

Stevenson, P. C.; D'Cuhna, R. F.; Gryzwacz, D. Inactivation of Baculovirus by Isoflavenoids on Chickpea (*Cicer arietinum*) Leaf Surface Reduces the Efficacy of Nucleopolyhedrovirus Against *Helicoverpa Armigera. J. Chem. Ecol.* **2010,** *36,* 227–235.

Strand. M. R.; Burke, G. R. Polydnavirus-Wasp Associations: Evolution, Genome Organization and Function. *Curr. Opin. Virol.* **2013,** *3*(5), 587–594.

Tanada, Y. A Granulosis Virus of the Codling Moth, *Carpocapsa pomonella* (Linnaeus) (Olethreutidae, Lepidoptera). *J. Insect. Pathol.* **1964,** *8,* 378–380.

Tanada, Y.; Reiner, C. The Use of Pathogens in the Control of the Corn Earworm Heliothis Zea (Boddie). *J. Invertebr. Pathol.* **1962,** *4,* 139–154.

Tinsley, T. W.; Kelly, D. C. Taxonomy and Nomenclature of Insect Pathogenic Viruses. In *Viral Insecticides for Biological Control;* Maramorosh, K., Sherman, K. E., Eds.; Academic Press: Orlando, FL, 1985; pp 3–25.

Vago, C. A New Type of Insect Virus. *J. Invert. Pathol.* **1963,** *5,* 275–276.

Valicente, F. H.; da Costa, E. F. Controle da Lagarta Do Cartucho *Spodoptera frugiperda* Com *Baculovirus spodoptera*, Aplicado Via água de Irrigação. *An. Soc. Entomol. Bras.* **1995,** *24,* 61–67.

van Oers, M. M.; Vlak, J. M. Baculovirus Genomics. *Curr. Drug Targ.* **2007,** *8,* 1051–1068.

van Regenmortel, M. H. V.; Fauquet, C. M.; Bishop, D. H. L.; Cartens, E. B.; Estes, M. K.; Lemon, S. M.; Maniloff, J.; Mayo, M. A.; McGeoch, D. J.; Pringle, X. R.; Wickner, R. B. *Virus Taxonomy;* Seventh Report of the International Committee of Taxonomy of Viruses: Academic Press: San Diego, 2000, p 1162.

Vásquez, J.; Zeddam, J. L.; Tresierra, A. A. Control Biológico Del "cogollero del maíz" *Spodoptera frugiperda* (Lepidoptera: Noctuidae) Con el Baculovirus SfVPN, en Iquitos-Perú. *Folia Amazon.* **2002,** *13,* 25–39.

Vega, F. E.; Kaya, H. K. *Insect Pathology;* Elsevier: Amsterdam, the Netherlands, 2012; p 508.

Virto, C.; Navarra, D.; Mar Tellez, M.; Herrero, S.; Williams, T.; Murillo, R.; Caballero, P. Natural Populations of *Spodoptera exigua* are Infected by Multiple Viruses that are Transmitted to Their Offspring. *J. Invertebr. Pathol.* **2014,** *122,* 22–27.

Vlak, J. M. The Biology of Baculovirus *in vivo* and in Cultured Insect Cells. In *Baculovirus and Recombinant Protein Production Processes;* Vlak, J. M., Schlaeger, E. J., Bernard, A. R., Eds.; Editiones Roche: Interlaken, Switzerland, 1992; pp 2–10.

White, G. F. Sacbrood, A. *USDA Bull.* **1917,** *431,* 1–55.

Xu, P.; Liu, Y.; Graham, R. I.; Wilson, K.; Wu, K. Densovirus is a Mutualistic Symbiont of a Global Crop Pest (*Helicoverpa armigera*) and Protects Against a Baculovirus and *Bt* Biopesticide. *PLoS Pathog.* **2014,** *10*(10), e1004490. doi:10.1371/journal.ppat.1004490

Yasem de Romero, M. G.; Romero, E.; Sosa Gómez, D.; Willink, E. Evaluación de Aislamientos de Baculovirus Para el Control de *Spodoptera frugiperda* (Smith) Lep.: Noctuidae, Plaga Clave del Maíz en el Noroeste Argentino. *Rev. Ind. Agríc. Tucumán.* **2009,** *86,* 7–15.

Yu, S. J. Detection and Biochemical Characterization of Insecticide Resistance in Fall Armyworm (Lepidoptera: Noctuidae). *J. Econ. Entomol.* **1992,** *85,* 675–691.

PART III
Parasitoids and Predators

CHAPTER 9

TRICHOGRAMMA: AN EGG PARASITOID IN INSECT PEST MANAGEMENT

TARAK NATH GOSWAMI[1*], ANIL[1], TARAK BRAMBHA MAJI[2], PRANAB BARMA[3], and SHREE NIWAS RAY[1]

[1]Department of Entomology, Bihar Agricultural University, Sabour, India

[2]Department of Agricultural Entomology, Bidhan Chandra Krishi Viswavidyalaya, Mohanpur 741252, Nadia, West Bengal, India

[3]Darjeeling Krishi Vigyan Kendra, Uttar Banga Krishi Viswavidyalaya, Kalimpong, Darjeeling 734301, West Bengal, India

*Corresponding author. E-mail: tarakento@gmail.com

CONTENTS

ABSTRACT

Trichogramma spp are very important egg parasitoids for many of the agricultural pests. In this chapter we have tried to gather and compile information available particularly from India. But in a few cases we had to collect it from outside India. In this chapter we have put the list of Trichogramma spp available in India, their rate of release as recommended by National Bureau of Agricultural Insect Resources (NBAIR). Further, identification and taxonomy, biology of Trichogramma spp., use of Trichogramma in pest management of various crops in India like Paddy, Maize, Sugarcane, Okra, Tomato, Brinjal, and Cotton have been discussed in this chapter. We also have incorporated the use of Trichogramma against a very important pest, Asian corn borer in China. Mass production technique of *Trichogramma* spp., *in vitro* rearing of *Trichogramma* spp. on artificial host eggs as well as the quality control guidelines for *Trichogramma* spp. have also been attempted to discuss in this chapter.

9.1 INTRODUCTION

The family *Trichogrammatidae* is one of the most poorly known groups of the superfamily Chalcidoidea (Order: Hymenoptera) mainly due to the small size and fragility of the insects belonging to it (Pinto & Stouthamer, 1994). This family represents ~80 genera and includes the smallest of insects, ranging in size from 0.2 to 1.5 mm. However, the genus *Trichogramma* is the best known in the family due to its use in the biological control of agricultural pests with over 200 described species.

Trichogramma and *Trichogrammatoidea* are cosmopolitan in distribution and parasitize more than 200 species of insect pests under the orders Lepidoptera, Neuroptera, Diptera, Hemiptera, Coleoptera, and Hymenoptera. Majority of them are primary eggs parasitoids of Lepidoptera (Tanwar et al., 2006). It is important for plant protection because of its wide spread natural occurrence and its success as biological control agent by mass releasing (Van Lenteren, 2003). According to Smith (1996), the *Trichogramma* has been considered as one of the most important parasitoids for more than 100 years. Since this parasitoid kills the pest in the egg stage itself before the pest could cause any damage to the crop and also that it is quite amenable to mass production in the laboratories, it has the distinction of being the highest produced and most utilized biological control agent in the world.

9.2   *TRICHOGRAMMA* SPECIES AVAILABLE IN INDIA

The details of *trichogramma* species available in India are provided in Table 9.1.

9.3   IDENTIFICATION AND TAXONOMY

Under the family *Trichogrammatidae, Trichogramma* is one of the 80 genera. All members of this family are parasites of insect eggs. *Trichogrammatidae* includes the smallest of insects, ranging in size from 0.2 to 1.5 mm. Within the genus *Trichogramma*, there are 145 described species worldwide (Olkowski & Zhang, 1990). Due to the importance of its use, to succeed in a biological control program with the *Trichogramma* species, correct identification is essential.

Trichogramma are difficult to identify because they are so small and have generally uniform morphological characters. Also, certain physical characteristics such as body color and the number and length of body hairs can vary with body size, season, rearing temperature, and the host on which the adult is reared (Pinto & Stouthamer, 1994).

A major advance in the systematics of *Trichogramma* was the discovery that characteristics of male genitalia can be used to identify species. This is the primary means of identification today, but body color, wing venation, and features of the antennae serve as supporting characteristics. Females cannot be identified with the same level of confidence, so collections submitted for identification must include males. In addition to physical characteristics, studies of reproductive compatibility and mode of reproduction also have been especially valuable in identifying species (Pinto & Stouthamer, 1994).

The discovery of the male genitalia as a morphological characteristic in the identification of *Trichogramma* species was a major breakthrough in the taxonomy of these parasitoids (Pinto & Stouthamer, 1994). However, in some cases the identification is difficult due to the absence of males in thelytokous species infected by bacteria of the genus Wolbachia, the

TABLE 9.1 *Trichogramma* Species Available in India.

Species	Stage supplied	Target pest	Recommended dosage	No. of releases recommended
Trichogramma chilonis Ishii (I) #	Parasitized egg cards	Sugarcane borers *Chilo infuscatellus, Chilo sacchariphagus indicus, Chilo auricilius, Acigona steniellus*; Cotton (Non Bt) bollworms *Helicoverpa armigera, Pectinophora gossypiella & Earias* spp.; Maize stem borer *Chilo partellus, Diamond back moth Plutella xylostella*; Tomato fruit borer *Helicoverpa armigera*	50,000/ha on sugarcane and vegetables; 100,000/ha on maize and 150,000/ha on cotton	Sugarcane: 4–6 releases at 10 days intervals for early shoot borer; 8–10 releases for stalk, internode and Gurdaspur borers Cotton (Non Bt) & Vegetables: six weekly releases Maize: three releases at five days intervals
Trichogramma japonicum (I)	Parasitized egg cards	Top shoot borer of sugarcane *Scirpophaga excerptalis* and paddy stem borer *Scirpophaga incertulas*	Sugarcane & paddy: 50,000/ha	Sugarcane: 4–6 releases at 10 days intervals on observing pest or from 60th day Paddy: 6 releases on appearance of pest or from 30th day after transplantation
Trichogramma achaeae (I)*		Cotton (Non Bt) bollworms and Bhendi Borer	150,000/ha on cotton (Non Bt) 50,000/ha on vegetables	Six releases at weekly intervals
Trichogramma pretiosum (E)*		Tomato fruit borer *Helicoverpa armigera*	50,000/ha	Six releases at weekly intervals on appearance of pest or from 45th day from transplantation
Trichogramma embryophagum (E)*		Apple Codling moth *Cydia pomonella*	2000 adults per tree or 100,000/ha	Releases starting from the first moth catch, continue at weekly intervals till pest egg availability in the field

TABLE 9.1 *(Continued)*

Species	Stage supplied	Target pest	Recommended dosage	No. of releases recommended
Trichogramma dendrolimi Matsumara (E)*		Targeted against tissue borers on maize and sugarcane for research work		
Trichogramma brassicae (E)*		Diamondback moth *Plutella xylostella* and Cabbage butterfly *Pieris brassicae* on cabbage and cauliflower	100,000/ha	Six releases at weekly intervals
Trichogramma evanescens Westwood		Targeted against tissue borers on maize and sugarcane for research work		
Trichogramma mwanzai (E)*		Targeted against *Helicoverpa armigera* for research work		
Trichogrammatoidea armigera (E)*		Targeted against *Helicoverpa armigera* for research work		
Trichogrammatoidea bactrae (E)*		Diamond back moth *Plutella xylostella* on cabbage	250,000/ha	Five releases at weekly intervals

*Only nucleus culture will be supplied, either for research work or for further scaling up for commercialization.

#Sugarcane, cotton, and vegetable strains of *T. chilonis* maintained; I: indigenous; E: exotic.

Source: Anonymous (2015).

presence of cryptic species and the small size of the individual (Almeida & Stouthamer, 2003; Borba et al., 2005). A more recent method, molecular biology, has been widely used in the taxonomic identification of diverse groups of insects (Borba et al., 2005).

In recent studies the internal transcribed region of space in ITS2 DNA has been used in sequencing for identification of *Trichogramma* species, differentiation of populations, as well as in the reconstruction of phylogenetic relationships between closely related species. This identification method is gaining recent importance in the study of cryptic species of *Trichogramma*, where the polymerase chain reaction is used to amplify these ITS spacers, via universal primers that bind to the highly conserved regions, it is possible to obtain the sequence and thus, the description of the species (Samara et al., 2008; Sumer et al., 2009).

Species particularly difficult to identify with the use of traditional taxonomic techniques may have their identities clarified using molecular sequencing techniques or sequence length polymorphism generated through the techniques of molecular markers as described by Borba et al. (2005).

Santos et al. (2015) identified and differentiated *Trichogramma exiguum* Pinto and Platner species, *Trichogramma pretiosum* Riley, and *Trichogramma galloi* Zucchi using sequences of the ITS2 region of ribosomal DNA. After extracting DNA from the studied species, a PCR reaction was performed by them, where the amplified samples were subjected to sequencing. The sequences obtained were submitted to a similarity search in GenBank (NCBI: National Center for Biotechnology Information) using the BLAST program, aiming to determine the similarity of these sequences with the species already deposited in the referenced database, and then multiple sequences were aligned using version 2.0 of the ClustalX program. According to the results of the multiple alignments of all sequences obtained by them, it was possible to observe the differences between the *T. pretiosum*, *T. galloi* and *T. exiguum* species. It was concluded by Santos et al. (2015) that using the sequences of the ITS2 region of the ribosomal DNA was efficient in the differentiation of the studied *Trichogramma* species, which suggested a strong inter-specific variation among species.

9.4 BIOLOGY OF *TRICHOGRAMMA* SPP. IN GENERAL

The development of all *Trichogramma* spp. is very similar. The egg (0.1 mm in length) is inserted into the host and doubles its size before hatching. The larva, which is reduced to a digestive sac provided with two crochets (mandibles) framing the mouth, feeds in the vitelline mass or embryo of the host, in the latter instance the embryo being destroyed by a process of lysis (Metcalfe & Breniere, 1969). There are three larval instars, all sacciform. These are followed by a prepupa, when the adult characters form, and a pupa. At the beginning of the third larval instar, the host egg turns black due to the deposition of black granules at the inner surface of the chorion, an invaluable diagnostic character for parasitized eggs (Brenière, 1965b; Flanders, 1937; Krishnamurti, 1938). The development time for *Trichogramma australicum* on *Corcyra cephalonica* staint at 28 °C is given in Table 9.2 (Brenière, 1965b).

TABLE 9.2 The Development Time for *Trichogramma australicum* on *Corcyra cephalonica* Staint at 28 °C.

Stages	Duration (hours)
Egg	24 hours
1st larval instar	21 hours
2nd larval instar	27 hours
3rd larval instar	48 hours
Prepupa	24 hours
Pupa	48 hours
Total	8 days

Fecundity varies between 20 and 120 eggs per female according to the species, the host, and the longevity of the adult. Longevity is related to food supply (sugar and water), availability of host eggs, temperature, humidity, and the activity of the female (Nayaranan & Mookherjee, 1955).

Under identical conditions with *C. cephalonica* as host, the fecundity of *T. australicum* was 43, *Trichogramma fasciatum* 67, and *Trichogramma minutum* 76 (Brenière, 1965d). The oviposition habit varies with species (Quednau, 1960). While *Trichogramma fasciatum* lays most of its eggs in the first 24 hours, *T. minutum* spreads them over 2–3 days, and *T. australicum* over 10–12 days (Metcalfe & Breniere, 1969).

If we consider bollworms, the adult female wasp uses chemical and visual clues to locate a bollworm egg. The chemical clues, called kairomones, are on the moth scales left near the egg by the female moth during oviposition (Nordlund et al., 1981). Some of these same chemicals are also bollworm sex pheromones. However, shape of the eggs and color also may be visual clues to the wasp for parasitization (Ruberson & Kring, 1993). After finding a bollworm egg, the female drills a hole through the chorion (egg shell) and inserts two to three eggs into the bollworm egg. The internal pressure of the bollworm egg forces a small drop of yolk out of the oviposition hole. Females feed on this yolk, which increases their longevity. Under laboratory conditions a female parasitizes from one to ten bollworm eggs per day or from 10 to 190 during her life span. Large females parasitize more eggs than smaller females. Females which are provided honey and young bollworm eggs to feed on live an average of 11 days, while females receiving only honey live only for three days (Ruberson & Kring, 1993). However, the study done by Suh (1998) revealed that the average adult life span was 24 days.

According to Ruberson and Kring (1993), bollworm eggs in the early stages of development are more suitable for parasite development. Older bollworm eggs, especially those in which the head capsule of the larva is visible, are not usually parasitized and if they are, parasite survival is much lower. The yolk and embryo of the parasitized bollworm egg are digested before the *Trichogramma* egg hatches. A venom injected by the female at the time of oviposition is believed to cause this predigestion of the egg's contents. Eggs hatch in about 24 hours and the parasite larvae develop very quickly. According to Strand (1986), two *T. pretiosum* larvae can consume the digested contents of a young budworm egg within 10 hours of hatching. Larvae develop through three instars. During the 3rd instar (3 to 4 days after the host egg was parasitized) dark melanin granules are deposited on the inner surface of the egg chorion, causing the bollworm egg to turn black. Larvae then transform to the inactive pupal stage. After about 4.5 days, the adult wasps emerge from the pupae and escape the bollworm egg by chewing a circular hole in the eggshell. The black layer inside the chorion and the exit hole are evidence of parasitism by *Trichogramma*. The life cycle from egg to adult requires about 9 days, but varies from 8 days when mid-summer temperatures are high (90 degrees

F) to as many as 17 days at 60 degrees F. Adults are most active at 75–85 degrees (Strand, 1986).

9.5 USE OF TRICHOGRAMMA IN PEST MANAGEMENT OF VARIOUS CROPS IN INDIA

9.5.1 PADDY

Kaur et al. (2008) carried out an experiment on the management of rice leaf folder and stem borer in two rice varieties, *viz.*, PR 116 and *Basmati* 386 at PAU, Ludhiana. Organic practices and integrated practices (seven releases of *Trichogramma chilonis* and *Trichogramma japonicum* @ 100,000 each at weekly interval starting 30 DAT) proved to be effective in the management of rice leaf folder and stem borer in both the rice varieties.

Mishra and Kumar (2009) evaluated the bio-efficacy of *T. japonicum* and *T. chilonis* against stem borer and leaf folder of rice at farmer's field. Different doses of the egg parasitoids, *T. japonicum* and *T. chilonis* were evaluated and the results revealed the reduction in the tiller damage caused by stem borer and folded leaves by rice leaf folder varied from 50.1 to 63.5% and from 60.8 to 70.5%, respectively, where the egg parasitoid were released at 50,000/ha. Similarly, reduction in the tiller damage and folded leaves varied from 68.2 to 78.3% and from 70.3 to 78.8%, respectively, where egg parasitoids were released at 75,000/ha during the season, when the third dose of both bio-agent applied at 100,000/ha during the crop season, the reduction in damage tiller and folded leaves varied from 78.1 to 82.1% and from 76.6 to 84.6%, respectively. All the doses used in the inundative releases were found effective over the control but the releases at 100,000/ha was found to be superior over the lower dose.

Chakraborty (2012) evaluated the efficacy of the integrated pest management (IPM) system against the yellow stem borer on rice (cv. Swarna Mashuri) in West Bengal, India. The IPM module which was having release of *Trichogramma* resulted in lower dead heart and white head incidence, higher populations of spiders and beetles, higher grain and fodder yields, and higher returns and cost benefit ratios.

Singh et al. (2013) studied on performance of bioagent *T. japonicum* against stem borer, *Scirpophaga incertulas* in paddy. It was observed that plots where *T. japonicum* were released had lower incidence of stem borer during the years under study. On an average, IPM exhibited lower dead hearts and white ears (3.9 and 4.8%) as compared to 5.0 and 6.6%, respectively, in non-IPM. An increase of 3.9% in yield was also recorded in IPM treatments where *Trichogramma* was included.

Pandey and Choubey (2012) studied on management of yellow stem borer, *S. incertulas* in "rice". They found that the release of bioagent, *T. japonicum* 50,000 eggs/ha was effective in reducing the incidence of *S. incertulas* which was observed to be superior over some of the treatments.

Mohapatra and Nayak (2015) conducted field experiments at the experimental farm of Regional Research and Technology Transfer Station, Bhawanipatna, Orissa University of Agriculture and Technology during *kharif* 2011–2012, 2012–2013 and 2013–2014 to assess the performance of IPM module against major insect pests of rice (var. Swarna) in comparison to the recommended insecticidal control schedule (Non-IPM module). Results revealed that inclusion of release of *T. japonicum* @1.0 lakhs/ha at 30 days after transplanting (DAT), 40DAT and 50DAT in the IPM module gave excellent result in managing yellow stem borer.

In a field experiment conducted by Shirke and Bade (1997) in Maharashtra, India, with rice cv. Ambemohar revealed that the parasitoid *T. japonicum* (released at the rate of 50 000 adults/ha) was far more effective in controlling *S. incertulas* than endosulfan (50 vs. 28%).

The efficacy of *T. chilonis* and/or *T. japonicum* in controlling stem borer (*S. incertulas*) and leaf folder (*Cnaphalocrocis medinalis*) infesting rice was determined in a field experiment conducted by Bade et al. (2006) in Maharashtra, India during the kharif season of 1998–2000. Six releases of *T. japonicum* at 50,000 adults/ha each at weekly intervals starting 25 DAT resulted in the highest control of stem borer, seed yield (29.90 q/ha) and highest cost: benefit ratio (1:12.42). Six releases of *T. chilonis* at 50,000 adults each resulted in the highest control of leaf folder (2.57%), which was at par with three alternate spraying of 7% endosulfan and 0.04% monocrotophos.

9.5.2 MAIZE

Field trials carried out by Rawat et al. (1994) in Himachal Pradesh, India, in 1992–1993 showed that inundative releases of *T. chilonis* were effective against *Chilo partellus* on maize.

Jalali and Singh (2003) while studying the efficacy of *T. chilonis* and *Cotesia flavipes*, released at different time intervals, in controlling *C. partellus* infesting fodder maize found that egg parasitism by *T. chilonis* ranged from 41.9 to 42.8% after the first release and it rose to 75.2–76.8% after the third release in plots where parasitoids were released at a three-day interval. In plots where parasitoids were released in a five-day interval, parasitism rates increased from 24.8–27.4% to 62.3–62.6%. Crop yield was higher when *T. chilonis* was released at three-day intervals compared to five-day intervals. *C. flavipes* parasitized up to 51.6–53.2% and 65.8–68.4% of the first and second generation of *C. partellus*, respectively. *T. chilonis* was more efficient in controlling the pest compared to *C. flavipes*.

Chhata et al. (2014) tested several modules for organic pest and disease management in maize. Of the six organic modules/treatments evaluated under sole maize and maize-black gram intercropping systems, the module comprising deep summer ploughing+neem cake @ 2 q ha^{-1} + seed treatment with *T. harzianum* at 6 g kg^{-1} + spray of neem seed kernel extract (NSKE) @ 5% at 3 and 30–35 DAG + release of *T. chilonis* at 7–14 and 50 DAG + spray of *B. thuringiensis* formulation at 10 and 60 DAG + burying of older leaves during earthing + removal of dead hearts formed/disease infected plants was found most effective, economical in managing the diseases and insect pests both under sole maize and maize-black gram intercropping systems.

In study, conducted by Chaudhary et al. (2012) the IPM strategy for maize was developed and validated in the four blocks of Hoshiarpur district of Punjab. *C. partellus* was recorded as the major pest of *kharif* maize causing considerable economic loss. Along with other components of IPM, *T. chilonis*, an egg parasitoid of *C. partellus* was released to decimate its population in the egg stage. The number of infested plants by *C. partellus* in IPM field did not exceed 4% whereas in farmer's field the number of infested plants reached up to 12%.

Studies were conducted by Kanta et al. (2008) in Punkab, India, for the control of maize stem borer, *C. partellus* with *T. chilonis*. Release of *T. chilonis* with or without release of *C. partellus* eggs, was evaluated along with the recommended insecticide (deltamethrin at 100 mL/ha) and untreated control. Single release of *T. chilonis* in the form of parasitized eggs of *C. cephalonica* at 100,000/ha on 13-day-old crop proved very effective for the control of the maize stem borer (65% egg parasitization).

9.5.3 SUGARCANE

Rajendran and Hanifa (1998) studied on different techniques for the release of *T. chilonis* against sugarcane internode borer, *Chilo sacchariphagus indicus* (Kapur). The release of the parasitoid through closed cup methods (ice-cream cup and plastic disposal cup) and loose egg exposure procedure recorded higher emergence rates (70.8, 70.3 and 74.1%) than the conventional practice of card tying (61.0%) or card stapling (60.7%) to a sugarcane plant.

Pandey et al. (1997) studied integrated control of *Scirpophaga excerptalis* on sugarcane at Seorahi, Uttar Pradesh, India. Control methods included egg mass collection during oviposition by the 1st and 2nd generations in March and May, and treatment with carbofuran at 30 kg a.i./ha and release of *T. chilonis* and *T. japonicum* at 50,000 adults/ha per week during the 3rd generation. They found promising result of releasing of the *Trichogramma* spp. against the sugarcane borer.

Shenhmar (1996) evaluated egg parasitoid *T. chilonis* for the control of the sugarcane pest *Chilo auricilius* in farmers' fields in Punjab, India. The parasitoid, when released at a rate of 50,000 parasitized eggs/ha during July–October at 10-day intervals, proved to be highly effective. The mean level of damage by *C. auricilius* was 12.8% in release fields as compared to 37.7% in the control.

Sharma et al. (1997) concluded from a field study that biological control using the egg parasitoid *T. chilonis* and cultural control were equally effective at reducing sugarcane shoot borer, *Chilo infuscatellus* and increasing sugarcane productivity.

Mann and Doomra (1996) conducted a trial of the biological control of *C. auricilius* on sugarcane using *T. chilonis*. Parasitoids were released at a rate of about 50,000/ha at 10–15 day intervals from July to December. Sugarcane was less damaged in areas into which the parasitoid was released. In further experiments, the effectiveness of releases of *T. japonicum*, with applications of carbofuran and some mechanical control, was also investigated. It was concluded that *T. chilonis* may be a useful tool in the control of *C. auricilius*.

9.5.4 OKRA

Yadav et al. (2009) found *T. chilonis* as the most important natural enemy of *E. vittella*. The percentage of parasitism of *T. chilonis* was from 10 to

12%. The data revealed a high positive relationship between the pest and parasitoids indicating an important role in suppressing pest population.

Sardana et al. (2008) conducted field experiment in Raispur and Harsawan (Uttar Pradesh, India) on okra, to study the validation and economic viability of adaptable IPM technology in a farmers' participatory approach. The IPM technology for okra crop comprising growing of yellow vein mosaic virus (YVMV)-resistant hybrids, namely Tulsi, Sun 40 and Makhmali, 2–3 sprays of NSKE at 5%, five releases of egg parasitoid *T. chilonis* based on pheromone monitoring, erection of delta traps at 5/ha and bird perches at 10/ha and periodical rouging of borer affected shoots and fruits and disease infected plants. The study revealed the release of *Trichogramma* to be very effective in reducing the incidence of fruit borer of okra.

Yadav et al. (2008) in an experiment conducted in Kanpur, Uttar Pradesh, India evaluated the efficacy of botanical and biological control treatments against pests of okra (cv. Azad-1). Neem (*Azadirachta indica*) formulation with azadirachtin-endosulfan applied at 15-day intervals reduced the jassid (*Empoasca* spp.) population to 0.68 per 5 plants, whereas the application of *Bacillus thuringiensis* (*Bt*)-neem (*Azadirachta indica*) formulation with azadirachtin-endosulfan-*Trichogramma* applied at 15-day intervals reduced fruit and shoot borer (*Earias vittella*) infestation to 1.93% and gave the highest yield (79.70 quintal/ha). All the treatments significantly reduced the jassid population, fruit, and shoot borer infestation (1.0 quintal = 100.0 kg).

9.5.5 TOMATO

Vijayalakshmi (2007) carried out field experiment to evaluate *T. pretiosum* alone and in combination with other biological control agents against tomato fruit borer, *Helicoverpa armigera* (Ha). The treatment of biointensive pest management module (BIPM) comprised five releases of *T. pretiosum* at 50,000 adults/ha from flower initiation period+2 sprays of *H. armigera* nuclear polyhedrosis virus (HaNPV) at 1.5×10^{12} POBs/ha+3 sprays of *B. thuringiensis* (Bt) at 1.0 kg/ha+*Rhynochoris marginatus* at 5000 adults/ha was found to be superior in reducing the larval population and in yield as compared to the other treatments.

Sushil et al. (2006) evaluated safer management tools against major insect pests of tomato and garden pea. In tomato, 4 releases of *T. chilonis* at 50,000 insects per ha per release at an interval of 10 days from flowering initiation stage against fruit borer, *H. armigera* was found to be the most promising to manage the fruit borer of tomato.

Karabhantanal et al. (2005) carried out an investigation to evaluate different IPM modules against the tomato fruit borer, *H. armigera*. The results revealed that the IPM module consisting of trap crop (15 rows of tomato: 1 row marigold) + *T. pretiosum* (45,000/ha) – NSKE (5%) – HaNPV (250 LE/ha) – endosulfan 35 EC (1250 ml/ha) was significantly superior over the rest of the modules tested in restricting the larval population (100% after the fourth spray).

Kumar et al. (2004) studied the efficacy of *T. chilonis*, *T. pretiosum* and *Trichogramma brasiliense* at 50,000, 75,000 and 100,000/ha in controlling *H. armigera* infesting tomato cv. Punjab Chhuhara. Significant difference was observed in the larval population and the lowest mean larval population (0.5 larvae/5 plants) was recorded when *T. chilonis* was released at one lakh/ha. Parasitism (41.07%) was highest in *T. chilonis* was released at one lakh/ha, and it was at par with releasing 75,000/ha of *T. chilonis* (40.00%). The lowest fruit damage (8.01%) was recorded when *T. chilonis* was released at one lakh/ha, which was followed by the release of *T. chilonis* at 75,000/ha (9.20%), one lakh/ha of *T. brasiliense* (11.66%), and *T. pretiosum* at 100,000 and 75,000/ha (10.88 and 11.82%, respectively). The highest (261.07 q/ha) yield was obtained with the release of *T. chilonis* at one lakh/ha, followed by *T. chilonis* at 75,000/ha (248.27 q/ha).

Brar et al. (2003) determined the efficacy of *T. pretiosum* (five releases weekly at 50,000 per ha), *H. armigera* nuclear polyhedrosis virus (HaNPV; 2, 3 or 5 sprays at 7-, 10- or 15-day intervals at 1.5×10^{12} polyhedral occlusion bodies per ha) and/or endosulfan (3 sprays at 15-day intervals at 700 g/ha) for the management of tomato fruit borer (*H. armigera*) in Punjab, India. They found that the egg parasitism was very high (36.32–61.00%) in plots where *T. pretiosum* was released. The mean egg parasitism was highest in the plot treated with *T. pretiosum* alone (49.33%). The mean egg parasitism was 7.45 and 14.85% in the endosulfan-treated, and control plots, respectively. Fruit damage was highest during 1999 and 2000. Among all the treatments, treatment with *T. pretiosum* + HaNPV + endosulfan resulted in the lowest fruit damage (13.07%) and the highest mean yield (243.86 q/ha). The control treatment had the highest borer incidence and fruit damage, and the lowest yield (163.31 q/ha) among all treatments.

The yield in endosulfan alone was 209.31 q/ha, which was significantly superior to three HaNPV sprays (184.15 q/ha). It is concluded that the treatment combination *T. pretiosum* + HaNPV + endosulfan was the most effective for *H. armigera* control.

9.5.6 BRINJAL

Sardana et al. (2008) analyzed the IPM technology for brinjal. It comprised raising healthy nursery using soil solarization and mixing of *Trichoderma* along with FYM; application of neem cake @ 250 kg ha^{-1} at 30 DAT, erection of bird perches @ 10 ha^{-1}, installation of delta traps @ 5 ha^{-1} periodical collection and destruction of borer affected shoots and fruits and diseased plants, two sprays of NSKE @ 5%, five releases of egg parasite *Trichogramma brasiliensis* and 1–2 sprays of chemical pesticides. They found this module containing *Trichogramma* as very effective in reducing the incidence of pests and minimizing the yield losses.

Mani et al. (2005) reviewed that augmentative release of *Trichogramma* spp. to be useful in controlling aubergine fruit borer *Leucinodes orbonalis*.

Yadav and Sharma (2005) conducted a field experiment in Bikaner, Rajasthan, India, to assess the efficacy of bioagents and neem products in relation to malathion against aubergine shoot and fruit borer, *L. orbonalis*, on aubergine. The bioagents and neem products were not superior to the malathion 50 EC (0.05%), however, *B. thuringiensis* subsp. *kurstaki* (Dipel 8L at 2.5 mL/L water) provided sufficient control of the pest. Nimbecidine, NSKE, neem seed kernel solution, and *T. chilonis* were statistically superior over control, but were less effective than malathion and *B. thuringiensis* subsp. *kurstaki*, in suppressing the infestation.

9.5.7 COTTON

A study was conducted by Rahman et al. (2003) in Andhra Pradesh, India to evaluate various biological and non-insecticidal components under biointensive insect pest management (BIPM) module used in cotton (cv. NA 1588) production. BIPM module comprised: hand picking pests and putting them in screen cages; sowing maize as intercrop 10 days after main crop sowing; single release of 14000 *Chrysoperla carnea* larvae ha^{-1}, synchronized with bollworm (*H. armigera*) occurrence; eight releases of 150,000 *T. chilonis* ha^{-1} week^{-1}, synchronized with bollworm egg appearance; need-based application of *H. armigera* nuclear polyhedrosis virus at 500 larval equivalents ha^{-1}; and insecticides for sucking pests. In farmers' practice (FP) module, need-based application of insecticides (including monocrotophos, chlorpyrifos, endosulfan, quinalphos, triazophos, and acephate) was performed. They observed promising result of using *T. chilonis* against American boll worm.

Butter and Kular (2002) studied the efficacy of various strategies for controlling cotton bollworms (*Pectinophora gossypiella, H. armigera* and *Earias* spp.) in Punjab, India using cotton cv. LH 886. The treatments comprised IPM, alternate sprays of gossyplure and organic pesticides (GSP) and seven recommended schedules of pesticide spraying (RSS). The IPM comprised spraying of a sex pheromone (gossyplure) and organic pesticides and release of the egg parasitoid *T. chilonis*. They found that the mean incidence of bollworm infestation was lowest with GSP (32.4%), whereas crop yield was highest with RSS (1701 kg/ha), although differences in the values of the parameters measured due to the treatments were not significant.

Brar et al. (2002) found that when *T. chilonis* (Bathinda strain) was integrated with insecticides, it reduced damage by cotton bollworms (*Earias* spp., *H. armigera*, and *P. gossypiella*) by 70.3% and increased yields by 44.5% over insecticides alone.

Bharpoda et al. (2000) after a study stated that an IPM module comprising intercropping, use of pheromone traps releases of *C. carnea* and *T. chilonis* and sprays of various microbial and synthetic insecticides proved significantly effective in managing the populations of Lepidoptera and Homoptera on cotton H-6 and giving higher seed cotton yield in comparison to the cotton crop protected with the recommended insecticide schedule and left unprotected.

Sakhare and Kadam (1999) stated that need based staggered applications (treating alternate row strips alternately) with deltamethrin 0.005% + *B. thuringiensis* at 0.003%, endosulfan 0.05% + HaNPV 0.025% L.E. × *T. chilonis* at 1 lakh/ha applied on single or double row strips, and fenvalerate at 0.0075% + HaNPV 0.025% L.E. monocrotophos 0.035% + azadirachtin 0.00015% × *T. chilonis* at 1 lakh/ha applied on single row strips, were the most effective treatments in managing bollworm pests of H$_4$ cotton.

9.6 USE OF TRICHOGRAMMA AGAINST ASIAN CORN BORER IN CHINA

At Xifeng and Xiuyan of Liaoning province in China, where *Trichogramma* releases for control of the first generation of Asian corn borer continued for 17 years in the 1970s and 1980s, the natural parasitization of *Trichogramma* on first generation Asian corn borer egg masses increased year by year and reached 70%. The parasitism rates of parasitoids attacking larval and pupal stages of Asian corn borer also increased year by year, and the combined effect of increased numbers of *Trichogramma* and other parasitoids decreased the damage rates of corn and stabilized it at about 10% since 1980. The population of Asian corn borer in these regions maintained at a low dynamic equilibrium level for a long period (Zhang et al., 1990b). In Miyun of Beijing, where *Trichogramma* had been released for more than 20 years since 1977, the populations of natural enemies in cornfields have increased. This helps keeping other insect pests under control without application of pesticides (Shi, 1996). Parasitism of first generation Asian corn borer egg masses by naturally occurring *Trichogramma* in non-release corn fields increased from 1% in 1980 to 33% in 1991 (Shi, 1996). In Xifeng county, Liaoning province, where *Trichogramma* was released continuously on a large scale for over 30 years, the number of overwintering larvae has been reduced to 5.6 larvae/per hundred stalks with a corresponding yield of 7500 kg/ha, compared to 193.6 larvae/hundred stalks and 5250 kg/ha in surrounding counties where an outbreak of the Asian corn borer was observed in 1997 (Cao & Sun, 2002). In Wafangdian city, Liaoning province, *T. dendrolimi* were mass released on a large area of corn during the first generation of Asian corn borer for three years. The results showed a mean parasitism rate of *T. dentrolimi* of 75.1% and a damage reduction of 62.3% for first generation of Asian corn borer. Without further releases, parasitism rate of second generation of Asian corn borer was 48.3% with the number of holes in the stem and the number of stalks broken decreasing by 44.6 and 51.9%, respectively. The average number of larvae, assessed at harvest decreased by 46.2%. This indicated that first as well as second generation of Asian corn borer could be controlled by only one *Trichogramma* release against first pest generation (Yang et al., 2011). In Gongzhuling City, Jilin province, mean number of holes, the number of larvae per hundred stalks and the percentage of damaged plants decreased between 63 and 73% in areas where the *Trichogramma* have been released from 1990 to 2008. Furthermore, average natural parasitism of corn borer eggs by *Trichogramma* increased from 14.8% in 1989 to 30.6% in 2007 for consecutive 19 years of *Trichogramma* releases in Gongzhuling, Jilin province (Yu et al., 2009). The release of *Trichogramma* for control of the Asian corn borer has become one of the key techniques of IPM of corn pests in China (Yang et al., 2011). It has been commonly adopted by the farmers in Northeastern provinces in China because of its easy use and good control efficacy. No insecticides are applied in fields where *Trichogramma* is applied. *Trichogramma* releases for control Asian corn borer comprise 1–1.3 million ha each year. With the Chinese government paying more attention to grain production and environmental protection, *Trichogramma* based biological control has been expanded to the Huang-Huai-Hai summer corn region and the Northwestern corn region in recent years. Jilin, Liaoning, and Heilongjiang provincial governments have provided some subsidies for controlling the Asian corn borer with *Trichogramma* since 2002. The total release area of *Trichogramma* for control of the Asian corn borer between the years 2009 and 2011 was 3.23 million ha in Jilin province (Wang et al., 2014).

9.7   MASS PRODUCTION OF *TRICHOGRAMMA* SPP.

9.7.1   HISTORY OF PRODUCTION OF TRICHOGRAMMA

According to DeBach (1964), the maximum number of individuals of any genus cultured so far by man belongs to *Trichogramma*. The practice of large scale production and utilization of *Trichogramma*, described by Enock (1895) as *Trichogramma* farming, became a practical method for control of sugarcane borers in Lousiana and Barbados since late 1920s following the discovery (Flanders, 1930) that the eggs of Angoumois grain moth, *Sitotroga cerealella* (Oliv.) can serve as an excellent factitious host and the moth could be mass cultured rather easily. Flanders (1930) described the first known mass production system for *Trichogramma* using eggs of *S. Cerealella*. In India, *Trichogramma* was multiplied on eggs of rice moth, *C. cephalonica* Staint., which is equally amenable to laboratory production. The steps involved in the production of host insect and the parasitoid were standardized and described (Kunhi Kannan, 1931; Subramaniam, 1937; Subramaniam & Seshagiri Rao, 1940). Subsequently several modifications and improvements were made leading to less laborious and more economical production of *Corcyra* and *Trichogramma* (Parshad, 1975). The host larvae hatching from the eggs that have escaped parasitism tend to be predaceous; this problem was resolved with the discovery of exposure off eggs to ultra-violate light before parasitization kills the embryo without affecting the nutritive value (Brenière, 1965a, 1965b). In the USA the Rincon Vitova Insectary at Ventura, California, has developed excellent facilities for massive production of *Trichogramma* on *Sitotroga*. In India, commercial production of *Trichogramma* has been undertaken since 1981 in Biocontrol Research Laboratories at Bangalore.

The ergs of S. *cerealella* is the traditionally used host for the production of *Trichogramma* in the USA, USSR, France, Germany, and so forth while in India and China, *C. cephalonica* is preferred. The eggs of eri silk worm and the oak silk worm are also utilized in China. In an egg of *Sitotroga* or *Corcyra* only one and rarely two individuals of *Trichogramma* can develop, while in an egg of silkworm up to 60 parasitoids can complete their development. The latest progress with regard to *Trichogramma* was production in the development of synthetic diet for the parasite. The progress will be described later in a separate subtopic.

9.7.2 TECHNIQUES OF MASS PRODUCTION OF TRICHOGRAMMA SPP.

A comprehensive review was carried out by Wang et al. (2014) on "Mass rearing and release of *Trichogramma* for biological control of insect pests of corn in China" where they lucidly described the mass rearing procedure of *Trichogramma* spp. *C. cephalonica*, an insect pest of stored grain, is commonly used as host for *Trichogramma* mass rearing and also for maintaining the *Trichogramma* species in laboratory, especially those species that cannot be reared on large eggs of *Antheraea pernyi*. For *Trichogramma* mass production, it is very important to produce high quality host eggs in huge numbers at competitive costs. Mass rearing techniques for *C. cephalonica* have been improved and a set of machines or devices for *C. cephalonica* production have been designed in Shanxi province in the late 1970s (Chen et al., 2000). The production line for *C. cephalonica* includes medium blender, moth collector, egg collector, egg cleaner, egg sterilizer with UV light, egg card machine, and a *Trichogramma* mass rearing facility (Shi et al., 1982). Zhang et al. (1991) compared the effects different media on egg production by *C. cephalonica*. Among the media used including rice bran, wheat bran, rice bran/wheat bran with corn flour (9:1), the rice bran is the best for mass rearing of *C. cephalonica*, however Bajra/Sorghum/Maize grains along with Groundnut powder are also used in India (Vijaykumar, 2009). At the rearing density of 4000 eggs/kg of rice bran infested, a harvest of about 2254 adults and 378,738 eggs/kg can be attained, with a multiplication rate of 94.7 times in one generation. In north China, wheat bran is much easier to obtain than rice bran, but the survival rate and the reproduction of *C. cephalonica* reared on wheat bran is much lower than that on rice bran. When 90% wheat bran mixed with 10% soybean flour or corn flour is used as the medium for *C. cephalonica* rearing, the emergence of moth was 134% and 78% higher, and the number of eggs produced was 231% and 146% higher, respectively (Zhou et al., 1988). In order to meet the mass rearing of *Trichogramma* with the release dates, the parasitized *C. cephalonica* eggs are usually stored in a cold room for a certain period. However, emergence rates from the parasitized eggs, the sex ratio and fecundity are affected by storage duration. To keep good quality levels in *Trichogramma* rearing, the storage period for parasitized *C. cephalonica* eggs at 4 °C should not exceed 15 days (Zhang et al., 2008). However, emergence rate and fecundity of the parasitoids are affected by the developmental stage of the parasite, the incubation temperature, and the period of the storage. When the freshly parasitized eggs are incubated for 3–6 days at alternating temperatures of 8 and 25 °C for 16 and 8 h, parasitized eggs at the late egg stage can be successfully stored at 8 °C for a longer period without compromising emergence rate and fecundity. Gou (1985) showed that parasitized *C. cephalonica* eggs treated with this method showed an average emergence rate of 87.5% while emerged adults parasitized 55.8 host eggs per female wasp when the host eggs were stored at 8 °C for 90 days. The rearing techniques for *C. cephalonica* have been studied in Taiwan, especially the thickness of diets and the rearing density (Cheng & Hung, 1996, 1997). The development of mass rearing and inundative release techniques of *Trichogramma ostriniae* for controlling of the Asian corn borer in Taiwan began in 1984 (Yu et al., 1992). Tsneg (1990) developed the egg card machine to decrease labor hours and the costs in the mass production of *T. ostriniae* by using *C. cephalonica* eggs. The egg card machine is 10 times faster in producing egg cards than human laborers and also produces a more standardized number of eggs on the paper. Egg cards irradiated by UV light for 1 h and treated with the preservative agent "Fuyolin", then stored at 4 °C was the most effective preservation. The parasitism rate on the stored egg cards stored at 4 °C for 15 days was 74.3%, not significantly different from that of egg cards not stored at low temperatures. The preservation of egg card could save the surplus of alternate host *C. cephalonica* eggs from being discarded each day and could also help adjusting the differences in time between producing and releasing egg cards in the field (Wang et al., 2014).

For obtaining the highest egg yield in a shorter period of time 0.25 mL of *C. cephalonica* eggs is recommended to inoculate 1 kg of crushed brown rice (Cheng & Hung, 1996). Quality control process for *T. ostriniae* rearing with *C. cephalonica* eggs has been set up, based on the following steps: good *C. cephalonica* egg cleaning, selection of *C. cephalonica* eggs, superior breeding wasp culturing, environment of mass rearing management, parasitized eggs collection and screening and inspection. The criteria for high quality of *T. ostriniae* adults reared on *C. cephalonica* eggs include a survival (>80% of adults should survive two days without access to water and food at 25 °C, RH 80–90%), the ability to parasitize 50 or more *C. cephalonica* eggs per female and a 3:1 sex ratio (females: males). Survival is based on 60 female wasps while fecundity is based on 10 mated females (Chen et al., 2000).

Schematic representation of mass rearing procedure of *Trichogramma* spp through mass production of *Corcyra cephalonica*as laboratory host has been given by Vijaykumar (2009) which is as following:

i.   Take 2.5 kg cleaned insecticide free, fresh bajra/sorghum/maize/ paddy grains, sterilized at 100 °C for 30 min and put into *Corcyra* rearing cages/plastic trays (one set used for 100 days)

ii. Incorporate powdered groundnut kernel weighing 100 g as a source of protein

iii. Adding yeast tablets (10 nos or 5 g) as nutritional supplement + 0.05 g streptomycin sulfate as antibiotic. Supplementing 0.01 sulfur as Acaricide

iv. Mixing one cc fresh viable eggs of *Corcyra* in each cage/tray. Maintenance of culture at 26 °C

v.   After 40–45 days, emergence of adult moth commences

vi. Regular collection of adult moths manually and mechanically using vacuum cleaner

vii. Allow adult moths in fecundity cage used for three days supplementing with food swab impregnating honey 15%, water, and vitamin E (2 cap)

viii. Daily collection of *Corcyra* eggs

ix. Pass the eggs through 15, 30, and 40 mesh sieves and run over a slope of paper to eliminate dust particles or separation of scales admixture in eggs mechanically by motorized egg separator

x. Harvest fresh eggs

xi. Storage of eggs at 10 °C up to seven days

xii. Freshly collected *Corcyra* eggs are subjected to UV treatment using 30 watt UV tube light for 45 min

xiii. Pasting UV treated eggs on to the "*Tricho*" cards measuring size 15 × 10 cm comprising six pieces (12 × 3 cm each)

xiv. Introduction of single egg card inside polythene balloon/parasitization chamber along with nucleus culture of *Trichogramma* strain maintaining in ratio of one female to 30 eggs for effective parasitization

xv. Parasitization takes place within a week time

xvi. About 12,000 *Trichogramma* adults emerge out from single card

9.7.3 *IN VITRO REARING OF TRICHOGRAMMA SPP. ON ARTIFICIAL HOST EGGS*

Wang et al. (2014) while reviewing mass rearing and release of *Trichogramma* for biological control of insect pests of corn in China, reviewed the *in vitro* rearing of *Trichogramma* on artificial host eggs. According to them mechanized production of *T. dendrolimi* and *T. chilonis* with artificial host eggs has been successful. A prototype for producing artificial host eggs was manufactured which automatically completes all five processes, including setting-up synthetic membrane, forming the "eggshells", injecting the artificial media into the shells, sealing the double-layered membrane, and separating into egg cards (Liu et al., 1988). Based on this prototype designed by Liu et al. (1988), the model GD-5 automatic machine for mass production of artificial host egg-cards was successfully assembled in 1995, and the *Trichogramma* mass rearing technology based on artificial host egg-cards developed. Operating rules for mechanized production of artificial host eggs for *Trichogramma* and techniques for

propagating parasitoids, quality control, and releasing have been consti-
tuted (Dai et al., 1996; Liu et al., 1996). One GD-5 could produce 5000 egg
cards/8 h, each with 140 eggs. Each egg could produce 50–60 wasps with
good quality (Liu et al., 1998a, 1998b). Two artificial host egg production lines for *Trichogramma* were set up in Guangzhou and Beijing in the
late 1990s. The standard of quality control for mass production of *Trichogramma in vitro* has been put forward, which include the storage, backup
and rearing of the seed wasps, and quality monitoring of the artificial eggs
(Wang et al., 2005). The parasitoids from *in vitro* rearing have been used
for control of Asian corn borer and cotton bollworm over 150,000 ha in
the 1990s (Wang, 2001). A releasing container for *Trichogramma* reared
on artificial host eggs was also designed and the emergence rate was over
90% (Feng et al., 1999). China was the first country worldwide to make
use of *in vitro* rearing of *Trichogramma* for commercial production and
use for insect pest control on a large scale (Feng et al., 1997; Wang, 2001).
However, according to Wang et al. (2014) the artificial host eggs can only
be used for mass rearing of *T. dendrolimi* and *T. chilonis*, and also the costs
of artificial host egg production is costly.

9.7.4 QUALITY CONTROL GUIDELINES FOR *TRICHOGRAMMA SPP.*

According to Van Lenteren et al. (2003) three species of *Trichogramma*
namely, *T brassicae*, *T. cacoecie* and *T. dendrolimi* are having the guidelines
to maintain the quality. We are presenting here the guidelines for quality
control of *Trichogramma brassicae* Bezd. (=*T. maidis*) as mentioned by
Van Lenteren et al. (2003) initially designed by Bigler (1994) and coordinated by Hassan and Zhang (2001).

Guidelines for *Trichogramma brassicae* Bezd. (=*T. maidis*):
Test conditions

Temperature   : 23 ± 2 °C
RH   : 75 ± 10%
Light regime   : 16L:8D
Rearing hosts : *Ephestia kuehniella*
                 Sitotroga cerealella

Species identification	: The species is specified on the lavel and verified by the producer. Molecular techniques are available at laboratory of Entomology, Wageningen University, the Netherlands. Test necessary once a year, sample size minimum 30 individuals.
Quality control criteria Sex ratio	: ≥50% females; 100 adults assesses on ten release units each or 5 × 100 adults of bulk material; at least weekly or batch-wise test if batches were exposed to special treatments (e.g., storage)
Number of females	: As indicated on label; determined as for sex ratio.
Fecundity and longevity	: ≥ Offspring per seven days per female; 80% of females should live at least seven days; $n = 30$; monthly or batch-wise test.
Natural host parasitism	: ≥10 parasitized hosts per 4 h per female
Description of testing methods Fecundity and longevity	: 30 females (age 24 h) are confined individually in glass tubes; at least 200 factitious host eggs (<24 h) are glued with water on to a small cardboard strip; a small droplet of honey and water are added directly to the wall of the vial. Eggs of E. kuehniella (<24 h old) are ultraviolet (UV)-irradiated and provided on day 1 and removed after day 7; fresh eggs of S. cerealella are provided on day 1, 3 and 5. The number of living adults is recorded after day 7. Egg cards are incubated and the number of black eggs is counted not earlier than day10. Minimum fecundity after day 7 is 40 offspring per female; mortality after day 7 is <20%; at least monthly test or batch-wise if batches were exposed to special treatments (e.g., storage procedures and long-range shipments).
Natural host parasitism	: 30 females (age 24 h) are confined individually in tubes; two fresh egg masses of at least 20 eggs per egg masse of O. nubilalis (<24 h old) are added for 4 h; honey and water are provided as described above; after separation of egg masses from the females they are incubated for three days; the number of black eggs

is counted; the mean number of black eggs is ≥ 10 per female. The host-cluster acceptance rate (=females parasitizing at least one host egg) should be ≥ 80%. This measure is important because parasitism drops drastically if a high proportion of female does not accept their hosts. This test is and indirect measure of acceptance and suitability of natural-host egg. The test should be performed two to four times per year depending on the rearing system (number of generations reared on the factitious hosts).

9.8 CONCLUSION AND FUTURE PROSPECTS

In Indian perspective, adoption of natural biological control was a tradition before the chemical pesticides came into action. In the present day too it should get the prime importance. Regarding *Trichogramma*, number of mass rearing laboratories should be increased to get better impact by releasing those. For that purpose the government private agencies need to take strong initiatives. Understanding the quantitative nutritional profile of *Trichogramma* for the purpose of mass rearing, and to improve the survival and successful establishment of *Trichogramma* under field condition is required to be studied. More investigations should be done on the tritrophic interaction and the role of kairomones influencing the searching behavior of *Trichogramma* which can help in improving the effectiveness of inundative release. There is also need to develop strains which will be compatible with application of pesticides. The role of macroclimate on population buildup of the parasitoid has been well reviewed by many workers. However, little information is available on impact of microclimatic factors like air, current, evaporation, and so forth on the parasitoid population. Therefore, research should be needed in this line too.

KEYWORDS

- Trichogramma
- morphological
- information
- parasitism rate
- monitoring

REFERENCES

Almeida, R. P.; Stouthamer, R. Molecular Identification of *Trichogramma cacoeciae* Marchal (Hymenoptera: Trichogrammatidae): A New Record for Peru. *Neotrop. Entomol.* **2003,** *32*(2), 269–272.

Anonymous. 2015. http://www.nbair.res.in (accessed Dec 30, 2015).

Bade, B. A.; Pokharkar, D. S.; Ghorpade, S. A. Evaluation of Trichogrammatids for the Management of Stem Borer and Leaf Folder Infesting Paddy. *J. Maharashtra Agr. Uni.* **2006,** *31*(3), 308–310.

Bharpoda, T. M.; Patel, H. P.; Patel, U. P.; Patel, G. P.; Patel, J. J.; Patel, J. R. Integrated Pest Management (IPM) in Cotton H-6 Cultivated in Middle Gujarat. *Indian J. Entomol..* **2000,** *62*(4), 327–331.

Bigler, F. Quality Control in *Trichogramma* Production. In *Biological Control with Egg Parasitoids;* Wajnberg, E., Hassan, S. A., eds.; CAB International: Wallingford, UK, 1994; pp 93–111.

Borba, R. S.; Garcia, M. S.; Kovaleski, A.; Oliveira, A. C.; Zimmer, P. D.; Castelo Branco, J. S.; Malone, G. Dissimilaridade Genética de Linhagens de *Trichogramma* Westwood (Hymenoptera: Trichogrammatidae) Através de Marcadores Moleculares ISSR. *Neotrop. Entomol..* **2005,** *34*(4), 565–569.

Brar, K. S.; Sekhon, B. S.; Singh, J.; Shenhmar, M.; Joginder Singh. Biocontrol-Based Management of Cotton Bollworms in the Punjab. *J. Biol. Control.* **2002,** *16*(2), 121–124.

Brar, K. S.; Singh, J.; Shenhmar, M.; Kaur, S.; Sanehdeep Kaur.; Joshi, N.; Singh, I.; Tandon, P. L.; Ballal, C. R.; Jalali, S. K.; Rabindra, R. J. In *Integrated Management of Helicoverpa armigera (Hübner) on Tomato,* Proceedings of the Symposium of Biological Control of Lepidopteran Pests, July 17–18, 2002; pp 271–274, 2003.

Brenière, J. Les Trichogrammes, Parasites de *Proceras Sacchariphag~is* Boj. Borer de la Canne à Sucre à Madagascar. II. Etude Biologique de *Trichogramma australiciim* Gir. *Ibid,* **1965b,** *10,* 99–117.

Brenière, J. Les Trichogrammes, Parasites de *Proceras sacchariphagiis* Boj. Borer de la Canne à Sucre à Madagascar. IV. Etude Comparée de Plusieurs Espèces de Trichogrammes. *Ibid,* **1965d,** *10,* 273–294.

Brenière, J. Les Trichogrammes, Parasites de *Proceras Saccltariphagus* Boj. Borer de la Canne à Sucre à Madagascar. I. Ecologie de *Triclzograntmn australicitm* Gir., Parasite Autochtone. Effet du Renforcement de la Population Parasite. *Entomophaga.* **1965a**, *10,* 83–96.

Butter, N. S.; Kular, J. S. Integrated Pest Management Strategy against Bollworm Complex of Cotton. *Indian J. Entomol.* **2002**, *64*(4), 531–534.

Cao, J. L.; Sun, H. J. Control Effect Analysis for Asian Corn Borer by Releasing *Trichogramma dendrolimi. Rain. Fed. Crops.* **2002**, *22,* 116.

Chakraborty, K. Evaluation of Integrated Pest Management Module against Paddy Yellow Stem Borer *Scirpophaga incertulas* Walk. *Karnataka J. Agr. Sci.* **2012**, *25*(2), 273–275.

Chaudhary, N.; Saharawat, Y. S.; Pradyumn Kumar. IPM: A Technology to Conserve Biological Control Agents in Maize. *Indian J. Entomol.* **2012**, *74*(4), 348–351.

Chen, H. Y.; Wang, S. Y.; Chen, C. F. Quality Control of Trichogramma Ostriniae (Hymenoptera: Trichogrammatidae) Mass Reared with Rice Moth, *Corcyra cephalonica* Eggs. *Nat. Enemies Insect.* **2000**, *22,* 145–1150.

Cheng, W. Y.; Hung, T. H.. Rearing Density for *Corcyra cephalonica* Stainton. *Rep. Taiwan Sugar Res. Inst.* **1996**, *153,* 39–58.

Cheng, W. Y.; Hung, T. H. Experiments on the Supplemental Inoculation in the Rearing of *Corcyra cephalonica. Rep. Taiwan Sugar Res. Inst.* **1997**, *155,* 23–39.

Chhata, L. K.; Verma, J. R.; Pokhar Rawal. Integrated Disease and Pest Management through Organic Farming Approaches in Maize. *J. Mycol. Plant Pathol.* **2014**, *44*(3), 264–267.

Dai, K. J.; Ma, Z. J.; Pan, D. S.; Xu, K. J.; Cao, A. H. In *Standardisation on Mechanized Production of Artificial Host Eggs of Trichogramma and Techniques for Propagating Wasps,* Proceedings of the National Symposium on IPM in China, Beijing, China, pp 1138–1139; Zhang, Z. L., Piao, Y. F., Wu, J. W. Eds.; China Agricultural Scientech Press: Beijing, 1996.

DeBach, P. Ed . *Biological Control of Insect Pests and Weeds;* Cambridge University Press: Cambridge, 1964; p 844.

Enock, F. Remarks on *Trichogramma evanescens* Westw. Recorded by the Secretary of the South London Entomological and Natural History Society. *Entomol.* **1895**, *28,* 283.

Feng, J. G.; Tao, X.; Zhang, A. S.; Yu, Y.; Zhang, C. W.; Cui, Y. Y. Studies on Using *Trichogramma* sp. Reared on Artificial Host Egg to Control Corn Pests. *Chin. J. Biol. Control.* **1999**, *15,* 97–99.

Feng, J. G.; Zhang, A. S.; Zhang, C. W. Study on Using Artificial Eggs of *Trichogramma chilonis* to Control *Helicoverpa armigera. Chin. J. Biol. Control.* **1997**, *13,* 6–9.

Flanders, S. E. Mass Production of Egg Parasites of the Genus *trichogramma. Hilgardia.* **1930**, *4,* 465–501.

Flanders, S. E. Notes on the Life History and Anatomy of *Trichogramma. Ann. ent. Soc. Am.* **1937**, *30,* 304–308.

Gou, X. Q. Cold Storage Test of Rice Moth Eggs Parasitized by *Trichogramma ostriniae. Chin. J. Biol. Control.* **1985**, *1,* 20–21.

Hassan, S. A.; Wen Quin, Z. Variability in Quality of *Trichogramma brassicae* (Hymenoptera: Trichogrammatidae) from Commercial Suppliers in Germany. *Biol. Control.* **2001**, *22,* 115–121.

Jalali, S. K.; Singh, S. P. Biological Control of *Chilo partellus* (Swinhoe) in Fodder Maize by Inundative Releases of Parasitoids. *Indian J. Pl. Prot.* **2003**, *31*(2), 93–95.

Kanta, U.; Snehdeep, K.; Singh, D. P.; Brar, K. S. Bio-suppression of *Chilo partellus* with *Trichogramma chilonis* on *Kharif* Maize. *J. Insect Sci.* **2008,** *21*(1), 87–89.

Karabhantanal, S. S.; Awaknavar, J. S.; Patil, R. K.; Patil, B. V. Integrated Management of the Tomato Fruit Borer, *Helicoverpa armigera* Hubner. *Karnataka J. Agri. Sci.* **2005,** *18*(4), 977–981.

Kaur, R.; Brar, K. S.; Rabinder, K. Management of Leaf Folder and Stem Borer on Coarse and *Basmati* Rice with Organic and Inorganic Practices. *J. Biol. Control.* **2008,** *22*(1), 137–141.

Krishnamurti, B. S. C. A Microscopical Study of the Development of *Trichogramma minutum* Riley (the Egg Parasite of the Sugar Cane Borers in Mysore). *Proc. Indian Acad. Sci.* **1938,** *7,* 3645.

Kumar, P.; Shenhmar, M.; Brar, K. S. Field Evaluation of Trichogrammatids for the Control of *Helicoverpa armigera* (Hübner) on Tomato. *J. Biol. Control.* **2004,** *18*(1), 5–50.

Kunhi Kannan, K. The Mass Rearing of the Egg Parasites of the Sugarcane Moth Borer in Mysore (Preliminary Experiments). *J. Mysore Agric. Expt. Union.* **1931,** *12,* 1–5.

Liu, Z. C.; Liu, J. F.; Yang, W. H.; Li, D. S.; Wang, C. X. Research on Technological Process of *Trichogramma* Produced with Artificial Host Egg and Quality Standards. *Nat. Enemies Insect.* **1996,** *18,* 23–25.

Liu, J. F.; Liu, Z. C.; Feng, X. X.; Li, D. S., Present Status of Mass-rearing *Trichogramma* with Artificial Host Eggs to Control Insect Pests. *Chin. J. Biol. Control.* **1998a,** *14,* 139–140.

Liu, S. S.; Zhangh, G. M.; Zhang, F. Factors Influencing Parasitism of *Trichogramma Dendrolini* on Eggs of Asian Corn Borer. *Biocontrol.* **1998b,** *43,* 273–287.

Liu, Z. C.; Sun, Y. R.; Yang, W. H.; Liu, J. F.; Wang, C. X. Some New Improvements in Mass Rearing of Egg Parasitoids on Artificial Host Eggs. *Chin. J. Biol. Control.* **1988,** *4,* 145–148.

Mani, M.; Krishnamoorthy, A.; Gopalakrishnan, C. Biological Control of Lepidopterous Pests of Horticultural Crops in India – A Review. *Agric. Rev.* **2005,** *26*(1), 39–49.

Mann, A. P. S.; Subhash Doomra. Biological Control and Management of Sugarcane Borers in Punjab. *Coop. Sugar.* **1996,** *28*(4), 288–292.

Metcalfe, J. R.; Breniere, J. Egg Parasites *(Trichogramma* spp.) for Control of Sugar Cane Moth Borers. In: *Pests of Sugarcane;* Williams, J. R., Metcalfe, J. R., Mongomery, R. W., Mathes, R. Eds.; Elsevier Publishing Company: Amsterdam, 1969; pp 81–115.

Mishra, D. N.; Kumar, K. Field Efficacy of Bio-agent *Trichogramma* spp. Against Stem Borer and Leaf Folder in Rice Crop Under Mid-western Plain Zone of UP. *Environ. Ecol.* **2009,** *27*(4A), 1885–1887.

Mohapatra, L. N.; Nayak, S. K. Evaluation of IPM Module against Insect Pest Complex of Rice. *Indian J. Entomol.* **2015,** *77*(1), 35–38.

Nayaranan, E. S; Mookherjee, P. B. Effect of Nutrition on the Longevity and Rate of Reproduction in *Trichogramma evanescens minutum* Ril. *Indian J. Ent.* **1955,** *17,* 376–382.

Nordlund, D. A.; Lewis, W. J.; Gross, Jr. H. R.; Beevers, M. Kairomones and Their Use for Management of Entomophagus Insects. XII. The Stimulatory Effects of Host Eggs and the Importance of Host-egg Density to the Effective Use of Kairomones for *Trichogramma pretiosum* Riley. *J. Chem. Ecol.* **1981,** *7,* 909–1017.

Olkowski, W.; Zhang, A. *Trichogramma*: a Modern Day Frontier in Biological Control. *IPM Pract.* **1990,** *12,* 1–15.

Pandey, K. P.; Singh, R. G; Singh, S. B. Integrated Control of Top Borer *Scirpophaga excerptalis* Wlk in Eastern U.P. *Indian Sugar.* **1997,** *47*(7)*,* 491–493.

Pandey, S.; Choubey, M. N. Management of Yellow Stem Borer, *Scirpophaga incertulas* in Rice. *Agric. Sci. Dig.* **2012,** *32*(1), 7–12.

Parshad, B. A New Method of Maintenance of Culture of *Corcyra cephalonica* with Least Manual Labor. *Indian J. Entomol.* **1975,** *37*(43)*,* 303–306.

Pinto, J. D.; Stouthamer, R. Systematics of the Trichogrammatidae with Emphasis on *Trichogramma,* pp 1–36. In *Biological Control with Egg Parasitoids;* Wajnberg, E., Hassan, S. A., Eds.; CAB International: IOBC, Wallingford, 1994, p 304.

Quednau, W. Ober die Identität der *Trichogramma*-Arten und Einiger Ihrer Okotypen (Hymenoptera, Chalcidoidea, Trichogrammatidae). *Mitt. biol. BundAnst. Ld-u. Forstw.* **1960,** *100,* 11–50.

Rahman, S. J.; Rao, A. G.; Reddy, P. S.; Tandon, P. L.; Ballal, C. R.; Jalali, S. K. In *Potential and Economics of Biointensive Insect Pest Management (BIPM) Module in Cotton for Sustainable Production,* Proceedings of the Symposium of Biological Control of Lepidopteran Pests, Bangalore, India, July 17–18, 2002; pp 279–283, 2003.

Rajendran, B.; Hanifa, A. M. Efficacy of Different Techniques for the Release of *Trichogramma chilonis* Ishii, Parasitising Sugarcane Internode Borer, *Chilo sacchariphagus indicus* (Kapur). *J. Entomol. Res.* **1998,** *22*(4)*,* 355–359.

Rawat, U. S.; Pawar, A. D.; Joshi, V. Impact of Inundative Releases of *Trichogramma Chilonis* Ishii in the Control of Maize Stem Borer, *Chilo partellus* (Swinhoe) in Himachal Pradesh. *Plant Protect. Bull.* **1994,** *46*(2–3)*,* 28–30.

Ruberson, J. R.; Kring, T. J. Parasitism of Developing Eggs by *Trichogramma pretiosum* (Hymenoptera: Trichogrammatidae): Host Age Preference and Suitability. Biol. Control. **1993,** *3,* 39–46.

Sakhare, M. V.; Kadam, J. R. Efficacy of Eco-friendly Staggered Application Technique in Management of Bollworm Pests of H_4-Cotton. *J. Maharashtra Agric. Univ.* **1999,** *24*(2)*,* 170–174.

Samara, R.; Monje, J. C.; Reineke, A; Zebitz, C. P. W. Genetic Divergence of *Trichogramma aurosum* Sugonjaev and Sorokina (Hymenoptera: Trichogrammatidae) Individuals Based on ITS2 and AFLP Analysis. *J. Appl. Entomol.* **2008,** *132*(3)*,* 230–238.

Santos, N. R.; Almeida, R. P.; Padilha, I. Q. M.; Araújo, D. A.; Creão-Duarte, A. J. Molecular Identification of *Trichogramma* Species from Regions in Brazil Using the Sequencing of the ITS2 Region of Ribosomal DNA. *Braz. J. Biol.* **2015,** 75, 391–395.

Sardana, H. R.; Bambawale, O. M.; Singh, D. K.; Kadu, L. N. Large Area Validation of Adaptable Integrated Pest Management Technology in Okra (*Hibiscus esculentus*) Through Farmers' Participation. *Indian J. Agri. Sci.* **2008,** *78*(12)*,* 1063–1066.

Sharma, B. K.; Nayak, N.; Das, P. K. Management of Sugarcane Shoot Borer, *Chilolinfuscatellus* Snellen in Nayagarh District of Orissa. *Indian Sugar.* **1997,** 46(11)*,* 879–881.

Shenhmar, M.; Brar, K. S. Evaluation of *Trichogramma chilonis* Ishii (Hymenoptera: Trichogrammatidae) for the Control of *Chilo auricilius* Dudgeon on Sugar Cane. *Indian J. Plant Prot.* **1996,** *24*(1–2)*,* 47–49.

Shi, G. R. In *Ecological Structure Becomes Better after Releasing Trichogramma Year by Year,* Proceeding of the National Symposium on IPM in China, Beijing, China, Zhang, Z. L., Piao, Y. F., Wu, J. W. Eds.; Agricultural Scientech Press: Beijing, 1996; p 488.

Shi, G. Z.; Zhou, Y. N.; Zhao, J. S. Studies on the Techniques for Rice Moth Mass Rearing and Application. *Nat. Enemies Insect.* **1982,** *4,* 1–8.

Shirke, M. S.; Bade, B. A. Efficacy of *Trichogramma japonicum* against Paddy Stem Borer. *J. Maharashtra Agri. Univ.* **1997,** *22*(3), 338–339.

Singh, D.; Praduman B.; Ahlawat, D. S.; Ahlawat, K. S. Performance of Bioagent *Trichogramma japonicum* Against Stem Borer *Scirpophaga incertulas* (Walker) in Basmati Rice Var CSR-30. *Environ. Ecol.* **2013,** *31*(1A), 293–295.

Smith S. M. Biological Control with *Trichogramma adances* Success and Potential of Their Use. *Ann. Rev. Entomol.* **1996,** *41,* 375–406.

Strand, M. R. Physiological Interactions of Parasitoids and Hosts. In *Insect Parasitoids,* Waage, J., Greathead, D., Eds.; Academic Press: London, 1986; pp 109–118.

Subramaniam, T. V.; Seshagiri Rao, D. "Biological Control of Sugarcane Borers in Mysore". *Mysore Dept. Agric. Ento. Ser. Bull.* **1940,** *12,* 2–10.

Subramaniam, T. V. Preliminary Experiments on' the Mass-Production of *Trichogramma* Parasites, for Control against Sugarcane Borer in Mysore. *Indian J. Asri. Sci.* **1937,** *7*(1), 149–155.

Suh, C. P. C. In *Reevaluation of Trichogramma Releases for Suppression of Heliothine Pests in Cotton,* Proceedings 1997. Beltwide Cotton Production Research Conference, San Diego, CA, Jan 5–9, 1998.

Sumer, F.; Tuncbilek, A. S.; Oztemiz, S.; Pintureau, B.; Rugman-jones, P.; Stouthamer, R. A Molecular Key to the Common Species of *Trichogramma* of the Mediterranean Region. *Bio. Control.* **2009,** *54*(5), 617–624.

Sushil, S. N.; Mohan, M.; Hooda, K. S.; Bhatt, J. C.; Gupta, H. S. Efficacy of Safer Management Tools against Major Insect Pests of Tomato and Garden Pea in Northwest Himalayas. *J. Biol. Control.* **2006,** *20*(2), 113–118.

Tanwar, R. K.; Bambawale, O. M.; Singh, S. K.; Singh, A. *A Handbook on Trichogramma: Production and Field Release;* National Centre for Integrated Pest Management: New Delhi, India, 2006.

Tsneg, C. T. The Improved Technique for Mass Production of *Trichogramma ostriniae.* Egg Card Machine and Preservation of Egg Cards. *Chin. J. Entomol.* **1990,** *10,* 101–107.

Van Lenteren, J. C.; Hale, A.; Klapwijk, J. N.; van Schelt, J.; Steinberg, S. Guidelines for Quality Control of Comercially Produced Natural Enemies. In *Quality Control and Production of Biological Control Agents Theory and Procedures*; Van Lenteren, J. C., Ed.; CAB International: Oxfordshire, UK, 2003; pp 300–303.

Vijayalakshmi, D. Evaluation of BIPM Module on Tomato Fruit Borer (*Helicoverpa armigera*) Larval Population. *Madras Agri. J.* **2007,** *94*(1–6), 130–133.

Vijaykumar, N. Improved Mass Production Technologies of Major Biocontrol Agents for Efficient Pest Management. In *Organic Pest Management: Potential and Applications*; Narayansamy, N. Ed.; Satish Serial Publishing House: New Delhi, India, 2009; pp 163–176.

Wang, S. Q. Research Progresss in *Trichogramma* Mass Rearing by Using Artificial Host Eggs. *Plant Prot. Techol. Extens.* **2001,** *21,* 40–41.

Wang, S. Q.; Luo, C.; Zhang, F.; Zheng, Z. L. In *The Quality Standard in Mass Rearing and Applied Technique in the Field of Artificial Host Eggs Trichogramma,* Proceedings of the Fifth International Conference on Biodiversity and Utilization, Beijing, China,

October 22–24, 2005; Luo, C., Ji, Y. S., Eds.; Beijing Science and Technology Press: Beijing, 2005; pp 132–136.

Wang, Z. Y.; He, K. L.; Zhang, F.; Lu, X.; Babendreier, D. Mass Rearing and Release of *Trichogramma* for Biological Control of Insect Pests of Corn in China. *Biol. Control.* **2014,** *68,* 136–144.

Yadav, D. S.; Sharma, M. M. Comparative Efficacy of Bioagents, Neem Products and Malathion against Brinjal Shoot and Fruit Borer, *Leucinodes orbonalis* Guenee. *Pest. Res. J.* **2005,** *17*(2)*,* 46–48.

Yadav, J. B.; Singh, R. S.; Tripathi, R. A. Evaluation of Bio-Pesticides against Pest Complex of Okra. *Ann. Plant Prot. Sci.* **2008,** *16*(1)*,* 58–61.

Yadav, J. B.; Singh, R. S.; Singh, H. P.; Singh, A. K. Effect of Abiotic and Biotic Factors on Jassid and Fruit and Shoot Borer in *Kharif* Okra Crop. *Int. J. Plant Prot.* **2009,** *2*(1)*,* 119–122.

Yang, C. C.; Wang, C. S.; Zheng, Y. N.; Fu, B.; Na, C. Y.; Su, X. M. Sustained Effects of *Trichogramma dendrolimi* on *Ostrinia furnacalis. J. Maize Sci.* **2011,** *19,* 139–142.

Yu, G. Q.; Zhang, X. G.; Zhang, Y. H. Extension and Research on Control Technique of Corn Borer by Releasing *Trichogramma Dendrolimi. Agric. Technol.* **2009,** *29,* 92–96.

Yu, J. Z.; Chen, C. C.; Chou, L. Y. Assessment of Field Releases of *Trichogramms ostrinae* (Hymenoptera: Trichogrammatidae) for Control of the Asian Corn Borer, *Ostrinia furnacalis* (Lepidoptera: Pyralidae). *J. Agric. Res. China.* **1992,** *41,* 295–309.

Zhang, G. H.; Lu, X.; Li, L. J.; Ding, Y.; Liu, H. W. Influence of *Corcyra cephalonica* Egg Storage on Trichogramma Fecundity. *J. Jilin Agric. Sci.* **2008,** *33,* 43.

Zhang, J.; Wang, J. L.; Yang, C. C.; Zhang, F.; Xu, G. M.; Zhao, D. S.; Li, T. H.; Zhao, X. G.; Wang, G. J. The Effect of Long-term Releasing of *Trichogramma* on Controlling *Ostrinia* in Large Area. *J. Shenyang Agric. Univ.* **1990b,** *21,* 285–290.

Zhang, Y. Z.; Cheng, M. Z.; Zhou, W. R.; Wang, C. X. Studies on the Efficiency of Rearing Rice Moth *Corcera cephalonica* (Lep.: Gelechiidae) with Rice and Wheat Bran. *J. Biol. Control.* **1991,** *7,* 71–73.

Zhou, Y. N.; Zhao, J. S.; Li, M. Selection Medium for Rice Moth Mass Rearing. *Nat. Enemies Insect.* **1988,** *10,* 191–193.

CHAPTER 10

COCCINELLIDS IN INSECT PEST MANAGEMENT: PROBLEMS AND PROSPECTS

PRANAB BARMA[1*], SUPRAKASH PAL[2], and
TARAK NATH GOSWAMI[3]

[1]*Darjeeling Krishi Vigyan Kendra, Uttar Banga Krishi Viswavidyalaya, Kalimpong, Darjeeling, West Bengal 734301, India*

[2]*Directorate of Research (RRS-TZ), Uttar Banga Krishi Viswavidyalaya, Pundibari, Cooch Behar, West Bengal 736165, India*

[3]*Department of Entomology, Bihar Agricultural University, Sabour 813210, Bihar, India*

Corresponding author. E-mail: pranab.barma@gmail.com

CONTENTS

ABSTRACT

The family Coccinellidae is a very important predatory family under the order coleoptera. In this chapter we discuss various aspects related to the predatory coccinellids like examples of classical biological control using *Rodolia cardinalis* against *Icerya purchase*, *Diomus hennesseyi* and *Hyperaspis notata* against *Phenacoccus manihoti*, *Hippodamia convergens* against *Schizaphis graminum*, *Cryptolaemus montrouzieri* against *Mealy Bugs*, *Cryptognatha nodiceps* against *Aspidiotus destructor*, *Rhyzobius lophanthae* against *Diaspidinae scales*, *Chilocorus bipustulatus* against armored scale insects, *Chilocorus nigritus* against variety of armored scales, *Chilocorus circumdatus* against *Citrus Snow Scale*, *Aphidecta obliterate* against balsam woolly adelgid, *Sasajiscymnus tsugae* against *Adelges tsugae*. We also have tried to discuss the techniques of conservation of coccinellid predators in different crop ecosystems and predominant coccinellid predators in different crop ecosystems like rice, wheat, maize, mustard, tea, and horticultural ecosystems in Indian context. Augmentation of coccinellid predators in crop ecosystems has also been discussed in this chapter keeping in view the limitations of using coccinellid predators.

10.1 INTRODUCTION

The coccinellids or ladybird beetles are conserved, imported, and augmented in different ecosystems, to achieve biological control of pests. They have been associated with good fortune in many myths and legends. These beetles have been honored through the centuries as their vernacular indicates that the term "Lady" is in reference to biblical Mother Mary (Roache, 1960). They are commonly known variously as ladybirds (English English, Australian English, and South African English), ladybugs (North American English), lady beetles, or coccinellid beetles (preferred by scientists). The family name comes from its type genus, *Coccinella*. Most of them are of bright shining colors with a pattern of spots or patches against a contrasting background. There are about 5200 species reported across the world. Four hundred species under 79 genera of coccinellid beetles were recorded from Indian sub-continent (Poorani, 2002).

They are of great economic importance as predaceous both in their larval and adult stages on various important crop pests such as aphids, coccids, and other soft-bodied insects (Hippa et al., 1978; Kring et al.,

1985). About 90% of coccinellid species are considered beneficial because of their predatory activity, mainly against Homopteran insects and phytophagous mites injurious to various agricultural and forest plants. When insect food is insufficient, these coccinellids feed on pollen and nectar. They overwinter in unfavorable seasons mostly in adult stage.

The predaceous coccinellids are linked to biological control more often than any other taxa of predatory organisms. The history of using cocci-nellid beetle in biological control dates back to nineteenth century when the vedalia beetle, *Rodolia cardinalis* (Mulsant), was used to manage the cottony cushion scale, *Icerya purchasi,* in citrus during 1888. The beetle was introduced in the United States from Australia. In southern India, this scale was controlled by introduction of these beetles. This is an example of remarkable success for classical biological control with introduction of any exotic coccinellid beetle. For more than 100 years since then, this ladybird beetle has kept cottony cushion scale populations below economic levels.

10.2 TECHNIQUES OF USING COCCINELLIDS IN BIOLOGICAL CONTROL

In the ecosystem, coccinellids are used as a predator of soft-bodied insects. The approaches involve three major techniques, *viz.* classical biological control, conservation, and augmentation.

10.2.1 CLASSICAL BIOLOGICAL CONTROL

10.2.1.1 RODOLIA CARDINALIS AGAINST ICERYA PURCHASI

The vedalia beetle, *R. cardinalis*, was introduced in 1887 from Australia to combat the cottony cushion scale *I. purchasi* in the nascent California citrus industry. The huge success of this project became an example of the great potential of classical biological control as a tactic for suppressing invasive pests (DeBach, 1964). Subsequently, the beetle was introduced to at least 29 other countries and control of the target was either complete or substantial in all locations. The success of the vedalia beetle cata-lyzed the introduction of many exotic coccinellids targeting other pests, although rarely with the same degree of success (Caltagirone & Doutt, 1989). The high degree of specificity of *R. cardinalis* for *Icerya* species, in combination with a very short generation time, highly efficient detection of isolated host patches, and the ability of a larva to complete development on a single mature female scale are all thought to be key factors in the effectiveness of the vedalia beetle (Prasad, 1990).

10.2.1.2   DIOMUS HENNESSEYI AND HYPERASPIS NOTATA AGAINST PHENACOCCUS MANIHOTI

Against the cassava mealybug, *Phenacoccus manihoti,* the neotropical ladybirds, *D. hennesseyi* and *Hyperaspis notate* were both established in central and eastern Africa but none of those gave significant success against the target pest (Neuenschwander, 2003).

10.2.1.3   HIPPODAMIA CONVERGENS AGAINST SCHIZAPHIS GRAMINUM

Introduction of a US strain of *H. convergens* to Kenya for control of *S. graminum* in 1911 failed despite a good history of the predator–prey association on wheat (Hunter, 1909). This was probably because the source material was not adapted to tropical conditions (Greathead, 1971).

10.2.1.4   CRYPTOLAEMUS MONTROUZIERI AGAINST MEALY BUG

Greathead (2003) reviewed classical programs in Africa and listed nine coccinellid species successfully established, of which eight were scale-feeding species and the other the mealy bug predator *C. montrouzieri.*

10.2.1.5   CRYPTOGNATHA NODICEPS AGAINST ASPIDIOTUS DESTRUCTOR

Successful control of the coconut scale, *A. destructor,* was attributed to the introduction and establishment of *C. nodiceps* from Trinidad to Fiji (Singh, 1976).

10.2.1.6   RHYZOBIUS LOPHANTHAE AGAINST DIASPIDINAE SCALES

Successes have been achieved by introducing *R. lophanthae* in controlling various scales under the subfamily Diaspidinae in Italy, the Black Sea coast of the Ukraine, and North Africa (Yakhontov, 1960). The species is now considered an important introduced predator of citrus scale insects both in the United States (Flint & Dreisdadt, 1998) and in Australia (Smith et al., 1997).

10.2.1.7   CHILOCORUS BIPUSTULATUS AGAINST ARMORED SCALE INSECTS

Successful introduction of *C. bipustulatus,* an effective predator of armored scale insects was achieved in the United States (Huffaker & Doutt, 1965), North Africa (Iperti & Laudeho, 1969), Central Africa, and Australia (Waterhouse & Sands, 2001).

10.2.1.8   CHILOCORUS NIGRITUS AGAINST VARIETY OF ARMORED SCALES

C. nigritus feeds on a variety of armored scales and it has been widely disseminated throughout tropical regions through both intentional and unintentional introductions (Samways, 1989). As early as 1940, it was introduced effectively against a complex of scale insects damaging coconuts in the Seychelles (Greathead, 2003).

10.2.1.9   CHILOCORUS CIRCUMDATUS AGAINST CITRUS SNOW SCALE, UNASPIS CITRI

C. circumdatus has been successfully introduced for controlling citrus snow scale, *Unaspis citri*, to Australia, South Africa, the United States, and elsewhere (Samways et al., 1999). It appears quite specific to its prey and is able to track it effectively at low densities, although in Hawaii it reportedly feeds on two additional species of Diaspididae (Funasaki et al.,

1988). *C. circumdatus* is now considered to be the most important predator of *U. citri* in Australia (Smith et al., 1997).

10.2.1.10 APHIDECTA OBLITERATE AGAINST BALSAM WOOLLY ADELGID, ADELGES PICEAE

The larch ladybird, *A. obliterate,* a species of European origin was introduced to South Carolina in 1960 (Amman, 1966) to control the balsam woolly adelgid, *Adelges piceae*, an invasive pest that arrived in North America around 1900. Later, *A. obliterate* was introduced to British Columbia, Canada (Harris & Dawson, 1979), where it has emerged as an important biological control of both *A. piceae* and the hemlock woolly adelgid, *Adelges tsugae* (Humble, 1994; Montgomery & Lyon, 1996).

10.2.1.11 SASAJISCYMNUS TSUGAE AGAINST ADELGES TSUGAE

Sasajiscymnus tsugae, an Asian was introduced from Japan to target *A. tsugae* and has been established in various regions of the Eastern United States due to release programs initiated during 1997 (Cheah et al., 2004). Although established, the impact of *S. tsugae* on the pest population was not as great it was expected.

10.2.2 CONSERVATION OF COCCINELLID PREDATORS IN CROP ECOSYSTEMS

For most agricultural systems, conservation techniques for coccinellids are lacking, even though they are abundant in these habitats. Conservation of coccinellids can be facilitated by modifying the environment so that minimal disruption can be caused to the coccinellids.

Pesticides are the most important factor with a detrimental effect on the survival of the coccinellids. Therefore, efforts should be given toward eliminating the deleterious effect of pesticides on the coccinellids.

Plants may directly or indirectly influence coccinellid effectiveness as a result of altered prey suitability or host-finding success of the predator. Prey reared on resistant host plants may have a significant impact on the fitness of coccinellids. However, because these effects are generally complex and interacting, generalities about the compatibility of plant resistance and biological control are lacking (Obrycki & Kring, 1998). Searching behavior of coccinellids is influenced by the complexity of the substrate searched (Obrycki, 1986). Variations in pea plant architecture significantly influenced *Coccinella septempunctata* and *Hippodamia variegate* foraging behavior (Kareiva & Shakian, 1990). So, resistant varieties should be used selectively for encouraging the activity of ladybird beetles.

Changes in cultural practices can be achieved through maintaining refugia for coccinellid beetles by adopting strip-planting, field borders, eco-feast cropping, and cover cropping. Agricultural practices that use conservation tillage and cover crops are promoted for their benefits to soil quality, weed suppression, and water conservation (Schomberg, 2003). In addition, these should be promoted for their ability to encourage biological control, with numerous cases where cover crops have been shown to increase beneficial species and reduce pest species (Olson et al., 2012).

The abundance and efficacy of coccinellid as biocontrol agents may depend on the structure, composition, and diversity of the surrounding landscape at different spatial scales. The worldwide trend toward simplification of agricultural landscapes, caused by planting annual crops and eliminating edges of less disturbed natural or semi-natural habitats, can alter the structure of natural enemy communities, favoring exotic over native species and lowering the diversity of natural enemies. This, in turn, could result in a lower biological control because of the loss of species with different roles in the agro-ecosystems (Grez et al., 2014).

The non-farmed areas surrounding agricultural lands are therefore an important component of the stability of agro-ecosystems. These areas can be putative refuges for predators and can act as an important source of biological control agents. Refuges can enhance the survival, fecundity, longevity, and behavior of natural enemies, increasing the colonization of crops by such enemies and contributing to the control of pest populations. Ecologically based research on biological control by natural-enemy populations requires an understanding of the population dynamics of these species and of their dispersal into the crop from a particular resource or habitat patch.

Being an important group of predator, Coccinellids should be conserved in the different crop ecosystems. The rate of predation and species diversity of coccinellid community is influenced by plant community structure and microclimate (Tooker & Hanks, 2000). This is because different species of ladybird exhibit a preference for specific vegetation types coupled with suitable food in sufficient abundance.

10.2.2.1   IMPORTANT COCCINELLID PREDATORS IN DIFFERENT ECOSYSTEMS

Rice

The coccinellid fauna found in the terai region of West Bengal were identified as *C. septempunctata* Linnaeus and *Micraspis* sp. (Ghosh et al., 2013). Numerous reports show coccinellids as opportunistic, feeding on a variety of food material in addition to their preferred prey. *Micraspis discolor* is the most abundant species of coccinellid in rice ecosystems and touted as a biocontrol option for brown planthopper (BPH), *Nilaparvata lugens* (Stal), a key pest of rice. However, it has been reported as both entomophagous and phytophagous. The abundance of adult *M. discolor* in rice at flowering phase does not correspond to prey abundance in the field but rather reflects an inclination to pollen feeding more than entomophagy (Shanker et al., 2013).

Ten genera under four tribes of the family Coccinellidae were collected and identified from Indo-Bangladesh border in northeast India. They are *Brumoides suturalis* (Fabricius), *C. nigritus* (Fabricius), *C. septempunctata* Linnaeus, *C. transversalis* (Fabricius), *Harmonia octomaculata* (Fabricius), *Illeis cincta* (Fabricius), *M. crocea* (Fabricius), *Propylea sp. nr. japonica* (Thunberg), *C. montrouzieri* Mulsant, and *Scymnus nubilus* Mulsant. Result also revealed that out of ten coccinellids, *Harmonia octomaculata* (Fabricius) (22.9%), *M. crocea* (Fabricius) (18.5%), *Brumoides suturalis* (Fabricius) (12.36%), and *C. nigritus* (Fabricius) (9.06%) were major predatory coccinellids beetles in rice ecosystems of Indo-Bangladesh border in the northeastern region of India.

Wheat and Maize

Coccinellid fauna found in wheat crop was *C. transversalis* Fabricius and *Cheilomenes sexmaculata* Fabricius (Megha et al., 2015). Jat et al. (2009) also noticed the population of *C. sexmaculata* in wheat crop. Coccinellid fauna found on maize crop was *Scymnus nubilus*, *Harmonia octomaculata,* and *Cheilomenes sexmaculatus* (Megha et al., 2015). The predatory activity of *Cheilomenes sexmaculata* in maize was noticed by Tank et al. (2006) and Joshi et al. (1999).

Mustard

The coccinellid species namely *C. septempunctata* Linnaeus, *C. transversalis* Fabricius, *Menochilus sexmaculatus* Fabricius, *M. discolor* Fabricius, *Coelophora unicolor* Fabricius, and *Oenopia luteopustulata* Mulsant have been recorded in the terai region of West Bengal predating on the mustard aphid (Choudhury & Pal, 2006).

Tea

Twenty species of coccinellid predators were observed in sub-Himalayan tea plantation of North Bengal. Of these, *M. discolor* (Fabricius) was dominant (42.5%) in the conventionally managed tea plantations. The abundance of *M. discolor* populations was positively correlated with the abundance of red spider mites (*Oligonychus coffeae* Nietner, Acarina: Tetranychidae) ($R^2 = 0.705$) and tea aphid (*Toxoptera aurantii* Boyer de Fons, Homoptera: Aphidae) ($R^2 = 0.893$).

Horticultural Ecosystems

The faunal complex of coccinellids found in the horticultural ecosystems of hilly region of Kalimpong was identified as *O. kirbyi* Mulsant, *O. sauzeti* Mulsant, *C. sexareata* Mulsant, *C. unicolor* (Fabricius), *Illeis* sp., *C. septempunctata* Linnaeus, *C. transversalis* Fabricius, *M. discolor* (Fabricius), and *Cheilomenes sexmaculata* (Fabricius) (Pal, 2007).

10.2.3 AUGMENTATION OF COCCINELLID PREDATORS IN CROP ECOSYSTEMS

Augmentation of Coccinellids is very much important nowadays because climate is changing and way of controlling pests is becoming more biobased. Augmentation may be either inoculative or inundative. The main motto of augmentation is to increase the effectiveness of natural enemies. Inconsistent release of inoculation, that is, once in a year for a new establishment of coccinellids species as a natural enemy for the deletion from the nature due to unfavorable weather condition. Recent reviews of such approach are rarely utilized in this program (Collier & Van Steenwyck, 2004; Powell & Pell, 2007).

Augmentative biological control (or "augmentation") is simply the release of large numbers of insectary-reared natural enemies with the goal of "augmenting" natural enemy populations or "inundating" pest populations with natural enemies. *C. montrouzieri* was the first species used to demonstrate an inoculative approach for augmentative biological control (Hagen, 1974). Augmentative biological control in greenhouses has been documented for mealy bugs using *C. montrouzieri* (Hussey & Scopes, 1985). Augmentative biological control of insect pests in outdoor cropping systems is an attractive option for integrated pest management (IPM) programs. Augmentative releases of biological control agents have promise as environmentally safe applications of biological control, and as an approach that should be compatible with the application of appropriate pest monitoring and economic injury levels. One of the major stimuli for investigating augmentative biological control has been the drive to reduce a historic reliance on broad-spectrum pesticides for pest control. Augmentation might be used as a substitute for pesticide applications if the pest is sufficiently suppressed by the released natural enemies.

However, the effectiveness and economic value of augmentative biological control options is questionable in many cases—64% of augmentative control projects are failures, and in many cases, the costs associated with these programs are as high as or higher than insecticides (Collier & Van Steenwyk, 2004). In direct comparisons, augmentation was typically less effective than pesticide applications, but not always. In the evaluation of economics, augmentative releases were frequently more expensive than pesticides, though there were cases where augmentation was clearly cost effective. Finally, 12 ecological factors were implicated as potential limits on the efficacy of augmentation. Unfavorable environmental conditions, enemy dispersal, mutual interference, and pest refuges from parasitism or predation were most often suggested as possible ecological limitations.

The many failed attempts at augmentative biological control are primarily attributable to a poor understanding of the natural enemy's ability to locate hosts in specific crops after release (Wright et al., 2005). A lack of knowledge of the expected dispersal behavior of a natural enemy influences the decision on release rates and the crop system targeted for augmentative biological control. Work on *Trichogramma ostriniae* has shown that low density, early-season releases are effective in corn (Wright et al., 2002), yet in solanaceous crops, release rates have to be orders of magnitude higher to achieve even moderate parasitism levels (Kuhar et al., 2004). This will clearly impact the benefit-to-cost ratio of using the same species in different crop systems. Collier and Van Steenwyk (2004) suggested that the burden of future research is to identify crop–pest systems in which augmentation can cost-effectively control arthropod pests using rigorous field experiments that compare augmentation, pesticide application, and control (no management) treatments.

10.3 LIMITATIONS OF USING COCCINELLID PREDATORS

The use of pesticides is the most important factor for killing the coccinellids in the agroecosystem. This is the most important impediment for successful implementation of conservation and augmentation program. Pesticide-treated ecosystems were found to support less number of lady beetle species. The sprayed orchards were found to have 14 species of as compared to 17 in unsprayed ones. Similarly, only 10 species were recovered from the sprayed vegetable fields as compared to 12 from unsprayed fields. The biodiversity indices indicated appreciable effect of pesticide application on the coccinellids assemblages (Shah & Khan, 2014). Pesticides vary widely in their effect on coccinellids, and similarly, coccinellids vary greatly in their susceptibility to pesticides. Standardized techniques for testing the impact of pesticides on natural enemies have been developed by a western Palearctic working group of the International Organization of Biological Control (IOBC/OILB) (Hassan, 1985), although they are not accepted worldwide.

10.4 CONCLUSION AND FUTURE PROSPECTS

Coccinellids are extremely diverse in their habitats; they live in all the terrestrial ecosystems (Skaife, 1979). They are also recognized as bio-indicators (Iperti, 1999). Coccinellids are density-dependent predators, that means their number increases or decreases as the prey number raises or decreases (Dixon, 2000). This insect migrates between various crop fields throughout the season depending upon the availability of prey and habitat disturbance (Maredia et al., 1992). For better IPM, it is necessary to deploy these coccinellids for its biological control potentiality, by reducing the pesticides in crop and orchard fields for minimal impact on the flora and fauna of the habitat. Therefore, it is important that beneficial ladybird species should be recognized and reared to release in the natural habitat.

Coccinellids will continue to play a role in naturally occurring and human-assisted biological control, and they will be considered as possible natural enemies for importation whenever a homopteran pest invades a new region (Obrycki & Kring, 1998). Methods to evaluate the effect of pesticides on coccinellids must be standardized to provide comparable results among species and locations. In importation programs, rather than maximizing the number of individuals released, focus is warranted on seasonal dispersal behaviors, overwintering requirements, suitability of alternate prey, and breeding structure. Genetic analysis of introduced coccinellids should be a part of all introduction programs. Biological control has been enhanced by conservation, importation, and augmentation of coccinellids in numerous agro-ecosystems, but further application and refinement of these technologies are needed.

KEYWORDS

- coccinellids
- ladybird beetles
- ecosystems
- biological control
- soft-bodied insects

REFERENCES

Amman, G. D. *Aphidecta obliterate* (Coleoptera, Coccinellidae), an Introduced Predator of the Balsam Woolly Aphid, *Chermespiceae* (Homoptera, Cherminae), Established in North Carolina. *J. Econ. Entomol.* **1966**, *59*, 506–508.

Caltagirone, L. E.; Doutt, R. L. The History of the Vedalia Beetle Importation to California and its Impact on the Development of Biological Control. *Annu. Rev. Entomol.* **1989**, *34*, 1–16.

Cheah, C. A. S. J.; Montgomery, M. S.; Salom, S. M. Biological Control of the Hemlock Woolly Adelgid. USDA For. Ser. FHTET-2004-04. FHTET: Fort Collins, 2004; p 22.

Choudhury, S.; Pal, S. Pest Complex and Their Succession in Mustard under Terai Ecological Conditions of West Bengal. *Indian J. Entomol.* **2006**, *68* (4), 387–395.

Collier, T.; Van Steenwyk, R. A Critical Evaluation of Augmentative Biological Control. *Biol. Control.* **2004**, *31* (2), 245–256.

DeBach, P. Ed.; *Biological Control of Insect Pest and Weedsl;* Chapman and Hall: London, 1964; p 844.

Dixon, A. F. G. *Insect Predator-Prey Dynamics Lady Birds Beetles and Biological Control;* University Press: New York, 2000; p 257.

Flint, M. L.; Dreisdadt, S. H. Eds.; *Natural Enemies Handbook: The Illustrated Guide to Biological Pest Control;* Publication No. 3386. University of California, Division of Agriculture and Natural Resources: CA, 1998; p 154.

Funasaki, G. Y.; Lai, P. Y.; Nakahara, L. M.; Beardsley, J. W.; Ota, A. K. A Review of Biological Control Introductions in Hawaii, 1890–1985. *Proc. Hawaiian Entomol. Soc.* **1988**, *28*, 105–160.

Ghosh, S. K.; Chakraborty, K.; Mandal, T. Bio-Ecology of Predatory Coccinellid Beetle, *Coccinella septempunctata* (Coleoptera: Coccinellidae) and its Dynamics in Rice Field of Terai Region of West Bengal, India. *Int. J. Bioresource Stress Manag.* **2013**, *4* (4), 571–575.

Greathead, D. J. A Review of Biological Control in the Ethiopian Region.In *Commonwealth Institute of Biological Control, Technical Communication No. 5.* CAB: Farnham Royal, UK, 1971; p 162.

Greathead, D. J. A Historical Overview of Biological Control in Africa. In *Biological Control in IPM Systems in Africa;* Neuenschwander, P., Borgemeisterand, C., Langewald, J., Eds.; CABI: Wallingford, UK, 2003; p 414.

Grez, A. A.; Zaviezo, T.; Hernandez, J.; Rodriguez-San Pedro, A.; Acuna, P. The Heterogeneity and Composition of Agricultural Landscapes Influence Native and Exotic Coccinellids in Alfalfa Fields. *Agri. Forest Entomol.* **2014**, *16*, 382–390.

Hagen, K. S. The Significance of Predaceous Coccinellidae in Biological and Integrated Control of Insects. *Entomophaga Mem. Hors. Ser.* **1974**, *7*, 25–44.

Harris, J. W. E.; A. F. Dawson. Predator Release Program for Balsam Wooly Adelgid, *Adelges piceae* (Homoptera, Adelgidae), in British Columbia, 1960–1969. *J. Entomol. Soc. B.C.* **1979**, *16*, 21–26.

Hassan, S. A. Standard Methods to Test Side-Effects of Pesticides on Natural Enemies of Insects and Mites Developed by the IOBC/WPRS Working Group "Pesticides and Beneficial Organisms." *Bull. OEPP/EPPO.* **1985**, *15*, 214–55.

Hippa, H.; Kepekenand, S. D.; Laine, T. On the Feeding Biology of *Coccinella hiero-glyphica* L. (Coleoptera: Coccinellidae). *Kevo-subaretitic Ras. Station.* **1978,** *14,* 18–20.

Huffaker, C. B.; Doutt, R. L. Establishment of the Coccinellid *Chilocorus bipustulatus* L. in California Olivegroves. *Pan-Pac. Entomol.* **1965,** *41,* 61–63.

Humble, L. M. Recovery of Additional Exotic Predators of Balsam Wooly Adelgid, *Adelges piceae* (Ratzeburg) (Homoptera, Adelgidae), in British Columbia, Canada. *Can. Entomol.* **1994,** *12,* 1101–1103.

Hunter, S. J. *The Green Bug and Its Enemies;* Kansas University Science Bulletin No. 137, 1909; p 163.

Hussey, N. W.; Scopes, N. Eds.; *Biological Pest Control: The Glasshouse Experience;* Cornell University Press: Ithaca, 1985.

Iperti, G. Biodiversity of Predaceous Coccinellidae in Relation to Bioindication and Economic Importance. *Agric. Ecosyst. Environ.* **1999,** *74,* 323–342.

Iperti, G.; Laudeho, Y. Les Entomophage de *Parlatoriablanchardi*Terg. dans les Palmerales de l'adrar Mauritanien. *Ann. Zool. Ecol. Anim.* **1969,** *1,* 17–30.

Jat, H.; Swaminathan, R.; Upadhyay, B. Bio-Ecology of Aphidophagous Coccinellids in Maize-Sorghum Based Cropping System. *Indian J. Entomol.* **2009,** *71* (2), 170–185.

Joshi, S.; Ballal, C. R.; Rao, N. S. Biotic Potential of Three Coccinellid Predators on Six Different Aphid Hosts. *J. Entomol. Res.* **1999,** *23* (1), 1–7.

Kareiva, P; Shakian, R. Tritrophic Effects of a Simple Architectural Mutation in Pea Plants. *Nature.* **1990,** *345,* 433–34.

Kring, T. J.; Gilstrap, F. E.; Michels, G. J. Role of Indigenous Coccinellid in Regulating Green Bugs (Homoptera: Aphididae) on Texas Grain Sorghum. *J. Econ. Entomol.* **1985,** *78* (1), 269–273.

Kuhar, T. P.; Hofmann, M. P.; Fleischer, S. J.; Groden, E.; Gardner, J.; Wright, M. G.; Pitcher, S. A.; Speese, J.; Westgate, P. Potential of *Trichogramma ostriniae* (Hymenoptera: Trichogrammatidae) As a Biological Control Agent of European Corn Borer (Lepidoptera: Crambidae) in Solanaceous Crops. *J. Econ. Entomol.* **2004,** *97,* 1209–1216.

Maredia, K. M.; Gage, S. H.; Landis, D. A.; Scriber, J. M. Habitat Use Patterns by the Seven Spotted Lady Beetle (Coleoptera: Coccinellidae) in a Diverse Agricultural Landscape. *Biol. Control.* **1992,** *2,* 159–65.

Megha, R. R.; Vastrad, A. S.; Kamanna, B. C.; Kulkarni, N. S. Species Complex of Coccinellids in Different Crops at Dharwad Region. *J. Exp. Zool. India.* **2015,** *18* (2), 931–935.

Montgomery, M. E.; Lyon, S. M. In Natural Enemies of Adelgids in North America: Their Prospect for Biological Control of *Adelges tsugae* (Homoptera, Adelgidae). Proceedings of the First Hemlock Woolly Adelgid Review, Charlottesville, Oct 12, 1995;Department of Agriculture, Forest Service: Charlottesville, VA,1996; pp 89–102.

Neuenschwander, P. Biological Control of Cassava and Mango Mealybugs in Africa. In *Biological Control in IPM Systems in Africa;* Neuenschwander, P.; Borgemeister, C.; Langewald, J., Eds.; CABI: Wallingford, UK, 2003; pp 45–594.

Obrycki, J. J. The Influence of Foliar Pubescence on Entomophagous Species. In *Interaction of Host Plant Resistance and Parasites and Predators of Insects;* Boethel, D. J.; Eikenbary, R. D., Eds.; Ellis Horwood Publ.: West Sussex, 1986; pp 61–83.

Obrycki, J. J.; Kring, T. J. Predaceous Coccinellidae in Biological Control. *Annu. Rev. Entomol.* **1998,** *43,* 295–321.

Olson, D. M.; Webster, T. M.; Scully, B. T.; Strickland, T. C.; Dadis, R. F.; Knoll, J. E.; Anderson, W. F. Use of Winter Legumes as Banker Plants for Beneficial Insect Species in a Sorghum and Cotton Rotation System. *J. Entomol. Sci.* **2012,** *47* (4), 350–359.

Pal, S. *Annual Report* 2007. Regional Research Station. UBKV: Kalimpong, 2007; p 48.

Poorani, J. An Annotated Checklist of the Coccinellidae (Coleoptera) (Excluding Epilachnae) of the Indian Sub Region. *Oriental Insects.* **2002,** *36,* 307–383.

Powell, W.; Pell, J. K. Biological Control. In *Aphids as Crop Pests.* van Emden, H. F., Harrington, R., Eds.; CABI: Wallingford, UK, 2007; pp 469–513.

Prasad, Y. K. Discovery of Isolated Patches of *Icerya purchase* by *Rhodolia cardenalis*: A Field Study. *Entomophaga.* **1990,** *35,* 421–429.

Roache, L. C. Ladybug, Ladybug, Ladybug: What's a Name? *Coleopt. Bull.* **1960,** *14,* 21–25.

Samways, M. J.; Osborn, R.; Hastings, H.; Hattingh, V. Global Climate Change and Accuracy of Prediction of Species' Geographical Ranges, Establishment Success of Introduced Ladybirds (Coccinellidae, *Chilocorus* spp.) Worldwide. *J. Biogeog.* **1999,** *26,* 795–812.

Samways, M. J. Climate Diagrams and Biological Control, an Example from the Areography of the Ladybird *Chilocorus nigritus* (F.) (Insecta, Coleoptera, Coccinellidae). *J. Biogeog.* **1989,** *16,* 345–351.

Schomberg, F. H.; Lewis, J.; Tillman, G.; Olsan, P.; Timper, P.; Wauchope, D.; Phatak, S.; Jay, M. Conceptual Model for Sustainable Cropping Systems in the Southeast: Cotton System. *J. Crop Prod.* **2003,** *8,* 307–327.

Shah, M. A.; Khan, A. A. Assessment of Coccinellid Biodiversity under Pesticide Pressure in Horticulture Ecosystems. *Indian J. Entomol.* **2014,** *76* (2), 107–116.

Shanker, C.; Mohan, M.; Sampathkumar, M.; Lydia, Ch.; Katti, G. Functional Significance of *Micraspis discolor* (F.) (Coccinellidae: Coleoptera) in Rice Ecosystem. *J. Appl. Entomol.* **2013,** *137,* 601–609. doi: 10.1111/jen.12035

Singh, S. Other Insect Pests of Coconuts. *Cocomunity.* **1976,** *16,* 18–24.

Skaife, S. H. *African Insect Life.* Struik Publishers: Cape Town, 1979; p 279.

Smith, D., Beattie, G. A. C.; Broadley, R. (Eds.). *Citrus Pests and their Natural Enemies: Integrated Pest Management in Australia.* Dept. of Primary Industries: Brisbane, 1997; p 272.

Tank, B. D. Carry-Over and Biology of Ladybird Beetle, *Cheilomenes sexmaculata* (Fab.) Under Middle Gujarat Conditions. M.Sc. (Agri.) Thesis, Anand Agricultural University, Anand, Gujarat, India, 2006.

Tooker, J. F.; Hanks, L. M. Influence of Plant Community Structure on Natural Enemies of Pine Needle Scale (Homoptera: Diaspididae) in Urban Landscapes. *Environ. Entomol.* **2000,** *29* (6), 1305–1311.

Waterhouse, D. F.; Sands, D. P. A. *Classical Biological Control of Arthropods in Australia;* ACIAR: Canberra, 2001; p 559.

Wright, M. G.; Kuhar, T. K.; Hoffmann, M. P.; Chenus, S. A. Effect of Inoculative Releases of *Trichogramma ostriniae* on Populations of *Ostrinia nubilalis* and Damage to Sweet Corn and Field Corn. *Biol. Control.* **2002,** *23,* 149–155.

Wright, M. G.; Kuhar, T. P.; Diez, J. M.; Hoffmann, M. P.In *Effective Augmentative Biological Control-Importance of Natural Enemy Dispersal, Host Location, and Post-Release Assessment*, Second International Symposium on Biological Control of Arthropods, Sept 12–16, 2005; Davos, Switzerland, University of California, Riverside USA USDA Forest Service Publication, FHTET: Davos , 2005–2008.

Yakhontov, V. V. *Utilization of Coccinellids in the Control of Agricultural Pests* Izd (In Russian.). AN Uzbek, SSR: Tashkent, 1960; p 85.

CHAPTER 11

CHRYSOPID: A POTENTIAL BIOCONTROL AGENT

NITHYA CHANDRAN[1*] and TAMOGHNA SAHA[2]

[1]Division of Entomology, Indian Agricultural Research Institute, New Delhi 110012, India

[2]Department of Entomology, Bihar Agricultural University, Sabour 813210, Bhagalpur, India

*Corresponding author. E-mail: nithyacr@yahoo.com

CONTENTS

ABSTRACT

Chrysoperla carnea, recognized as the common green lacewing, is an insect in the Chrysopidae family. The family Chrysopidae contains many species that could be considered as important biological control agent. Chrysopid larvae and the adults of certain species are polyphagous predators and feed on several pests of economic importance. The use of chrysopids in biological control has been enabled by the development of efficient mass rearing facilities for species bearing certain characteristics, such as *Chrysoperla* species. As we know, *Chrysoperla* sp. is a potential biological control agent; so advance studying the performance of *Chrysoperla* species under field condition could enhance their role in integrated pest management (IPM) program.

11.1   INTRODUCTION

The green lacewing *C. carnea* (Stephens) (Neuroptera:Chrysopidae) is a cosmopolitan predator found in a wide range of agricultural habitats, with high relative frequency of occurrence, good searching ability, and easy rearing in the laboratory. The larva of Green lacewing, *C. carnea* (Stephens), generally known as aphid-lion, is a generalist predator of a wide range of pest species such as mealybugs, aphids, thrips, whiteflies, mites, and eggs of insect pests (Carrillo & Elanov, 2004). They can inhabit many diverse agroecosystems, and they can be easily mass reared (Rajakulendran & Plapp, 1982). Adult lacewings are free living and feed only on nectar, pollen, and honey dew (El Serafi et al., 2000). One larva may devour as many as 500 aphids in its life and there is no doubt that they play an important part in the natural control of many small homopterous pests. It is their excellent predatory potential on other soft-bodied insect pests of crops, which suppresses the insect pests' density below the economic injury level (McEwen et al., 1999). Commercial products contain one growth stage (eggs, larvae, or pupae) or mixed stages, dependent on producer (Van Lenteren, 2003).

Green lacewing larvae are effective for inundative biological control of several pests of greenhouse and field crops (Harbaugh & Mattson, 1973; Sattar et al., 2007). *C. carnea* has effectively been used in Pakistan and thus has been proved to be a voracious predator of cotton mealybug (Sattar et al., 2007). Its single larva, consumed 487.2 aphids and 510.8 whitefly pupae in its entire life span (Afzal & Khan, 1978). Gurbanov (1984) used *C. carnea* against the thrips and aphids, the population of thrips fell down to 95.6% and those of aphids to 98.5%. Jin (1998) has evaluated the effectiveness of *C. carnea* against the *Heliothis armigera* (Hubner) and observed that pest infestation was reduced from 1.6 to 0.1%.

The biology of *C. carnea* on its developmental and reproductive parameters is very essential and vital for its survival in a set of environmental conditions. Chrysoperla presents in nature in many crop habitats and predates its target pests in areas where humidity is high like in greenhouses and irrigated crops. It is active in larval and adult forms present on leaves of crops, vegetables, fruit plants, and weeds. Its adult has green color with delicate soft wings. Because of its appearance it is commonly called green lace wing. They have a short life cycle and a vast diversity of host; and they can easily be mass reared (Sattar et al., 2007; Whitcomb & Bell, 1964; Ridgway et al., 1970). They can search efficiently and resist some pesticides (Sattar & Abro, 2011). However, the commercialization of natural enemies such as *C. carnea* and their increased use in pest management requires reducing the cost of their mass-rearing and improving the success rate.

11.2 ORIGINAL AREA AND DISTRIBUTION OF THE SPECIES

The *C. carnea* complex is cosmopolitan with Holarctic distribution. The Holarctic is the terrestrial ecozone that encompasses the majority of habitats found throughout the northern continents of the world. This region is divided into the Palearctic, consisting of North Africa, all of Eurasia, with the exception of Southeast Asia and the Indian subcontinent, and the Nearctic, consisting of North America and north of southern Mexico (Malais & Ravensberg, 2003).

11.3 TAXONOMIC CHALLENGES

C. carnea was originally considered to be a single species with a Holarctic distribution, but it has now been shown to be a complex of many cryptic, sibling species (Lourenço et al., 2006). These are indistinguishable from each other morphologically but can be recognized by variations in the vibrational songs the insects use to communicate with each other, especially during courtship (Henry et al., 2002) and to a certain extent by molecular (DNA) analyses (Lourenço et al., 2006). A more recent and extensive study by Henry et al. (2012) shows that many nodes on the phylogenetic tree of the *C. carnea* complex are well supported, but several taxa appear on more than one branch.

As of now, the *C. carnea* complex comprises 21 valid described species (Table 11.1). In addition, several new cryptic taxa have been identified but not yet described. Poor sampling of the *C. carnea* complex across much of Asia, northern and central Africa, and western North America makes it very likely that many new species remain to be discovered (Henry et al., 2013).

TABLE 11.1 Recognized Species and Subspecies of the Carnea-Group, with Citations to Literature Where Their Duet Songs are Described (Henry et al., 2013).

C. adamsi (Henry et al., 1993)	Henry et al. (1993), and Wells and Henry (1992)
C. downesi (Smith, 1932)	Henry (1980)
C. downesi-Kyrgyzstan	Wells and Henry (2004)
C. carnea-Kyrgyzstan(="motorboat KR")	Wells and Henry (1998)
C.carnea (Stephens)	Henry et al. (2002)
C. mediterranea (Holzel, 1972)	Henry et al. (1999)
C. agilis Henry et al. (2003)	Henry et al. (2003, 2011)

11.4   DESCRIPTION

The green lacewing eggs are oval and secured to the plant by long slender stalks. They are pale green when first laid but become gray later. The larvae are about 1 mm long when they first hatch. They are brown and resemble small alligators, crawling actively around in search of prey. They have a pair of pincer-like mandibles on their head with which they grasp their prey, sometimes lifting the victim off the leaf surface to prevent its escape. The larvae inject enzymes into the bodies of their victims which digest the internal organs, after which they suck out the liquidated body fluids. The larvae grow to about 8 mm long before they spin circular cocoons and pupate (Syngenta-bioline.co.uk., 2011).

Adult green lacewings are a pale green color with long, threadlike antennae and glossy, golden, compound eyes. They have a delicate appearance and are from 12 to 20 mm long with large, membranous, pale green wings which they fold tent-wise above their abdomens (Syngenta-bioline. co.uk., 2011).

11.5   HABITAT RANGE

The *C. carnea* complex is associated with herbaceous vegetation (Henry et al., 2013). It has been found that the larvae do not establish well in tall growing crops. They establish better in lower-growing crops since they easily fall from the leaves, and in tall crops they are unable to reach the growing tips again where their prey congregate. Adults feed on pollen, nectar, and honeydew (Malais & Ravensberg, 2003). According to documentation (Wikipedia printout) from the applicant, the gardeners can attract lacewings by using certain plants and therefore ensure a steady supply of larvae. Adults Chrysopidae can be attracted mainly to Asteraceae (e.g. *Coreopsis, Cosmos, Helianthus*, and *Taraxacum*)—and Apiaceae such as dill (*Anethum*) or angelica (*Angelica*). Volatiles from eggplant, okra, and peppers are also attractive to lacewings, while odors from tomato are not (Reddy, 2002).

11.6   HOST RANGE

Lacewings prefer to feed on aphids, but will also prey on whitefly, spider mites, and thrips. If present, other insects may also be consumed, like Lepidopteran eggs, small larvae (Klingen et al., 1996), and mealy bugs. When prey is scarce, the larvae can resort to cannibalism, with the older larvae eating the younger. Lacewing eggs are often ignored because they stand on stalks (Malais & Ravensberg, 2003). Within the Neuroptera order, cannibalism seems to be correlated with polyphagous feeding habits (Duelli, 1981). Chrysophid larvae are active hunters, characterized by swift movements, aggressive behavior, and fast growth. These larvae are voracious and usually polyphagous predators; therefore their diet also includes prey of the same species (Canard, 1984).

Mochizuki et al. (2006) observed 100% cannibalism among green lacewings larvae in the absence of aphids; but if the aphids were present, the abundance of the cannibalism was negligible. In most cases the larger, elder, and consequently better fed larvae consumed the smaller, starving, and the younger larvae (Bar & Gerling, 1985). Arzet (1973) reported that after 72 h of starvation the cannibalism occurred between larvae, which were the same size and age. Bar and Gerling (1985) found out that *C. Carnea* pupae were eaten only by 3rd-instar larvae. Cannibalism is a disadvantage in mass production of green lacewings larvae for biological control of plants (Van Lenteren, 2003). But in natural conditions when the food is in shortage, the cannibalism is only option for survival of individuals and it is the only possibility to prevent local extinction (McEwen et al., 2001).

11.7   BIONOMICS

Understanding lacewing biology will help you attract and maintain healthy populations of these beneficial insects. Lacewings pass through four developmental stages: egg, larva, pupa, and adult (Fig. 11.1). It takes a month or two for them to mature from egg to adult lacewing. The egg stage is several days to a week or two long, the larval stage about 14 to 28 days, and the pupal stage about two weeks to a month (Coppel & Mertins, 1977). Fully mature adults live from 20 to 40 days. Lacewings lay their eggs on plant leaves and stems (www.ces.uga.edu/pubcd/b1140/html). Green lacewing eggs are laid individually, supported by long filaments (Coppel & Mertins, 1977). The eggs are pale green, turning gray before they hatch. Lacewing eggs hatch into flat, greyish brown alligator-shaped larvae. All larvae have a pair of tusks on the front of their heads. They use these tusks to pierce their prey, inject digestive enzymes, and suck out body fluids (Olkowski et al., 1991). Lacewings are highly predacious and cannibalistic as larvae (Olkowski et al., 1991; Nordlund, 1993). They will eat the eggs and larvae of the Colorado potato beetle, most caterpillars, corn borers, spider mites, scales, psylla, mealybugs, whiteflies, thrips, leafhoppers, aphids, and other soft bodied prey. A single larva can consume up to 250 leafhopper nymphs, 300–400 aphids, 11,200 spider mites, 3780 scales, or 6500 scale eggs. After 14–21 days of feeding, the larva weaves itself a silken cocoon. The cocoon is white and pearl-like. Lacewings pupate for up to two weeks in this cocoon, depending on the temperature (Olkowski et al., 1991). Lacewings then emerge as adults. Green lacewing adults are green or yellow-green, approximately 1/2–3/4 inches long with golden eyes and two pairs of intricately veined, translucent wings. As adults, most lacewings feed solely on nectar, pollen, and honeydew, while some are predatory (Coppel & Mertins, 1977; IPM Laboratories, 1991).

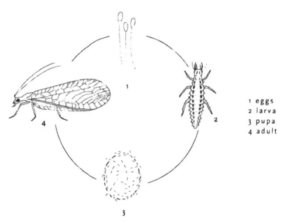

1 eggs
2 larva
3 pupa
4 adult

FIGURE 11.1   Lifecycle and appearance of green lacewings (Malais & Ravensberg, 2003).

11.8   CLIMATIC CONDITIONS

Population growth is dependent on temperature, the species of prey, and on atmospheric humidity. Adult lacewings do not like extremes of temperature, and in warm periods they leave the greenhouse. The development from egg to adult takes an average of 69 days at 16°C, 35 days at 21°C, and 25 days at 28°C. At constant temperatures below 10°C development is not completed. When low temperatures (even sub-zero temperatures) alternate with higher temperatures, development can take place to maturity. Temperature above 35°C is lethal (Malais & Ravensberg, 2003).

11.9   MASS REARING TECHNIQUE OF *CHRYSOPERLA CARNEA*

i.   2.5 kg cleaned insecticide free, fresh Bajra/sorghum/maize/paddy grains, sterilized at 100°C for 30 min are put into Corcyra rearing cages/plastic trays.

ii.   Incorporate powdered groundnut kernel weighing 100 g as source of protein.

iii.   Add yeast tablets (10 numbers or 5 g) as nutritional supplement +0.05 g streptomycin sulfate as antibiotic. Supplementing 0.01 g sulfur as acaricide.

iv.   Mix one cc fresh viable eggs of Corcyra in each cage/tray. Maintenance of culture at 26°C.

v.   After 40–45 days, emergence of adult moth commences.

vi.   Perform regular collection of adult moths manually and mechanically using vacuum cleaner.

vii.   Allow adult moths in fecundity cage used for three days supplementing with food swab impregnating honey 15%, water, and vitamin E.

viii.   Perform daily collection of Corcyra eggs.

ix.   Pass the eggs through 15, 30, and 40 mesh sieves and run over a slope of paper to eliminate dust particles or separation of scales admixture in eggs mechanically by motorized egg separator.

x.   Harvest fresh eggs.

xi.   Store eggs at 10°C up to 7 days.

xii.   Expose eggs of Corcyra to freshly emerged Chrysopid larvae for three days in group.

xiii.   Transfer Chrysopid larvae to plastic louvers for individual rearing on Corcyra eggs till cocoon formation.

xiv.   Collect cocoons.

xv.   Observe adult emergence.

xvi.   Collect adults and transfer to oviposition cage for rearing with adult diet.

xvii.   Collecct eggs for 25–30 days.

xviii. Store eggs at 10°C for 15 days.

xix.   Eggs may be shifted to room temperature for hatching.

xx.   Hatched larvae are provided Corcyra eggs again to repeat the process for more production (Vijayakumar, 2009).

11.10   MECHANISMS FOR DISPERSAL

The larvae of the *C. carnea* complex can be very mobile when hungry, but compared to the effect of adult flight activities their movement is irrelevant with regard to the dispersal of a population. Adults are good fliers and manage to disperse into and colonize other plants and fields. In the first two nights after emergence, the adult lacewings perform straight down-wind dispersal flights. Take-off behavior is elicited by the decrease in illumination at sunset. In the Central Valley (California) an average initial flight distance of 40 km per night has been estimated. Lacewings (eggs, larvae, pupae, and adults) can also disperse with human activity (handling and transportation of plant materials, etc.).

11.11   RELEASING TECHNIQUE OF *CHRYSOPERLA*

Experiments in vineyards suggest that the best way to ensure success with lacewings is to release them before pest populations' peak. Very low pest populations as well as high populations are difficult for lacewings to control (Daane & Yokata, 1997). Monitor pests to know when to begin *Chrysoperla* release. After release, continue to monitor both pest and lacewing populations.

Eggs are the most economical, but do not begin feeding immediately. Larvae have been the most effective in reducing pest populations in vineyard experiments (Daane & Yokata, 1997). Adults are nomadic and disperse well into trees and orchards.

Keep the purchased eggs at a temperature between 80 and 90°C. This can be done by placing them in a cooler with a hot water bottle. Chilling lacewing eggs reduces the hatch rate. Once eggs begin hatching, sprinkle them near your pest infestation. It is a good idea to mist foliage with water before your release, as lacewings thrive in high humidity (Raupp et al., 1994). Lacewing larvae can be gently tapped out of their packaging onto plants or placed individually on plants with a paint brush. If their shipping containers are opened and placed in the interior of the target plants, larvae will crawl from the packaging onto the plant (Raupp et al., 1994). Several experiments indicate that effective control requires releasing relatively large numbers of larvae. Researchers have used up to between 1 and 16 lacewing per plant (Daane & Yokata, 1997). You may need to do a follow-up release one or two weeks later. Continue to release lacewings until their populations are easily detectable or your pest populations decrease (Coppel & Mertins, 1977).

11.12　*CHRYSOPERLA* AS POTENTIAL BIOCONTROL AGENT

The efficacy of chrysopids as biological control agents against aphids was studied in 1742 for the first time (Senior & MCEwen, 2001). These species are considered to be important biological control agents in certain agroecosystems worldwide and they are most commonly released as commercially available chrysopids (Tauber et al., 2000). Besides *Chrysoperla* sp., there is also some scattered information concerning other important chrysopid species tested for use in biological control belonging to other genera, as well, such as *Chrysopa, Mallada* and *Ceraeochrysa* (Table 11.2).

TABLE 11.2　Some Chrysopid Species Which Have Been Studied for Biological Control Under Field and Laboratory Condition.

Chrysopa stricto	*Chrysoperla* sp
Chrysopa Formosa Brauer	*Chrysoperla carnea*
Chrysopa kulingensis Navas	*Chrysoperla rufilbaris* Burmeister
Chrysopa oculata Say	*Chrysoperla externa* (Hagen)
Chrysopa pallens Rambur	*Chrysoperla sinica* (Tjder)
Chrysopa perla (L.)	*Chrysoperla comanche* Banks
Chrysopa nigricornis Burmeister	*Chrysoperla lucasina* (Laeroix)
Ceraeochrysa spp	*Mallada* sp.
Ceraeochrysa cubana (Hagen)	*Mallada signata* (Schneider)

Reported by Albuquerque et al. (2001), Daane and Haegen (2001), Maisonneuve (2001), and Senior and McEwen (2001).

In greenhouse, *C. carnea, Chrysopa septempuctata* Wesmael, and *C. formosa* Brauer have been successfully used for aphid control in several crops, such as pepper, cucumber, eggplant, and lettuce (Tulisalo, 1984). On crops, the *Chrysoperla* larvae have been reported as attacking several species of aphids, red spider mites, thrips, whitefly, the eggs of leafhoppers, leaf miners, psyllids, small moths and caterpillars, beetle larvae, and the tobacco budworm. They are considered to be important predators of the long-tailed mealybug under glass (Tulisalo, 1984). Some examples of the earliest till the latest attempts to use *Chrysoperla* species in biological control both in field crops and in greenhouses are shown in Table 11.3.

TABLE 11.3   Pest Species Which Have Been Successfully Controlled by *Chrysoperla* Species.

Pest	References
Aphis gossypii Glover	Zaki et al. (1999)
Aphis pomi (DeGeer)	Hagley (1989)
Bemisia argentifolii Bellows and Perring	Legaspi et al. (1996)
Bemisia tabaci Gennadius	Breene et al. (1992)
Erythroneura elegantula Osborne	Daane and Yokota (1997)
Heliothis sp.	Ridgway and Jones (1968); Stark and Whitford (1987)
Myzus persicae (Sulzer)	Shands et al. (1972)
Pectinophora gossypiella (Saunders)	Irwin et al. (1974)
Pesudococcus maritimus Ehrhorn	Doutt and Hagen (1949, 1950)
Pesudococcus obscures Essig	Doutt and Hagen (1949 , 1950)
Tetranychus ludeni Zacher	Reddy (2001)
Tetranychus urticae Koch	Gomaa and Eid (2008)

Regarding field application is concerned, when considering the different biological control tactics used to control pests, four different methods could be used when releasing Chrysopid species:

• classical biological control, by importation of a new natural enemy;
• augmentation by means of inoculative release;
• augmentation by means of inundative release;
• conservation.

In general, Chrysopids have not been widely used in classical biological control, and most work has been focused on their augmentation and conservation in the agro ecosystems (Hagen et al., 1999). Due to the great ability of chrysopid adults to disperse and the occasional need of immediate response in pest control, augmentation of chrysopids by means of inundative releases (mainly based on the ability of the individuals released to suppress pest populations) has been most commonly used in cases where release cost is not restrictive. A few researchers have also documented the efficacy of chrysopids in augmentation biological control (Ridgway & Jones, 1969; Ehler et al., 1997; Knutson & Tedders, 2002).

Conservation techniques in biological control aim to the establishment of increased numbers of a chrysopid species within the field through the enhancement and manipulation of their habitat (e.g. crop fields). Research related with chrysopids has been mainly focused on the evaluation of certain chemicals or blends that could be sprayed on plants and act as attractants for the chrysopids and on the use of food supplements, such as artificial honeydews or pollen (Tauber et al., 2000; Tooth et al., 2009; Yu et al., 2008). Other conservation methods that have application in the manipulation of chrysopids are the use of hibernation shelters for the protection of overwintering adults (Wennemann, 2003; Weihrauch, 2008), cultural methods, such as intercropping (Senior & McEwen, 2001; Weihrauch, 2008), and the flowering plant (Shrewabury et al., 2004).

11.13 CONCLUSION AND FUTURE PROSPECTS

Chrysoperla is an important predator of many insect pests. It has a wider geographic distribution. Low cost mass production techniques have been standardized. It can be conserved or augmented in the field. So this predator has a greater role to play in the future IPM programs.

KEYWORDS

- *Chrysoperla carnea*
- chrysopids
- lacewings
- biocontrol agent
- integrated pest management

REFERENCES

Afzal, M.; Khan, M. R. Life History and Feeding Behaviour of Green Lacewing, *C. carnea* (Stephens) (Neuroptera: Chrysopidae). *Pak. J. Zool.* **1978**, *10* (1), 83–90.

Albuquerque, G. S.; Tauber, C. A.; Tauber, M. J. *Chrysoperla externa* and *Ceraeochrysa* spp.: Potential use in Biological Control in the New World Tropics and Subtropics. In

Lacewings in the Crop Environment; McEwen, P. K., New, T. R., Whittington, A. E., Eds.; Cambridge University Press: Cambridge, 2001; pp 408–423.

Arzet, H. R. Suchverhalten der Larven von *Chrysopa carnea* Steph. (Neuroptera: Chrysopidae). *Z. Angew. Entomol.* **1973,** *74,* 64–79.

Bar, D.; Gerling, D. Cannibalism in *Chrysoperla carnea* (Stephens) (Neuroptera, Chrysopidae). *Isr. J. Entomol.* **1985,** *19,* 13–22.

Breene, R. G.; Meagher, Jr. R. L.; Nordlund, D. A.; Wang, Y. T. Biological Control of *Bemisia tabaci* (Homoptera: Aleyrodidae) in a Greenhouse Using *Chrysoperla rufilabris*. *Biol. Control.* **1992,** *2,* 9–14.

Carrillo, M.; Elanov, P. The Potential of *C. carnea* as a Biological Control Agent of *Myzus persicae* in Glass Houses. *Ann. Appl. Biol.* **2004,** *32,* 433–439.

Coppel, H. C.; Mertins, J. W. *Biological Insect Pest Suppression.* Berlin, Germany: Springer-Verlag, 1977.

Daane, K. M.; Haegen, K. S. An Evaluation of Lacewing Release in North America. In *Lacewing in the Crop Environment,* McEwen, P. K., New, T. R., Whittington, A. E., Eds.; Cambridge University Press: Cambridge, 2001; pp 398–407.

Daane, K. M.; Yokata, G. Y. Release Strategies Affect Survival and Distribution of Green Lacewings (Neuroptera: Chrysopidae) in Augmentation Programs. *Environ. Entomol.* **1997,** *26* (2), 455–464.

Doutt, R. L.; Hagen, K. S. Biological Control Measures Applied against *Pseudococcus martimus* on Pears. *J. Econ. Entomol.* **1950,** *43,* 94–96.

Doutt, R. L; Hagen, K. S. Periodic Colonization of *Chrysopa californica* as a Possible Control of Mealybugs. *J. Econ. Entomol.* **1949,** *42,* 560–561.

Duelli, P. Is Larval Cannibalism in Lacewings Adaptive? (Neuroptera: Chrysopidae). *Res. Popul. Ecol.* **1981,** *23,* 193–209.

Ehler, L. E.; Long, R. F.; Kinsey, M. G.; Kelley, S. K. Potential for Augmentive Biological Control of Black Bean Aphid in California Sugarbeet. *Entomophaga.* **1997,** *42,* 241–256.

El Serafi, H. A. K.; Salam, A.; Bakey, N. F. Effect of Four Aphid Species on Certain Biological Characteristics and Life Table Parameters of *C. carnea* (Stephens) and *Chrysoperla septempunctata* Wesmael (Neuroptera: Chrysopidae) Under Laboratory Conditions. *Pak. J. Biol. Sci.* **2000,** *3,* 239–245.

Gomaa, W. O.; Eid, F. M. H. Release of *Chrysoperla carnea* (Stephens) (Neuroptera: Chrysopidae) and the Predaceous Mite, *Phytoseiulus macropilis* (Acari: Phytoseiidae) to Control *Tetranychus urticae* in Greenhouse in Egypt. *Egypt J. Biol. Pest.* **2008,** *18,* 381–384.

Gurbanov, G. G. Effectiveness and Use of Common Green Lacewing (*Chrysoperla carnea*) in Control of Sucking Pests and Cotton Moths on Cotton. *Biol. Nauk.* **1984,** *2,* 92–96.

Hagen, K. S.; Mills, N. J.; Gordh, G.; McMurtry, J. A. Terrestrial Arthropod Predator of Insects and Mites, Chrysopidae. In *Handbook of Biological Control: Principal and Application of Biological Control;* Bellows T. S., Fisher, T. W., Eds.; Academic Press: San Diego, CA, 1999; pp 415–423.

Hagley, E. A. C. Release of *Chrysoperla carnea* Stephens (Neuroptera: Chrysopidae) for Control of the Green Apple Aphid, *Aphis pomi* Dageer (Homoptera: Aphididae). *Can. Entomol.* **1989,** *121,* 309–314.

Harbaugh, B. K.; Mattson, R. H. Lacewing Larvae Control Aphids on Greenhouse Snap-Dragons. *J. Am. Soc. Hort. Sci.* **1973,** *98,* 306–309.

Henry, C. S.; Wells, M. M; Pupedis, R. J. Hidden Taxonomic Diversity within *Chrysoperla plorabunda* (Neuroptera:Chrysopidae): Two New Species Based on Courtship Songs. *Annal. Entomol. Soc. Am.* 1993, *86*, 1–13.

Henry, C. S.; Brooks, S. J.; Duelli, P.; Johnson, J. B.; Wells, M. M.; Mochizuki, A. Parallel Evolution in Courtship Songs of North American and European Green Lacewings (Neuroptera: Chrysopidae). *Biol. J. Linnean Soc.* **2012**, *105*, 776–796.

Henry, C. S. The Courtship Call of *Chrysopa downesi* Banks [sic] (Neuroptera: Chrysopidae): Its Evolutionary Significance. *Psyche.* **1980**, *86*, 291–297.

Henry, C. S.; Brooks, S. J.; Duelli, P.; Johnson, J. B. A Lacewing with the Wanderlust: The European Song Species 'Maltese', *Chrysoperla agilis* sp.n., of the Carnea Group of *Chrysoperla* (Neuroptera: Chrysopidae). *Syst. Entomol.* **2003**, *28*, 131–148.

Henry, C. S.; Brooks, S. J.; Duelli, P.; Johnson, J. B. Discovering the True *Chrysoperla carnea* (Insecta: Neuroptera: Chrysopidae) Using Song Analysis, Morphology, and Ecology. *Ann. Entomol. Soc. Am.* **2002**, *95* (2), 172.

Henry, C. S.; Brooks, S. J.; Duelli, P.; Johnson, J. B.; Wells, M. M.; Mochizuki, A. Obligatory Duetting Behaviour in the *Chrysoperla carnea*-Group of Cryptic Species (Neuroptera: Chrysopidae): Its Role in Shaping Evolutionary History. *Biol. Rev.* **2013**, *88*, 787–808.

Henry, C. S.; Brooks, S. J.; Johnson, J. B.; Wells, M. L. M.; Duelli, P. Song Analysis Reveals a Permanent Population of the Mediterranean Green Lacewing *Chrysoperla agilis* Henry (Neuroptera: Chrysopidae) Living in Central Alaska. *Annal. Entomol. Soc. Am.* **2011**, *104*, 649–657.

Henry, C. S.; Wells, M .L. M. Adaptation or Random Change? The Evolutionary Response of Songs to Substrate Properties in Lacewings (Neuroptera:Chrysopidae:Chrysoperla). *Anim. Behav.* **2004**, *68*, 879–895.

Henry, C. S.; Wells, M. L. M.; Simon, C. M. Convergent Evolution of Courtship Songs among Cryptic Species of the Carnea Group of Green Lacewings (Neuroptera: Chrysopidae:*Chysoperla*). Evolution. **1999**, *53* (4), 1165–1179.

Hoffmann, M. P.; Frodsham, A. C. *Natural Enemies of Vegetable Insect Pests. Cooperative Extension;* Cornell University: Ithaca, NY, 1993; p 63.

Holzel, H. *Anisochysa (Chrysoperla) mediterránea* n. sp. eine neue europaische Chrysopiden- Spezies (Planipennia, Chysopidae). *Nachricht. Bayer. Ent.* **1972**, *21* (5), 81–83.

IPM Laboratories, Inc. Lacewings for Aphid Control. *IPM Lab. Quart.* **1991**, *3* (1), 1–2.

Irwin, M. E.; Gill, R. W.; Gonzalez, D. Field Cage Studies of Native Egg Predators of the Pink Boll Worm in Southern California Cotton. *J. Econ. Entomol.* **1974**, *67*, 193–196.

Jin, Z. S. Integrated Control of Insect Pests on Cotton for Years. *Nat. Enem. Ins. China.* **1998**, *8*, 25–28.

Klingen, I.; Johansen, N. S.; Hofsvang, T. The Predation of *Crysoperla carnea* (Neuroptera: Chrysopidae) on Eggs and Larvae of *Mamestra brassicae* (Lepidoptera: Noctuidae). *J. Appl. Entomol.* **1996**, *120*, 363–367.

Knutson, A. E.; Tedders, L. Augmentation of Green Lacewing, Chrysoperla rufilabris, in Cotton in Texas. *Southwest. Entomol.* **2002**, *27*, 231–239.

Legaspi, J. C.; Nordlund, D. A.; Legaspi, B. C. Jr. Tri-Trophic Interactions and Predation Rates in *Chrysoperla* Spp. Attacking the Silverleaf Whitefly. *Southwest. Entomol.* **1996**, *21*, 33–42.

Lourenço, P.; Brito, C.; Backeljau, T.; Thierry, D.; Ventura, M. A. Molecular Systematics of the Chrysoperla Carnea Group (Neuroptera: Chrysopidae) in Europe. *J. Zool. Syst. Evol. Res.* **2006**, *44* (2), 180–184.

Maisonneuve, J. C. Biological Control with *Chrysoperla lucasina* against *Aphis fabae* on Artichoke in Brittany (France). In *Lacewings in the Crop Environment;* McEwen, P., New, T. R., Whittington, A., Eds.; Cambridge University Press: Cambridge, UK, 2001; pp. 513–517.

Malais, M. H.; Ravensberg, W. J. Knowing and Recognizing. In *The Biology of Glasshouse Pests and Their Natural Enemies;* Koppert B. V., Ed.; Berkel en Rodenrijs: The Netherlands, 2003; p 288.

McEwen, P.; Kidd, N. A. C.; Bailey, E.; Eccleston, L. Small-Scale Production of the Common Green Lacewing *Chrysoperla carnea* (Stephens) (Neuroptera, Chrysopidae): Minimizing Costs and Maximizing Output. *J. Appl. Entomol.* **1999**, *123*, 303–306.

McEwen, P.; New, T. R.; Whittington, A. E. *Lacewings in the Crop Environment;* Cambridge University Press: UK, 2001; p 564.

Mochizuki, A.; Naka, H.; Hamasaki, K.; Mitsunaga, T. Larval Cannibalism and Intraguild Predation between the Introduced Green Lacewing, *Chrysoperla carnea*, and the Indigenous Trash-Carrying Green Lacewing, *Mallada desjardinsi* (Neuroptera: Chrysopidae), as a Case Study of Potential Non target Effect Assessment. *Environ. Entomol.* **2006**, *35*, 1298–1303.

Nordlund, D. A. Improvements in the Production System for Green Lacewings: A Hot Melt Glue System for Preparation of Larval Rearing Units. *J. Entomol. Sci.* **1993**, *28* (4), 338–342.

Olkowski, W.; Daar, S.; Olkowski. H. *Common Sense Pest Control;* The Taunton Press: Newtown, CT, 1991; pp 66–68.

Rajakulendran, S. V.; Plapp, F. W. Comparative Toxicities of Five Synthetic Pyrethroids to the Tobacco Budworm (Lepidoptera: Noctuidae), an Ichneumonid Parasite, *Campoletis sonorensis*, and a Predator, *Chrysopa carnea*. *J. Econ. Entomol.* **1982**, *75*, 769–772.

Raupp, M. J.; Hardin, R. M.; Braxton, S. M.; Brenda, B. B. Augmentative Release for Aphid Control on Landscape Plants. *J. Arbor.* **1994**, *20* (5), 241–249.

Reddy, G. V. P. Comarative Effectiveness of an Integrated Pest Management System and Other Control Tactics for Managing the Spider Mite *Tetranychus ludeni* (Acari: Tetranychidae) on Egg Plant. *Appl. Acarol.* **2001**, *25*, 985–992.

Reddy, G. V. P. Plant Volatiles Mediate Orientation and Plant Preference by the Predator *Chrysoperla carnea* Stephens (Neuroptera: Chrysopidae). *Biol. Control.* **2002**, *25*, 49–55.

Ridgway, R. L.; Jones, S. L. Field-Cage Releases of *Chrysopa Carnea* for Suppression of Populations of the Bollworm on Cotton. *J. Econ. Entomol.* **1968**, *61*, 895–898.

Ridgway, R. L.; Jones, S. L. Inundative Release of *Chrysopa carnea* for Control of Heliothis on Cotton. *J. Econ. Entomol.* **1969**, *62*, 177–180.

Ridgway, R. L.; Morrison, R. K.; Badgley, M. Mass Rearing of Green Lacewing. *J. Econ. Entomol.* **1970**, *63*, 834–836.

Sattar, M.; Abro, G. H. Mass Rearing of *C. carnea* (Stephens) (Neuroptera: Chrysopidae) Adults for Integrated Pest Management Programmes. *Pak. J. Zool.* **2011**, *43* (3), 483–487.

Sattar, M.; Hamed, M.; Nadeem, S. Predatory Potential of *C. carnea* (Stephens) (Neuroptera: Chrysopidae) Against Cotton Mealy Bug. *Pak. Entomol.* **2007**, *29* (2), 103–106.

Senior, L. J.; MCEwen, P. K. The Use of Lacewings in Biological Control. In *Lacewing in the Crop Environment;* McEwen, P. K., New, T. R., Whittington, A. E., Eds.; Cambridge University Press: Cambridge, 2001; pp 296–302.

Shands W. A.; Simpson, S. W.; Brunson, M. H. Insect Predators for Controlling Aphids on Potatoes. 1. In Small Plots. *J. Econ. Entomol.* **1972**, *65,* 511–514.

Shrewabury, P. M.; Lashomb, J. H.; Hamilton, J.; Zhang, J.; Patts, J. M.; Casargrande, R. A. The Influence of Flowering Plants on Herbivore and Natural Enemy Abundance in Ornamental Landscapes. *Int. J. Environ. Sci.* **2004**, *30,* 23–33.

Smith, R C. The Chrysopidae (Neuroptera) of Canada. Ann. Entomol. Soc. Am. **1932**, *25,* 579–601.

Stark, S.; Whitford, F. Functional Response of *Chrysopa carnea* (Neuroptera: Chrysopidae) Larvae Feeding on *Heliothis virescens* (Lep.: Noctuidae) Eggs on Cotton in Field Cages. *Entomophaga.* **1987**, *12* (5), 521–527.

Syngenta-bioline.co.uk. Aphid Control by *Chrysoline carnea.* **2011**. Syngenta-bioline.co.uk.

Tauber, M. J.; Tauber, C. A.; Daane, K. M.; Hagen, K. S. Commercialization of Predators: Recent Lessons from Green Lacewings (Neuroptera: Chrysopidae). *Ann. Entomol.* **2000**, *46,* 26–37.

Tooth, M.; Szentkiraalyi, F.; Vuts, J.; Letardi, A.; Kundsen, G. K. Optimization of a Phenylacetaldehyde-Based Attractant for Common Green Lacewings (*C. carnea*). *J. Chem. Ecol.* **2009**, *35,* 449–458.

Tulisalo, U. Biological Control in the Greenhouse. In *Biology of Chrysopidae;* Canard, M., Semeria, Y., New, T. R., Eds.; Dr. W. Junk: The Hague, The Netherlands, 1984; pp 228–233.

Van Lenteren, J. C. *Quality Control and Production of Biological Control Agents: Theory and Testing Procedures;* CABI Publishing: UK, 2003; p 327.

Vijayakumar, N. Improved Mass Production Technologies of Major Biocontrol Agents for Efficient Pest Management. In *Organic Pest Management: Potentials and Applications;* Narayanasamy, P., Ed.; 2009; pp 164–176.

Weihrauch, F. Overwintering of Common Green Lacewings in Hibernation Shelters in the Hallertau Hop Growing Area. *Bull. Insectol.* **2008**, *61,* 67–71.

Wells, M. M.; Henry, C. S. Songs, Reproductive Isolation and Speciation in Cryptic Species of Insects: A Case Study Using Green Lacewings. In *Endless Forms: Species and Speciation;* Howard, D., Berlocher, S., Eds.; Oxford University Press: New York, 1998; pp. 217–233.

Wells, M. M; Henry, C. S. The Role of Courtship Songs in Reproductive Isolation among Populations of Green Lacewings of the Genus *Chrysoperla* (Neuroptera: Chrysopidae). *Evolution.* **1992**, *46* (1), 31–42.

Wennemann, L. Innovative Crop Protection. *Gesunde Pflanz.* **2003**, *55,* 236–243.

Whitcomb, W. H.; Bell, K. Predaceous Insects, Spiders and Mites of Arkansas Cotton Fields. *Univ. Ark. Agric. Exptl. Stn. Bull.* **1964**, *680,* 84.

Yu, H.; Zhang, Y.; Wu, K.; Gao, X. W.; Guo, Y. Y. Field-Testing of Synthetic Herbivore-Introduced Plant Volatiles as Attractants for Beneficial Insects. *Entomol.* **2008**, *37,* 1410–1415.

Zaki, F. N.; El-Shaarawy, M. F.; Farag, N. A. Release of Two Predators and Two Parasitoids to Control Aphids and Whiteflies. *Anz. Schadlingsk.* **1999**, *72,* 19–20.

PART IV
Genetically Modified Crops, *Bacillus thuringiensis* and Phytochemicals in Biconrol

CHAPTER 12

GENETICALLY MODIFIED CROPS FOR INSECT PESTS AND DISEASE RESISTANCE

TUSHAR RANJAN[1], SANGITA SAHNI[2], BISHUN DEO PRASAD[3*], and VIJAY KUMAR JHA[4]

[1]Department of Basic Science and Humanities Genetics, Bihar Agricultural University, Sabour 813210, Bhagalpur, Bihar, India

[2]Department of Plant Pathology, Tirhut College of Agriculture, Dholi, RAU, Pusa 843105, Bihar, India

[3]Department of Molecular Biology and Genetic Engineering, Bihar Agricultural University, Sabour 813210, Bhagalpur, Bihar, India

[4]Department of Botany, Patna University, Patna 800005, Bihar, India

*Corresponding author. E-mail: dev.bishnu@gmail.com

CONTENTS

ABSTRACT

Genetically modified crops, whose genetic material/DNA is either altered or modified using recombinant DNA technology, are popularly used in agriculture now these days. The whole idea behind generating generally modified (GM) crops is to add one or sometimes more genes to an organism's genome. They give the crops new characteristics, like insect resistance, larger yields, and faster growing traits. The use of generally modified organisms (GMOs) is hardly new, but many believe that sufficient research on the long-term effects has not been conducted. With an ever-increasing global population, massive world hunger, and with an estimation that a child dies for every two seconds worldwide from starvation or undernourished, there is a great promise in the use of this technology to benefit not only the farmers but also societies worldwide.

12.1 INTRODUCTION

Genetically modified crops (also known as GMCs or GM crops or sometimes biotech crops) are plants used in agriculture whose genetic material/ DNA has been altered or modified using recombinant DNA technology or genetic-engineering techniques. The most common modification is to add one or sometimes more genes to an organism's genome. Less commonly, genes are removed or their expression is increased or silenced or the number of copies of a gene is increased or decreased. The idea behind developing GM crops is to introduce a trait that does not naturally occur in the plant to help improve, and protect the crop. Genetic engineering within food crops is done to create resistant to disease, pest, environmental stress condition, reduction of spoilage, resistant to chemical treatment (such as herbicides), and improving the nutrient profiling. The non-food crops are utilized for the production of pharmaceutical agents, biofuel and bioremediation as well (Conner et al., 2003; Meagher, 2000; Roush, 1997). DNA transfers naturally between organisms on large scale (Bock, 2010). Several natural mechanisms allow gene flow across species. For example, development of antibiotic resistance in bacteria is facilitated by mobile genetic elements such as transposons and proviruses that naturally translocate DNA to new loci in the genome (Monroe, 2006). Since early, the introduction of foreign germplasm into the crops has been achieved by traditional crop breeders by overcoming species barriers. A hybrid cereal grain was created in 1875 for the first time by crossing wheat and rye (Chen, 2010). By that time, important including genes for dwarfing and rust resistance have been introduced. Recent advancements in plant tissue culture and deliberate mutations as well have enabled us to alter the genomic makeup of plants according to our convenience (Predieri, 2001; Duncan, 1996). Farmers have widely adopted GM technology worldwide, but India is still very far in this race. The total surface area of land cultivated with GM crops increased by a factor of 100, from 17,000 km^2 (4,200,000 acres) to 1,750,000 km^2 (432 million acres) in between 1996and 2013. In the United States, by 2014, 94% of the planted area of soybeans, 96% of cotton and 93% of corn were genetically modified varieties. Approximately, 18 million of farmers account for 54% of worldwide GM crops in the developing countries. According to the *United States Department of Agriculture* (USDA), the number of field releases for genetically engineered organisms has grown from four in 1985 to an average of about 800 per year. Cumulatively, more than 17,000 releases had been approved through September 2013 (ISAAA, 2013).

Once satisfactory strains are produced, the producer applies for "field testing" or "field release." Field-testing involves cultivating the plants on farm fields in a controlled environment. If these field tests are successful, the producer applies for regulatory approval to grow and market the crop. Once approved, specimens (seeds, cuttings, breeding pairs, etc.) are cultivated and sold to farmers. The farmers cultivate and market the new strain. In some cases, the approval covers marketing but not cultivation. GM crops provide a number of ecological benefits and a recent study also indicates no human health risk after utilizing these crops. However, opponents have objected to accept GM crops on the grounds of environmental concerns, whether food produced from GM crops is safe and economic concerns raised by the fact these organisms are subject to intellectual property law (Ronald, 2011). The first GM crop produced was an antibiotic-resistant tobacco plant in 1982 and the first field trials were conducted in France and the United States in 1986, when tobacco plants were engineered for herbicide resistance. Plant Genetic Systems (Belgium), founded by Montagu and Schell, was the first company to produce genetically engineered insect-resistant (tobacco) plants by incorporating genes that produced insecticidal proteins from *Bacillus thuringiensis* (Bt) in 1987 (Fraley et al., 1983; Vaeck et al., 1987). China was the first country to allow commercialized transgenic virus-resistant tobacco plant. But, GM crop was approved for sale for the first time in the United States, in 1994, which was the *Flavr*

Savr tomato. It had a longer shelf life, because it took longer to soften after ripening. In 1994, Europe approved engineered tobacco which was resistant to the herbicide bromoxynil. In 1995, Bt Potato, Bt maize, glyphosate-resistant soybeans, virus-resistant squash, and additional delayed ripening tomatoes were approved. In 2000, Vitamin A-enriched golden rice, was the first food with increased nutrient value (Bruening & Lyons, 2000; Conner et al., 2003). Plants engineered to tolerate non-biological stressors such as drought, frost, high soil salinity, and nitrogen starvation were in development. In 2011, Monsanto's DroughtGard maize became the first drought-resistant GM crop to receive US marketing approval (Lundmark, 2006; Manoj et al., 2012). In 2012, the Food and Drug Administration (FDA) approved the first plant-produced pharmaceutical, a treatment for Gaucher's disease. Till date, several plants including tobacco plants have been modified to produce therapeutic antibodies.

In 2005, about 13% of the Zucchini (a form of squash) grown in the United States was genetically modified to resist three viruses. In 2011, the potato was made resistant to late blight by adding resistant genes *blb1* and *blb2* that originate from the Mexican wild potato *Solanum bulbocastanum*. In 2013, the USDA approved the import of a GM pineapple that is pink in color and that "overexpresses" a gene derived from tangerines and suppress other genes, increasing production of lycopene. In 2013, Robert Fraley, Marc Van Montagu, and Mary-Dell Chilton were awarded the World Food Prize for improving the "quality, quantity or availability" of food in the world for applying genetic engineering technology in the crops. Figure 12.1 represents global area of GM crops worldwide. In 2014, the USDA approved a genetically modified potato developed by J.R. Simplot Company that contained ten genetic modifications that prevent bruising and produce less acrylamide when fried. The modifications eliminate specific proteins from the potatoes, via RNA interference, rather than introducing novel proteins. In February 2015, Arctic Apples were approved by the USDA and became the first genetically modified apple approved for sale in the United States. Gene silencing was used to reduce the expression of polyphenol oxidase (PPO), thus preventing the fruit from browning (Conner et al., 2003). Cereals such as corn basically used for the food and ethanol production has been genetically modified to tolerate various herbicides by expressing a protein from *Bacillus thuringiensis* (Bt) that kills certain insects. In 2015, 81% of corn acreage contained the Bt trait and 89% of corn acreage contained the glyphosate-tolerant trait in United States. Corn can be processed into grits, meal, and flour as an ingredient in pancakes, muffins, doughnuts, breadings and batters, baby foods, meat products, cereals, and some fermented products as well. Corn-based masa flour and masa dough are used in the production of taco shells, corn chips, and tortillas (Conner et al., 2003).

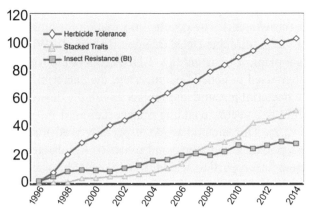

FIGURE 12.1   Global area of biotech crops, 1996–2007, by trait (million hectares) (Reproduced from James, Clive. 2009. Global Status of Commercialized Biotech/GM Crops: 2009. ISAAA Brief No. 41. ISAAA: Ithaca, NY. With permission.)

12.2   METHODS FOR GM CROPS DEVELOPMENT

Often used methods for developing GM crops are gene guns, electro-poration, microinjection, *Agrobacterium*, CRISPER, and TALEN which aids in either addition or removal of gene. Gene guns shoot target genes (DNA is bound to tiny particles of gold or tungsten) into plant tissue or single plant cells under high pressure (Fig. 12.2). The accelerated particles penetrate both the cell wall and membranes. The DNA separates from the metal and is integrated into plant DNA inside the nucleus. This method has been applied successfully for many cultivated crops, especially mono-cots like wheat or maize, for which transformation using *Agrobacterium tumefaciens* has been less successful. This procedure could cause a serious damage to the cellular tissue (Shrawat & Lorz, 2006). Electroporation is used when the plant tissue does not contain cell walls. In this technique, DNA enters the plant cells through miniature pores which are temporarily caused by electric pulse (Fig. 12.2). Microinjection basically injects the gene into the host directly. Indirect method of gene transfer is another common technique which involves *Agrobacterium tumefaciens* as living host (Fig. 12.2). *Agrobacterium* are natural plant parasites, and their natural ability to transfer genes provides another engineering method. To create a suitable environment for them, these Agrobacteria insert their genes into plant hosts, resulting in a proliferation of modified plant cells near the soil level (crown gall). The genetic information for tumor growth is encoded on a mobile, circular DNA fragment (plasmid). When *Agrobacterium* infects a plant, it transfers this T-DNA to a random site in the plant genome. When used in genetic engineering, the bacterial T-DNA is removed from the bacterial plasmid and replaced with the desired foreign gene. The bacterium is a vector, enabling transportation of foreign genes into plants (Fig. 12.2). This method works especially well for dicotyledonous plants like potatoes, tomatoes, and tobacco. Agrobacteria infection is less efficient in crops like wheat and maize (Shrawat & Lorz, 2006). Tobacco and *Arabidopsis thaliana* serve as model organisms for other plant species due to well-developed transformation methods, easy

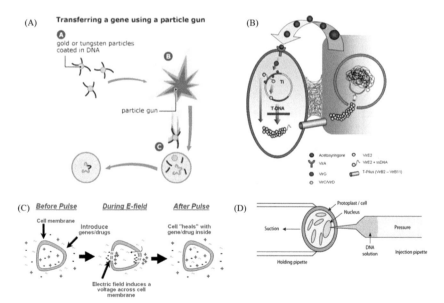

FIGURE 12.2 Represents different methodology used in the development of GM crops. Steps involving in gene guns (A), *Agrobacterium* mediated gene transfer (B), electroporation (C), and microinjection (D) are illustrated in details.

propagation, and well-studied genomes. Introducing new genes into plants requires a promoter specific to the area where the gene is to be expressed. For example, to express a gene only in rice grains and not in leaves, an endosperm-specific promoter is used. The codons of the gene of interest must be optimized for the organism due to codon usage bias. In addition, transgenic gene products should be able to be denatured by heat so that they are destroyed during cooking (Koornneef & Meinke, 2010).

Transgenic plants have genes inserted into them that are derived from another species. The inserted genes can come from species within the same kingdom (plant to plant) or between kingdoms (e.g., bacteria to plant). In many cases, the inserted DNA has to be modified slightly in order to correctly and efficiently express in the host organism. Transgenic plants are used to express proteins like the cry toxins from *B. thuringiensis*, herbicide-resistant genes, antibodies, and antigens for vaccinations (Tables 12.1 and 12.2). Transgenic carrots have been used to produce the drug Taliglucerase alfa which is used to treat Gaucher's disease. In the laboratory, transgenic plants have been modified to increase photosynthesis. This is possible by

TABLE 12.1   Different Methods for Gene Transfer.

Methods	Description
Indirect gene transfer	
Agrobacterium mediated	Delivery of foreign DNA inside the nucleus by using Ti-plasmid as a vector.
Direct gene transfer	
Particle bombardment	Delivery of DNA into cells using microscopic gold or tungsten particles coated with DNA as carriers accelerated into target cells by gunpowder, gas or air pressure or by electrical discharge.
Protoplast transformation	DNA introduction into protoplasts using PEG-mediated DNA uptake and electroporation, liposome (containing plasmid DNA) fusion.
Tissue electroporation	Transformation of plant organs or regenerable cell cultures.
Ultrasound-induced transformation	DNA uptake into protoplasts, suspension cells, and tissues induced by ultrasound waves.
Silicon-carbide fiber or whisker transformation	The fibers perforate cell walls and allow DNA to penetrate the cells.
Laser-mediated transformation	Laser beams are used to create openings in cell components and organelles allowing DNA insertion.
Microinjection	Direct delivery of DNA into plant cells using a micro-syringe.

changing the rubisco enzyme (i.e. changing C3 plants into C4 plants), by placing the rubisco in a carboxysome, by adding CO_2 pumps in the cell wall (*Walmsley & Arntzen, 2000*). In between 1995 and 2014, GM crop has made a great impact on agronomic and economic status. The study found that herbicide-tolerant crops have lower production costs, while for insect-resistant crops the reduced pesticide use was offset by higher seed prices, leaving overall production costs about the same. Yields increased 9% for herbicide tolerance and 25% for insect resistance. Farmers who adopted GM crops made 69% higher profits than those who did not. It was observed that GM crops help farmers in developing countries, increasing yields by 14% points (Klümper & Qaim, 2014). Table 12.2 lists important GMCs with advantageous traits.

TABLE 12.2 Milestones for the Development of Transgenic Crops with Several Agronomical Important Traits.

Traits	Modification of gene	Name of crops
Tolerance of glyphosate or glufosinate	Gene added	Alfalfa
Delayed browning	Genes added for reduced polyphenol oxidase (PPO) production from other apples	Apple
Tolerance of glyphosate or glufosinate high laurate canola, oleic acid canola	Gene added	Canola
Tolerance of herbicides glyphosate glufosinate and 2,4-D insect resistance. Added enzyme, alpha amylase, that converts starch into sugar to facilitate ethanol production. Viral resistance	Gene from Bt added	Corn
Insect resistance	Gene from Bt added	Cotton
Insect resistance	Gene from Bt added	Eggplant
Resistance to the papaya ring spot virus	Gene added	Papaya
Resistance to Colorado beetle Resistance to potato leaf roll virus and Potato virus Y. Reduced acrylamide when fried and reduced bruising	Bt cry3A, coat protein from PVY "Innate" potatoes added genetic material coding for mRNA for RNA interference	Potato (Food)

TABLE 12.2 *(Continued)*

Traits	Modification of gene	Name of crops
Antibiotic resistance gene used for selection	Antibiotic resistance gene from bacteria	Potato (Starch)
Better starch production	Modifications to endogenous starch-producing enzymes	
Enriched with beta-carotene (a source of vitamin A)	Genes from maize and a common soil microorganism	Rice
Tolerance of glyphosate or glufosinate. Reduced saturated fats.	Herbicide resistant gene taken from bacteria added	Soybeans
Kills susceptible insect pests.	Knocked out native genes that catalyze saturation.	
Viral resistance.	Gene for one or more Bt crystal proteins added.	
Resistance to watermelon, cucumber and zucchini/courgette yellow mosaic viruses	Viral coat protein gene	Squash
Tolerance of glyphosate, glufosinate	Genes added	Sugar beet
Pesticide tolerance High sucrose content	Genes added	Sugarcane
Resistance to cucumber mosaic virus	Viral coat protein gene	Sweet Peppers
Suppression of the enzyme polygalactu-ronase (PG)	Antisense gene of the gene responsible for PG enzyme production added	Tomato

Three types of modifications are needed for the development of GM crops, which includes transgenic, cisgenic, and subgenic. Transgenic crops have genes inserted into them that are derived from another species. The inserted genes can come from species within the same kingdom (e.g., plant to plant) or between kingdoms (e.g., bacteria to plant). In many cases, the inserted DNA has to be modified slightly in order to correctly and efficiently express in the host organism (codon bias). Transgenic plants are used to express proteins like the cry toxins (from *B. thuringiensis),* herbicide-resistant genes, antibodies, and antigens for vaccination. Transgenic carrots have been used to produce the drug Taliglucerase alfa which is used to treat Gaucher's disease (Podevina & Jardin, 2012; Walmsley & Amtzen, 2000). Cisgenic plants are basically made using genes found within the same species or a closely related one, where conventional plant breeding can occur. Some breeders and scientists argue that cisgenic modification is useful for plants that are difficult to crossbreed by conventional means (such as potatoes), and that plants in the cisgenic category should not require the same regulatory scrutiny as transgenics. In 2014, Chinese researcher Gao Caixia filed patents on the creation of a strain of wheat that is resistant to powdery mildew. The strain lacks genes that encode proteins that repress defenses against the mildew. The researchers deleted all three copies of the genes from wheat's hexaploid genome. The strain promises to reduce or eliminate the heavy use of fungicides to control the disease. Gao used the TALENs and CRISPR gene editing tools without adding or changing any other genes. Such kind of modification is known as subgenic. No field trials were immediately planned (Yanpeng, 2014).

12.3 WHY ARE THE CROPS BEING GENETICALLY MODIFIED?

Agricultural productivity is highly influenced by pest and disease, the most harmful factor concerning the growth and productivity of the crops worldwide. Conventional breeding methods are being used to develop the varieties more resistant to the biotic stresses. At the same time, these methods are tedious, time and resource consuming, germplasm dependent as well. On the other hand, pest management by chemicals obviously made considerable protection to the crops yields over the past few decades. Regrettably, extensive and very frequent usage of chemical pesticides has resulted in adverse effect on environment, human health, eradication of beneficial microorganism and insects, and development of pest-resistant insects too. At this situation, recently, biotechnology has been targeted for the enhancement of crop production in term of their efficiency that is locally important in developing countries. GM crops grown today, or under development, have been modified with various useful traits. These traits basically include improved shelf life, disease resistance, stress resistance, herbicide, pest resistance, production of useful goods such as biofuel or drugs, and ability to absorb toxins and for use in bioremediation of pollution (Table 12.1) (Wahab, 2009). The DNA transmission capabilities of *Agrobacterium* have been vastly explored in biotechnology as a means of inserting foreign genes into plants. Marc Van Montagu and Jeff Schell discovered the gene transfer mechanism between *Agrobacterium* and plants, which resulted in the development of methods to alter the bacterium into an efficient delivery system for genetic engineering in plants.

The plasmid T-DNA that is transferred to the plant is an ideal vehicle for genetic engineering. This is done by cloning a desired gene sequence (beneficial traits) into the T-DNA that will be inserted into the host DNA (Fig. 12.3). This process has been performed using firefly luciferase gene to produce glowing plants. This luminescence has been a useful device in the study of plant "chloroplast" function and as a reporter gene. It is also possible to transform *Arabidopsis thaliana* by dipping flowers into a broth of *Agrobacterium*: the seed produced will be transgenic. Under laboratory conditions, the T-DNA has also been transferred to human cells, demonstrating the diversity of insertion application (Schell & Montagu, 1997).

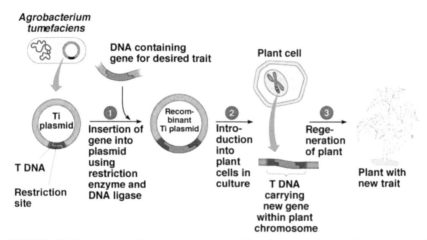

FIGURE 12.3 A model for *Agrobacterium* mediated gene transfer technology http://www.whatisthebiotechnology.com/blog/plant-transformation/

12.4 CROP PROTECTION THROUGH INSECT PESTS RESISTANT GENE

Insect attack is a serious agricultural problem leading to yield losses and reduced product quality. Insects can cause damage both in the field and during storage in silos. Each year, insects destroy about 25% of food crops worldwide. The larvae of *Ostrinia nubilalis*, the European corn borer, can destroy up to 20% of a maize crop. Pest and disease resistance is expressed in the ability of an organism to exclude or overcome, completely or to some extent, the effect of pathogen/pest/other damaging factors. Pests, weeds, and disease are most difficult and costly to deal with once resistance is acquired. Therefore, it is paramount that management of resistance programs exists for preventing resistance from developing, which possesses genetic bases, and is dependent on environmental and ecological factors. Transgenic plants should only be released into the field once well-planned integrated pest management programs exist. This entails implementation of polygenic traits with integrated chemical and biological control methods, which should ensure the stability of transgenic plants to pest and disease resistance. Theoretically, the most effective pest-resisting transgenic lines or conventionally bred cultivars are those that do not cause a selection or adaptation of the target organism to the resistant trait. However, this goal is elusive and rarely achieved. Plant pests (insects, mites nematodes), diseases (viral, fungal, and bacterial), and weeds are the cause of major crop damage resulting in economic losses to farmers worldwide. Pests, diseases, and weeds are managed mainly by the widespread application of insecticides, fungicides, and herbicides. However, these synthetic chemicals have a detrimental effect on human health, the environment, and non-target organisms. Specifically, acute problems which have arisen are due to pest resistance, chemical residues in agricultural products, detrimental effect on beneficial organisms, reduction of biodiversity of species, and pollution of surface and groundwater.

12.5   MECHANISM OF PLANT DEFENSE

Plants need to protect themselves against biotic stress such as pests (insects, mites, and nematodes), diseases (viral, fungal, and bacterial), and even weeds. Plants do encode innate immunity to fight against several diseases enabling them to survive in unfavorable environment. At many occasions, a deadly combination of detrimental organisms may coexist while attacking a plant and making plants more vulnerable for biotic attack. For instance, herbivorous insects are capable not only of causing mechanical damage but may serve also as vectors for numerous microbial pathogens such as virus. In a typical non-host interaction, the invading microbial pathogen or even pest is incapable of colonizing plant host tissue and sometimes no response is evident by such host plants. Alternatively, the host plant may resist invasion by activating several complex defense mechanisms preventing multiplication of pathogen through its replication and their spread between the cells either through plasmodesmata or through some other mechanisms. This kind of defense mechanism is frequently governed directly or indirectly by very interesting events known as gene-to-gene interaction among the products of a plant resistant (R) gene and a pathogen avirulence (*avr*) gene. Study suggests that knockout of either R-gene or *avr* gene makes plant more prone to many diseases, indicating critical role of these gene(s) interactions in innate immunity. Lacking of these genes leads to the suppression of defense mechanism and finally plant colonization takes place. During altered or incompatible interaction among R-gene or *avr* gene, the pathogen enters the first sub-cuticular cell and activates a cascade of host defense response which results in the collapse or apoptosis of the plant cells surrounding the infection site thereby "walling off" the pathogen. During the earlier stages of an incompatible reaction, the plant recognizes specific pathogen-generated molecules through a specific receptor protein (transmembrane protein) impinge on the plasma membrane, which causes the activation of cellular signaling transduction pathways, and rapid accumulation of specific gene transcripts or transcription factors, which subsequently activates the synthesis of pathogenesis-related (PR) proteins. Thereafter, these PR proteins activate a series of complex biochemical pathways (or cascades) inducing a resistant reaction which are usually accompanied by cell death, termed as hypersensitive response (HR). Various plant compounds such as phytoalexins, proteases, peroxidases, free radicals, lignin, callose, and several PR proteins have been found to be associated with HR and plant resistance against many agronomically important diseases. Subsequently, a long-enduring broad-spectrum resistance develops, termed as systemic acquired resistance (SAR). Interestingly, while a compatible host interaction, plant death occurs as the pathogen emerges from the first infected cell, without activating the plant host defense systems. After spreading of the pathogen through host-infected tissue, the early infected cells disintegrate and release several plant metabolites which triggers a defense response at that location. Interestingly, synthesis of PR proteins was observed to be delayed during a compatible interaction between a pathogen and its susceptible host. The major differences between compatible and incompatible interactions appear to be the timing and magnitude of defense system activation and enhanced PR protein synthesis (Van Loon, 1997; Maagd et al., 1999).

12.6   INSECT PEST RESISTANCE

In the recent year, chemical control method is generally preferred over continuous plant breeding efforts for the management of insect pests. Many fascinating new genetic methods for insect control are being used these days, which could substantially reduce expenditures and crop losses, and to be precise, these methods are less detrimental to the environment. A very successful molecular approach to engineering resistance has involved generating chimeric plants with the capability of synthesizing antimicrobial or insecticidal products. These products are usually constitutively produced in plants. This means that these insecticidal genes are put up under the control of strong constitutive promoters. In many cases, an inducer or activator is necessary for the gene expression in order to activate synthesis of these engineered chemicals if the presence of a pest is detected (e.g., the tetracycline-inducible promoter system). One of the drawbacks of this system is requirement of large quantities of tetracycline for induction, which makes this system not to be ideal. Therefore, alternative inducers are currently being developed by the researchers.

Btδ-**endotoxins:** *Bacillus thuringiensis* (Bt), a Gram-positive soil bacterium, produces Btδ-endotoxins, crystalline inclusions during sporulation. These inclusions contain several insecticidal proteins out of which over 100 different Bt toxins have been identified till date. The δ-endotoxins are processed inside the insect midgut to form the active form of toxin. Numerous plant species have now been transformed with Bt δ-endotoxin, a bacterial gene expressing the insecticidal proteins, making the plants tissues toxic to several insect pests. The Btδ-endotoxin gene was cloned in 1981 for the first time and the source of gene was from *B. thuringiensis*. Moreover, a research article on transgenic plants protected from insects by δ-endotoxins was published in 1987. Commercial seeds are available for several important crops including corn, potato, and cotton, expressing different synthetic Bt genes that show significant protection against the European corn borer, Colorado potato beetle, cotton bollworm, and pink bollworm infestations. Bt δ-endotoxin expression is currently under development stage in crops such as Alfalfa, apples, cranberry, eggplant, rice, and other plant species. One of the major challenges for scientific community is the durability and stability of produced Bt δ-endotoxins, due to the fact that certain pests have shown resistance to some of the toxins (Maagd et al., 1999).

Apart from Bt gene, several other insecticidal proteins are derived from plants which include chitinases, peroxidases, β-amylase inhibitors, proteinase inhibitors, trypsin inhibitors, and lectins. These proteins have also very significant effect on pest resistivity by the crops harboring these genes. Transgenic plants expressing these compounds have been generated and evaluated for control of various pests worldwide. Recently, transgenic tobacco has been used for expressing proteinase inhibitors and peroxidase as well, for control of *Manducasexta* larvae and *Helicoverpa zea*. Interestingly, transgenic pea seedlings expressing alpha-amylase inhibitor showed significant resistance to bruchid beetles. *Streptomyces*-derived cholesterol oxidase proteins have also recently been reported to have insecticidal activity worldwide against the cotton boll weevil, which is very difficult to control using conventional pesticide sprays. These compounds are being widely expressed in transgenic tobacco cells and may prove very useful alternative in the coming future (Kuc, 1990).

12.7   DISEASE RESISTANCE

Plants can suffer from infections caused by fungi, bacteria, viruses, nematodes, and other pathogens. Various high-tech approaches have been proposed to protect plants from harmful afflictions. To date, most interest has been focused on virus-resistant transgenic plants, but using biotechnology to confer resistance to fungi, bacteria, or nematodes has also been gaining attention (Table 12.1). Fungi are responsible for a range of serious plant diseases such as blight, grey mould, bunts, powdery mildew, and downy mildew. Crops of all kinds often suffer heavy losses. Fungal plant diseases are usually managed with applications of chemical fungicides or heavy metals. In some cases, conventional breeding has provided fungus-resistant cultivars. Besides combatting yield losses, preventing fungal infection keeps crops free of toxic compounds produced by some pathogenic fungi. These compounds, often referred to as mycotoxins, can affect the immune system, and disrupt hormone balances. Some mycotoxins are carcinogenic.

Genetic engineering enables new ways of managing fungal infections. Several approaches have been taken: (a) Introducing genes from other plants or bacteria encoding enzymes like chitinase or glucanase. These enzymes break down chitin or glucan, respectively, which are essential components of fungal cell walls; (b) introducing plant genes to enhance innate plant defense mechanisms (e.g., activating phytoalexins, proteinase inhibitors, or toxic proteins); and (c) invoking the hypersensitive reaction: Plants varieties that are naturally resistant to specific types of fungal diseases are often programmed to have individual cells quickly die at the site of fungal infection. This response, known as the hypersensitive reaction, effectively stops an infection in its tracks. Genetic engineering can help plant cells "know" when a fungus is attacking.

12.7.1   VIRAL DISEASE

Viruses cause many economically important plant diseases. For example, the *Beet necrotic yellow vein virus* (BNYVV) causes sugar beets to have smaller, hairier roots, reducing yields by up to 50%. The spread of most viruses is very difficult to control. Once infection sets in, no chemical treatment methods are available. Losses are usually very high and require longer rotation intervals and modified cropping systems. This translates into considerable losses. Viruses are often transmitted from plant to plant by insects. Insecticides are sometimes used to control viral infections, but success is very limited. The most effective ways of managing viruses are cultural controls (e.g., removing diseased plants) and using resistant cultivars. Although conventional methods of breeding have been able to provide some virus-resistant or tolerant cultivars, they are not available for most crops. In some cases, biotechnology can be used to make virus-resistant crops. The most common way of doing this is by giving a plant a viral gene encoding the virus' "coat protein" (Table 12.1). The plant can then produce this viral protein before the virus infects the plant. If the virus arrives, it is not able to reproduce. The explanation for this is called cosuppression or virus-induced gene silencing (VIGS) (Fig. 12.4). The plant has ways of knowing that the viral coat protein should not be produced, and it has ways of eventually shutting down the protein's expression. When the virus tries to infect the plant, the production of its essential coat protein is already blocked. Figure 12.4 represents RNA-induced gene silencing in plants. All genetically modified virus resistant plants on the market (e.g., papayas and squash) have coat protein mediated resistance. It may also be possible to confer resistance by taking a resistance gene naturally found in one plant and then transferring it to an important crop (Baulcombe, 2004).

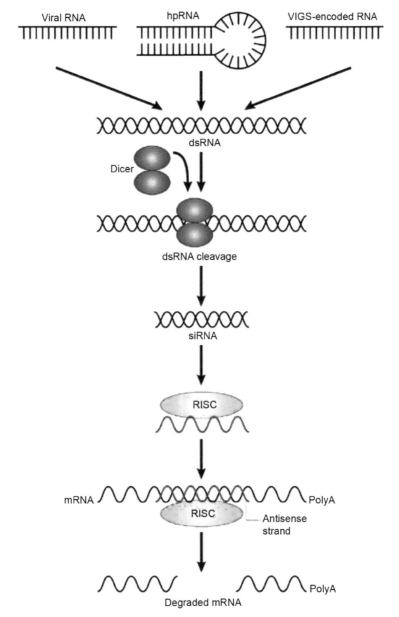

FIGURE 12.4   Overall mechanism of RNA interference or virus-induced gene silencing (VIGS) technology operates in plants (Adapted by permission from Macmillan Publishers Ltd: [Waterhouse, P. M.; Helliwell, W. C. Exploring Plant Genomes by RNA-Induced Gene Silencing. Nature Rev. Genet. 2003, 4 (1), 29–38. © 2003.)

Virus-induced gene silencing (VIGS) is a virus vector technology that exploits an RNA-mediated antiviral defense mechanism. In plants infected with unmodified viruses, the mechanism is specifically targeted against the viral genome. However, with virus vectors carrying inserts derived from host genes the process can be additionally targeted against the corresponding mRNAs. VIGS has been used widely in plants for analysis of gene function and has been adapted for high-throughput functional genomics. Until now, most applications of VIGS have been studied in *Nicotiana benthamiana*. However, new vector systems and methods are being developed that could be used in other plants, including Arabidopsis (Tumage et al., 2002). VIGS also helps in the identification of genes required for disease resistance in plants. These methods and the underlying general principles also apply when VIGS is used in the analysis of other aspects of plant biology. Table 12.3 summarizes the range of plant viruses with their suitable hosts used in RNA interference technology till date (Table 12.3). When a plant virus infects a host cell, it activates an RNA-based defense that is targeted against the viral genome. The double-stranded RNA (dsRNA) in virus-infected cells is thought to be the replication intermediate that causes the small interfering RNA/ribonuclease (siRNA/RNase) complex to target the viral single-stranded RNA (ssRNA). In the initially infected cell, the viral ssRNA would not be a target of the siRNA/RNase complex because this replication intermediate would not have accumulated to a high level. However, in the later stages of the infection, as the rate of viral RNA replication increases, the viral dsRNA and siRNA would become more abundant. Eventually, the viral ssRNA would be targeted intensively and virus accumulation would slow down (Voinnet, 2001). Many plant viruses encode proteins that are suppressors of this RNA silencing process. These suppressor proteins would not be produced until after the virus had started to replicate in the infected cell so they would not cause complete suppression of the RNA-based defense mechanism. However, these proteins would influence the final steady-state level of virus accumulation. Strong suppressors would allow virus accumulation to be prolonged and at a high level. Conversely, if a virus accumulates at a low level, it could be due to weak suppressor activity. The dsRNA replication intermediate would be processed so that the siRNA in the infected cell would correspond to parts of the viral vector genome, including any non-viral insert. Thus, if the insert is from a host gene, the siRNAs would target the RNase complex to the corresponding host mRNA and the symptoms in the infected plant would reflect the loss of the function in the encoded protein (Peter et al., 2003).

TABLE 12.3 Range of Plant Viruses Used for Silencing of Target Genes and Their Hosts with Targeted Genes.

Virus	Silencing host	Gene silenced	References
Tobacco mosaic virus	*Nicotiana benthamiana* *Nicotiana tabacum*	pds	Kumagai et al. (1995)
Potato virus X	*Nicotiana benthamiana* *Arabidopsis*	pds	Faivre-Rampant et al. (2004)
Tobacco rattle virus	*Nicotiana benthamiana,* Tomato, Solanum, Chili pepper	Rar1, EDS1, NPR1/NIMI, pds, rbcS	Ratcliff et al. (2001), Hileman et al. (2005)
Barley stripe mosaic virus	Barley	Pds, Lr21, Rar1, Sgt I, Hsp90	Holzberg et al. (2002), Scofield et al. (2005)
Bean pod mottle virus	Glycine max	pds	Zhang et al. (2006)
Cabbage leaf curl virus	Arabidopsis	CH42, pds	Turnage et al. (2002)
Pea early browning virus	*Pisum sativum*	Pspds, uni, kor, pds	Constantin et al. (2004)
Tomato yellow leaf curl china virus	*Nicotiana benthamiana*	Pcna, pds, su, gfp	Kjemtrup et al. (1998)
African cassava mosaic virus	*Nicotiana benthamiana*	Pds, su, cyp79d2	Fofana et al. (2004)

There are several examples that strongly support this approach to suppression of gene expression. Thus, when tobacco mosaic virus (TMV) or potato virus X (PVX) vectors were modified to carry inserts from the plant phytoene desaturase gene (*pds*) the photobleaching symptoms on the infected plant reflected the absence of photoprotective carotenoid pigments that require phytoene desaturase. Similarly, when the virus carried inserts of a chlorophyll biosynthetic enzyme, there were chlorotic symptoms and, with a cellulose synthase insert, the infected plant had modified cell walls (Kjemtrup et al., 1998). Genes other than those encoding metabolic enzymes can also be targeted by VIGS. For example, if the viral insert corresponded to genes required for disease resistance, the plant exhibited enhanced pathogen susceptibility. In one such example, the insert in a tobacco rattle virus (TRV) vector was from a gene (EDS1) that is required for N-mediated resistance to TMV. The virus vector-infected N-genotype plant exhibited compromised TMV resistance. The symptoms of a TRV vector carrying a *leafy* insert demonstrate how VIGS can be used to target genes that regulate development. *Leafy* is a gene required for flower development. Loss-of-function *leafy* mutants produce modified flowers that are phenocopied in the TRV-leafy infected plants. Similarly, the effects of tomato golden mosaic virus vectors carrying parts of the gene for a cofactor of DNA polymerase illustrate how VIGS can be used to target essential genes. The plants infected with this geminivirus vector were suppressed for division growth in and around meristematic zones of the shoot (Peele et al., 2001).

To exploit the ability to knock down, in essence, any gene of interest, RNAi via siRNAs has generated a great deal of interest in both basic and applied biology. Increasing number of large-scale RNAi screens are designed to identify the important genes in various biological pathways. Because disease processes also depend on the combined activity of multiple genes, it is expected that turning off the activity of a gene with specific siRNA could produce a therapeutic benefit to humanity. Based on the siRNAs-mediated RNA silencing (RNAi) mechanism, several transgenic plants has been designed to trigger RNA silencing by targeting pathogen genomes. Diverse targeting approaches have been developed based on the difference in precursor RNA for siRNA production, including sense/antisense RNA, small/long hairpin RNA, and artificial miRNA precursors. Virologists have designed many transgenic plants expressing viral coat protein (CP), movement protein (MP), and replication-associated proteins, showing resistance against infection by the homologous virus. This type of pathogen-derived resistance (PDR) has been reported in diverse viruses including tobamo-, potex-, cucumo-, tobra-, carla-, poty-, and Alfalfa mosaic virus groups as well as the luteovirus group (Abel et al., 1986; Ding, 2010). Figure 12.3 represents preparation of VIGS construct and *Agrobacterium*-mediated gene transfer for the viral disease assessment in plants. Transgene RNA silencing-mediated resistance is a process that is highly associated with the accumulation of viral transgene-derived siRNAs. One of the drawbacks of the sense/antisense transgene approach is that the resistance is unstable, and the mechanism often results in delayed resistance or low efficacy/resistance. This may be due to the low accumulations of transgene-derived siRNA in post-transcriptional gene silencing (PTGS) due to defense mechanism encoded by plants. Moreover, numerous viruses, including potyviruses, cucumoviruses, and tobamoviruses, are able to counteract these mechanisms by inhibiting this type of PTGS. Therefore, the abundant expression of the dsRNA to trigger efficient RNA silencing becomes crucial for effective resistance. To achieve resistance, inverse repeat sequences from viral genomes were widely used to form hairpin dsRNA *in vivo*, including small hairpin RNA (shRNA), self-complementary hpRNA, and intron-spliced hpRNA. Among these methods, self-complementary hairpin RNAs separated by an intron likely elicit PTGS with the highest efficiency. The presence of inverted repeats of dsRNA-induced PTGS (IR-PTGS) in plants also showed high resistance against viruses. IR-PTGS is not required for the formation of dsRNA for the processing of primary siRNAs, but the plant RNA-dependent RNA polymerases (RDRs) are responsible for the generation of secondary siRNAs derived from non-transgene viral genome, which further intensify the efficacy of RNA silencing induced by hpRNA, a process named RNA silencing transitivity. Among them, the sequence similarity between the transgene sequence and the challenging virus sequence is the most important. Scientists have engineered several transgenic plants with multiple hpRNA constructs from different viral sources, or with a single hpRNA construct combining different viral sequence. Thus, multiple viruses can be simultaneously targeted, and the resulting transgenic plants show a broader resistance with high efficacy. In addition to the sequence similarity, the length of the transgene sequence also contributes to high resistance. In general, an average length of 100–800 nt of transgene sequence confers effective resistance (Bucher et al., 2006; Himber et al., 2003; Peele et al., 2001).

KEYWORDS

- genetically modified crops
- genetic material
- genetic-engineering
- plant tissue culture
- insect-resistant

REFERENCES

Abel, P. P.; Nelson, R. S.; De, B.; Hoffmann, N.; Rogers, S. G.; Fraley, R. T.; Beachy, R. N. Delay of Disease Development in Transgenic Plants that Express the Tobacco Mosaic Virus Coat Protein Gene. *Science.* 1986, *232* (3), 738–743.

Baulcombe, D. RNA Silencing in Plants. *Nature.* 2004, *431* (2), 356–363.

Bock, R. "The Give-and-Take of DNA: Horizontal Gene Transfer in Plants". Trends Plant Sci. **2010,** *15 (1), 11–22.*

Bruening, G.; Lyons, J. M. "The Case of the FLAVR SAVR Tomato." *Calif. Agric.* 2000, *54* (4), 6–7.

Bucher, E.; Lohuis, D.; Van Poppel, P. M.; Geerts-Dimitriadou, C.; Goldbach, R.; Prins, M. Multiple Virus Resistance at a High Frequency Using a Single Transgene Construct. *J. Gen. Virol.* 2006, *87* (1), 3697–3701.

Chen, Z. "Molecular Mechanisms of Polyploidy and Hybrid Vigor." *Trends Plant Sci.* **2010,** *15 (2), 57–71.*

Conner, A. J.; Glare, T. R.; Nap, J. P. The Release of Genetically Modified Crops into the Environment. *Plant J.* **2003,** *33* (1), 19–46.

Constantin, G. D.; Krath, B. N.; MacFarlane, S. A.; Nicolaisen, M.; Johansen, I. E.; Lund, O. S. Virus-Induced Gene Silencing as a Tool for Functional Genomics in a Legume Species. *Plant J.* 2004, *40* (2), 622–631.

Ding, S. W. RNA-Based Antiviral Immunity. *Nat. Rev. Immunol.* 2010, *10* (9), 632–644.

Duncan, R. "Tissue Culture-Induced Variation and Crop Improvement". *Adv. Agron.* 1996, *58* (8), 201–200.

Faivre-Rampant, O.; Gilroy, E. M.; Hrubikova, K.; Hein, I.; Millam, S.; Loake, G. J.; Birch P; Taylor, M.; Lacomme, C. Potato Virus X-Induced Gene Silencing in Leaves and Tubers of Potato. *Plant Physiol.* 2004, *134* (4), 1308–1316.

Fofana, I. B.; Sangaré, A.; Collier, R.; Taylor, C.; Fauquet, C. M. A Geminivirus-Induced Gene Silencing System for Gene Function Validation in Cassava. *Plant Mol. Biol.* 2004, *56* (4), 613–624.

Fraley, R. T.; Rogers, S. G.; Horsch, R. B.; Sanders, P. R.; Flick, J. S.; Adams, S. P.; Bittner, M. L.; Brand, L. A.; Fink, C. L.; Fry, J. S.; Galluppi, G. R.; Goldberg, S. B.; Hoffmann,

N. L.; Woo, S. C. Expression of Bacterial Genes in Plant Cells. *Proc. Nat. Acad. Sci. USA.* 1983, *80* (15), 4803–4807.

Himber, C.; Dunoyer, P.; Moissiard, G.; Ritzenthaler, C.; Voinnet, O. Transitivity Dependent Andindependent Cell-To-Cell Movement of RNA Silencing. *EMBO J.* 2003, *22* (1), 4523–4533.

Holzberg, S.; Brosio, P.; Gross, C.; Pogue, G. P. Barley Stripe Mosaic Virus-Induced Gene Silencing in a Monocot Plant. *Plant J.* 2002, *30* (3), 315–327.

ISAAA. Annual Report Executive Summary, Global Status of Commercialized Biotech/ GM Crops: 2013.

James, C. Global Status of Commercialized Biotech/GM Crops: 2009. *ISAAA Brief No. 41,* Ithaca, New York, USA, **2009.**

Kjemtrup, S.; Sampson, K. S.; Peele, C. G.; Nguyen, L. V.; Conkling, M. V.; Thompson, W. F.; Robertson, D. Gene Silencing from Plant DNA Carried by a Geminivirus. *Plant J.* 1998, *14* (1), 91–100.

Klümper, W.; Qaim, M. A Meta-Analysis of the Impacts of Genetically Modified Crops. *PLoS One.* 2014, *9* (11), e111629.

Koornneef, M.; Meinke, D. The Development of Arabidopsis as a Model Plant. *Plant J.* 2010, *61* (6), 909–921.

Kuc, J. A Case for Self Defense in Plants Against Disease. *Phytoparasitica.* 1990, *18* (1), 3–7.

Kumagai, M. H.; Donson, J.; Della-Cioppa, G.; Harvey, D.; Hanley, K.; Grill, L. K. Cytoplasmic Inhibition of Carotenoid Biosynthesis with Virus-Derived RNA. *PNAS.* 1995, *92* (5), 1679–1683.

Lundmark, C. "Searching Evolutionary Pathways: Antifreeze Genes from Antarctic Hairgrass". BioScience. **2006,** *56 (6), 552.*

Maagd, R. A.; Bosch, D.; Stiekema, W. *Bacillus thuringiensis*toxin-Mediated Insect Resistance in Plants. *Trends Plant Sci.* 1999, *4* (1), 9–13.

Manoj, B.; Zhu, L.; Shen, G.; Payton, P.; Zhang, H. "Expression of an Arabidopsis Sodium/Proton Antiporter Gene (AtNHX1) in Peanut to Improve Salt Tolerance". *Plant Biotechnol. Rep.* 2012, *6,* 59–67.

Meagher, R. B. Phytoremediation of Toxic Elemental and Organic Pollutants. *Curr. Opin. Plant Biol.* 2000, *3*(2), 153–162.

Monroe, D. Jumping Genes Cross Plant Species Boundaries. *PLoS Biology.* 2006, *4* (1), e35.

Peele, C.; Jordan, C.V.; Muangsan, N.; Turnage, M.; Egelkrout, E.; Eagle, P.; Hanley-Bowdoin, L.; Robertson, D. Silencing of a Meristematic Gene Using Geminivirus-Derived Vectors. *Plant J.* 2001, *27* (4), 357–366.

Waterhouse, P. M.; Helliwell, W. C. Exploring Plant Genomes by RNA-Induced Gene Silencing. *Nature Rev. Genet.* **2003,** *4* (1), 29 -38.

Podevina, N.; Jardin, P. "Possible Consequences of the Overlap between the CaMV 35S Promoter Regions in Plant Transformation Vectors Used and the Viral Gene VI in Transgenic Plants". *GM Crops Food.* **2012,** *3, 296–300.*

Predieri, S. Mutation Induction and Tissue Culture in Improving Fruits. *Plant Cell Tiss. Org. Cult.* 2001, *64* (3), 185–210.

Ratcliff, F.; Martin-Hernandez, A. M.; Baulcombe, D. C. Technical Advance: Tobacco Rattle Virus as a Vector for Analysis of Gene Function by Silencing. *Plant J.* 2001, *25* (2), 237–235.

Ronald, P. Plant Genetics, Sustainable Agriculture and Global Food Security. *Genetics.* 2011, *188* (1), 11–20.

Roush, R. T. Bt-Transgenic Crops: Just another Pretty Insecticide or a Chance for a New Start in the Resistance Management. *Pestic. Sci.* 1997, *51* (3), 328–334.

Schell, J.; Van Montagu, M. "The Ti-Plasmid of Agrobacterium Tumefaciens, a Natural Vector for the Introduction of nif Genes in Plants?" Basic Life Sci. **1977,** *9 (1), 159–179.*

Scofield, S. R.; Huang, L.; Brandt, A. S.; Gill, B. S. Development of a Virus-Induced Gene-Silencing System for Hexaploid Wheat and Its Use in Functional Analysis of the Lr21-Mediated Leaf Rust Resistance Pathway. *Plant Physiol.* 2005, *138* (2), 2165–2173.

Shrawat, A.; Lörz, H. *Agrobacterium*-Mediated Transformation of Cereals: A Promising Approach Crossing Barriers. *Plant Biotechnol. J.* 2006, *4* (6), 575–603.

Turnage, M. A.; Muangsan, N.; Peele, C. G.; Robertson, D. Geminivirus-Based Vectors for Gene Silencing in Arabidopsis. *Plant J.* 2002, *30* (1), 107–114.

Vaeck, M.; Reynaerts, A.; Hofte, H.; Jansens, S.; Beukeleer, M. D.; Dean, C.; Zabeaug, M.; Montagu, M. V.; Leemans, J. Transgenic Plants Protected From Insect Attack. *Nature.* 1987, *328* (1), 33–37.

Van Loon, L. C. Induced Resistance in Plants and the Role of Pathogenesis-Related Proteins. *Eur. J. Plant Pathol.* 1997, *103* (2), 753–765.

Voinnet, O. RNA Silencing as a Plant Immune System against Viruses. *Trends Genet.* 2001, *17* (8), 449–459.

Wahab, S. Biotechnology Approaches in the Management of Plant Pests, Diseases, and Weeds for the Sustainable Agriculture. J. Biopesticides. **2009,** *2 (2), 115–134.*

Walmsley, A.; Arntzen, C. "Plants for Delivery of Edible Vaccines". *Curr. Opin. Biotechnol.* 2000, *11* (2), 126–129.

Yanpeng, W. "Simultaneous Editing Of Three Homoeoalleles in Hexaploid Bread Wheat Confers Heritable Resistance to Powdery Mildew". *Nat. Biotechnol.* 2014, *32* (2), 947–951.

Zhang, C.; Ghabrial, S. A. Development of *Bean Pod Mottle Virus*-Based Vectors for Stable Protein Expression and Sequence-Specific Virus-Induced Gene Silencing in Soybean. *Virology.* 2006, *344* (2), 401–411.

CHAPTER 13

BACILLUS THURINGIENSIS AND INSECT PEST MANAGEMENT

ANIL[1*], LOKENDER KASHYAP[2], TARAK NATH GOSWAMI[1], VIKAS KUMAR PATEL[1], and RAMESH KUMAR SHARMA[3]

[1]*Department of Entomology, Bihar Agricultural University, Sabour 813210, Bhagalpur, Bihar, India*

[2]*Department of Plant Protection, Lovely Professional University, Jalandhar 144410, Punjab, India*

[3]*Department of Horticulture (Veg. & Flori.), Bihar Agricultural University, Sabour 813210, Bhagalpur, Bihar, India*

Corresponding author. E-mail: aniljakhad@gmail.com

CONTENTS

ABSTRACT

Bacillus thuringiensis, a motile, gram positive, rod shaped, spore forming soil bacterium characteristically synthesizes parasporal crystalline inclusions containing crystal (Cry) and cytolytic (Cyt) proteins possessing pesticidal properties against wide range of insect orders and nematodes. During vegetative growth stage, it also secretes insecticidal proteins, that is, vegetative insecticidal proteins (Vip) which hold insecticidal activity against Lepidopteran, Coleopteran and some Homopteran insects. Recently, a less well-characterized secretory protein which has no amino acid similarity with Vip has also been found to possess insecticidal activity against Coleopteran insect and termed as secreted insecticidal protein (Sip). Being ubiquitous, *B. thuringiensis* has been isolated from a diverse source of habitat which led to the discovery of various new strains. These strains are characterized into 71 serotypes and 85 subspecies based on flagellar-serotyping. Different strains produce different types of toxins, each of which affects a narrow taxonomic group of insects and this has led to the development of over 100 types of insecticides using natural and genetically engineered strains for topical application. In addition, the genes possessing insecticidal properties have been successfully transferred into cotton, corn, soybean, and rice, which have led to significant economic benefits. However, prolonged and continuous use of *B. thuringiensis* based bio-pesticide and transgenic crops have created the problem of insect resistance in only few insect species. In this chapter, we discuss about *B. thuringiensis* on the historical developments, prevalence and ecology, general characteristics, classification and nomenclature, various insecticidal proteins, and the topical and transgenic products. Finally, this chapter focuses on the developments in insect resistance management and the future prospects of *B. thuringiensis* in insect pest management.

13.1 INTRODUCTION

Bio-pesticides especially the microbial pesticides is one of the suitable alternatives to non-chemical methods employed for the management of insect pests. Since microbial pesticides are natural disease-causing microorganisms which infect and inhibit the growth of specific pest group. These pesticides either contain microbes as a whole or their fermented products and some may contain the combination of both. Various microorganism *viz.,* bacteria, fungi, protozoa, viruses, nematodes, and so forth, pathogenic to insects and referred as entomopathogens, are being successfully exploited for protecting the crops from the ravages of insect pests. Among these, entomopathogenic bacteria gained immense importance and have become the most successful commercial biopesticide in recent decades. These include various spore-forming and non-spore-forming bacteria belonging to the families—Bacillaceae (*Bacillus*), Enterobacteriaceae (*Aerobacter, Enterobacter, Proteus,* and *Serratia*), Micrococcaceae (*Micrococcus*), Pseudomonadaceae (*Pseudomonas*), and Streptococcaceae (*Streptococcus*) (Tanda & Kaya, 1993). Bacteria belonging to these families except Bacillaceae, are non-spore formers and considered as potential pathogens when they get entry into haemocoel through various means. Although, non-spore forming bacteria play a vital role in natural disease outbreak of pest population, even then spore formers have been largely exploited in the suppression of pests. Since spore is a resistant form and its formation enables them to survive for longer period outside the body of the host. *B. thuringiensis (Bt), Bacillus papillae, Bacillus Lentimorbus,* and *Bacillus sphaericus* were identified as important spore forming entomopathogenic bacteria. Among these, *B. thuringiensis* has been used worldwide as bio-control agents against insect pests (Fernando et al., 2010) and represents about 78% of the all biological preparation and 4% of all insecticides by sales volume (Sanchis, 2011) due to its narrow host range, non-toxicity to animals and plants, and being environment friendly (Khetan, 2001).

The molecular potency of *Bt*-toxins is high as compared to the chemical pesticides, that is, 300 times higher than synthetic pyrethroids and 80,000 times stronger than organophosphates (Feitelson et al., 1992). More than 3000 insect species included in 16 orders have been found to be susceptible to different Cry proteins (Lin & Xiong, 2004) in addition to non-insect organisms such as nematodes, mites, protozoa, and plathelmintes (Feitelson, 1993). Its specificity for the control of insect pests such as lepidoptera, diptera, and coleoptera has led to the development of over 100 types of insecticides using natural and genetically engineered strains (Roh et al., 2007). In addition, the genes that code for the insecticidal crystal proteins (ICPs) have been successfully transferred into cotton, corn, soybean, and rice, which have led to significant economic benefits (Crickmore, 2006).

13.2 HISTORICAL DEVELOPMENTS

The insecticidal properties of *B. thuringiensis* were recognized many years before the bacterium was identified and its spores were already in use in ancient Egypt (Sanahuja et al., 2011). In late nineteenth and early twentieth centuries, scientists began to observe various insect pathogens including bacteria using commercially available microscope. During late nineteenth century, the silk industry observed sudden death of caterpillars (*Bombyx mori*) in Japan which led to the discovery of a spore forming bacteria in 1898 by Japanese biologist Sigetane Ishiwata (Aizawa, 2001). This bacterium was named *Bacillus sotto* to describe "sudden death" which occurred within hours of feeding by caterpillars (Ishiwata, 1901). The description of "sotto-kin" was published in Japanese and did not get much attraction among contemporary insect pathologist. In Japanese, this disease was called "sotto-byo-kin" or "sotto-kin" which means collapse-disease-microorganism. During experimentation, this bacterium survived for many years and showed maximum mortality from one-week-old culture (Ishiwata, 1905). Ishiwata already realized that toxins were involved in the pathogenic process, as larvae became rapidly paralyzed before multiplication of the bacillus. These discoveries were later confirmed by other Japanese Scientists (Aoki & Chigasaki, 1915, 1916; Mitani & Watarai, 1916) and they found that the filtrate of a culture dissolved in alkaline solution was lethal to silkworm (Beegle & Yamamoto, 1992; Aizawa, 2001). Across the world in 1909, a German scientist Ernst Berliner discovered the similar bacterium from dead Mediterranean flour moth, *Ephestia kuehniella* in Thuringia, Germany. Berliner formally described and named the bacteria *B. thuringiensis* (Berliner, 1915). This name has been retained for the species to the present and includes *B. sotto* and certain strains originally called *Bacillus cereus* (Beegle & Yamamoto, 1992).

The direct contact between the spores or Crys and healthy caterpillars had no effects whereas on feeding by insects, this bacterium was found lethal and thus Berliner suggested that it can be employed in insect pest management. However, those strains of *B. thuringiensis* were lost. After recognizing the potential of *B. thuringiensis* as an insecticide, German scientist Mattes (1927), isolated another strain of *B. thuringiensis* from flour moth and found promising results against European corn borer, *Ostrinia nubilalis* (Husz, 1929). These findings eventually led to the development of first commercial product "Sporeine" in 1938 (Beegle & Yamamoto,

1992; Milner, 1994). Berliner had noted the presence of a parasporal body or crystalline inclusion and the relation between the pathogenic effect of the bacterium and its crystalline inclusions was discovered by Edward Steinhaus in early 1950s (Steinhaus, 1951). The parasporal inclusions in sporulating cultures were first described by Hannay (1953) who later demonstrated that these inclusions were proteinaceous and might have insecticidal entity (Hannay & Fitz-James, 1955). Angus (1954) revealed that the inclusions were soluble in the alkaline digestive fluids of lepidopteran larvae. Later on, Heimpel and Angus (1959) demonstrated that the parasporal inclusions were responsible for the insecticidal activities. Production of such parasporal inclusions was considered as the unique phenotypic trait to distinguish *B. thuringiensis* from other *Bacillus* species (Vilas-Bôas et al., 2007). In 1981, the strong correlation between the loss of crystalline toxin production and the specific plasmid together evidenced that the gene in plasmids are responsible for the production of these toxins (González & Carton, 1984). During the same year, the work on cloning and expression of one of the genes from *B. thuringiensis* subsp. *kurstaki* HD1 encoding Cry proteins demonstrated toxicity against the larvae of *Manduca sexta* (Schnepf & Whiteley, 1981). During mid-1980s, these findings led to the development of earliest *B. thuringiensis* transgenic tobacco and tomato plants through the introduction of *cry* genes and their direct expression in plant tissues.

13.3 PREVALENCE AND ECOLOGY

B. thuringiensis is a ubiquitous soil microorganism and it has been frequently isolated from multiple environments *viz*., soil, insect-cadavers, insect rich environment, foliar plant surface, aquatic environment, animal faeces, flour mills, and grain storage in different parts of the world (Deluca et al., 1981; Padua et al., 1982; Hunag & Hunag, 1988; Martin & Travers, 1989; Smith & Couche, 1991; Lambert & Peferoen, 1992; Thanabalu et al., 1992; Bernhard et al., 1997; Chaufaux et al., 1997; Damgaard et al., 1997). In addition, *B. thuringiensis* has recently been isolated from marine sediments (Maeda et al., 2000) and soils of Antarctica (Forsty & Logan, 2000). This shows its presence and dominance everywhere in nature with its normal habitat in soil. *B. thuringiensis* has been isolated from different parts of India including Punjab, Haryana, Gujarat, Odisha, Maharashtra, Goa, Andhra Pradesh, Karnataka, Tamil Nadu, West Bengal, Sikkim, and

Himachal Pradesh (Manonmani & Balaraman, 2001; Chatterjee et al., 2007; Das et al., 2008; Patel et al., 2009; Randhawa et al., 2009; Eswarapriya et al., 2010; Nethravathi et al., 2010a, 2010b; Ramalakshmi & Udayasuriyan, 2010; Anitha et al., 2011; Patel et al., 2011; Anil, 2013).

     B. thuringiensis, although being soil bacterium, is not capable of multiplying in soil or water at levels that would allow favorable competition with other bacteria (Yara et al., 1997; Furlaneto et al., 2000) therefore; the insects are considered as the ideal site for *B. thuringiensis* multiplication and exchange of genetic material (Jarrett & Stephenson, 1990). Due to the fact, all the commercially relevant *B. thuringiensis* subspecies used in biopesticides were originally isolated from dead insects, including *kurstaki* (Kurstak, 1964), *israelensis* (Goldberg & Margalit, 1977), and *tenebrionis* (Krieg et al., 1983). It indicates that the soil may act as a reservoir for bacterial spores rather as a site of multiplication, especially as most known *B. thuringiensis* hosts are foliar feeders that would preferentially come into contact with soil bacteria during the non-feeding pupal stage. In fact, higher proportions of entomopathogenic serovars are identified from phylloplane compared to soil samples, denoting an evolutionary advantage for pathogenic *B. thuringiensis* on foliar surfaces and *B. thuringiensis* as a phylloplane epiphyte (Smith & Couche, 1991). Meadows (1993) proposed three prevailing hypothetical niches of *B. thuringiensis* as: an entomopathogen, a phylloplane inhabitant, and a soil microorganism. However, the true role of *B. thuringiensis* is not clear because some of the *B. thuringiensis* strains produced parasporal Cry inclusions that are toxic to many orders of insects whereas some strains showed no insecticidal activity (Iriarte et al., 2000).

13.4   GENERAL CHARACTERISTICS

B. thuringiensis is a gram-positive, rod-shaped, aerobic and spore forming soil bacterium which characteristically produces parasporal crystalline proteins during sporulation. In general, its colonies are irregular in form, cloudy in color, dry in texture, flat, and having rough surface (Fig. 13.1). During vegetative state, the rod-shaped bacterium has the dimensions of $1.0–1.2 \times 3.0–5.0$ microns and occurs in long to short tangled chains. When nutrients or oxygen is insufficient for vegetative growth, the bacterium sporulates, producing a spore and a parasporal body containing one or more ICPs (Federici & Siegel, 2008). During sporulation, the ellipsoidal spore measures $0.5–1.2 \times 1.3–2.5$ microns and resides in the sporangium with parasporal crystalline toxins. The parasporal crystalline insecticidal proteins are as large as or larger than the *B. thuringiensis* spores (Lingren & Green, 1984). The vegetative cell divides into two uniform daughter cells by formation of a division septum initiated midway along the plasma membrane. However, sporulation involves asymmetric cell division and it is characterized by seven stages shown in Figure 13.2. The genome size of *B. thuringiensis* strains varied from 2.4 to 5.7 million base pairs.

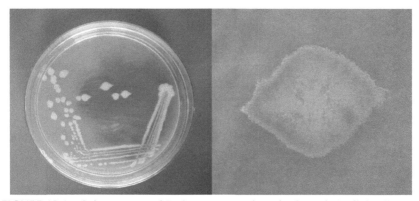

FIGURE 13.1   Colony pattern of *B. thuringiensis* subspecies *kurstaki* (Anil, 2013).

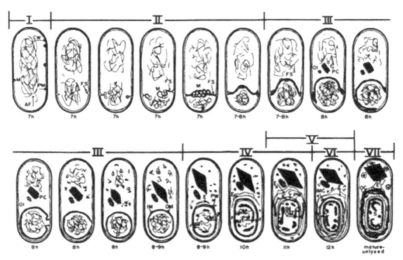

FIGURE 13.2   Diagrammatic scheme of sporulation in *B. thuringiensis* (Reprinted from Bechtel, D. B.; Bulla, L. A. Ultrastructural Analysis of Membrane Development during Bacillus thuringiensis Sporulation. J. Ultrastruct. Res. 1982, 79(2), 121–132. © 1982 with permission from Elsevier.)

B. thuringiensis strains synthesize various toxins namely exotoxins (diffused through the cell wall into culture medium) and endotoxins (contained in Cry inside bacterial cell as δ-endotoxins). The exotoxins like α-exotoxin (phospholipase C), β-exotoxin or thuringeinsin (thermostable exotoxin), γ-exotoxin (toxic to sawflies), louse factor exotoxin (active only against lice), mouse factor exotoxin (toxic to mice and Lepidopteran insects), Vip, and Sip are synthesized during vegetative phase of the bacterial growth. However, δ-endotoxins are synthesized at the onset of sporulation and during the stationary growth phase as parasporal crystalline inclusions. The endotoxins contain entomocidal proteins and referred as ICPs (Höfte & Whiteley, 1989). ICPs are classified into two major groups of toxins, namely, the Cry and the Cyt toxins. Both Cry and Cyt toxins are parasporal inclusion proteins, except that Cyt proteins exhibit haemolytic activity (Schnepf et al., 1998). Some of these proteins are highly toxic to certain insects, but they are harmless to most other organisms including vertebrates and beneficial insects (Gould, 1998; Christou et al., 2006). However, few Cry proteins did not show any activity against invertebrate target and exhibited strong and specific cytocidal activity against human cancer cells of various origins. These Cry proteins are called parasporin, as an alternative name. Moreover, *B. thuringiensis* isolates also synthesize other insecticidal proteins during the vegetative growth phase which are subsequently secreted into the culture medium and include Vip (Estruch et al., 1996) and Sip (Donovan et al., 2006).

13.5   CLASSIFICATION AND NOMENCLATURE

B. thuringiensis is a member of *B. cereus* group which includes other five species namely, *B. cereus, B. anthracis, Bacillus mycoides, Bacillus Pseudomycoides,* and *Bacillus weihenstephanensis.* Among these species, *B. thuringiensis* is the most diverse and has been classified into a large number of subspecies, strains, varieties, serotypes, serovars, pathovars, or crystovar. Initially, identification of *B. thuringiensis* is mainly based on the presence of crystalline inclusions that are detected by phase-contrast microscopy. This method is very useful to screen novel environmental isolates (Bravo et al., 1998) and it may be the best way to observe parasporal bodies (Ammons & Rampersad, 2002).

At subspecies level, many researchers attempted morphological, biochemical, antigenic methods, phase susceptibility, and plasmid profiling to introduce taxonomic classification system for *B. thuringiensis*. The initial efforts were concentrated on biochemical and serological methods including morphological and biochemical typing (Heimpel & Angus, 1958), esterase patterns of vegetative cells (Norris, 1964), and serotyping of the H flagellar antigen in vegetative cells (de Barjac & Bonnefoi, 1968). In addition to biochemical and serological methods, the insecticidal activities were also used to catalogue *B. thuringiensis* into different subspecies and differentiate among closely related subspecies (de Barjac et al., 1980). Lecadet et al. (1999) published the most prevalent and recent revision of the H-antigen classification which included 69 antigenic groups and 13 subgroups. Later on, four more serovars namely, *sinensis* (Li et al., 2000), *jordanica* (Khyami-Horani et al., 2003) and *mogi* (Roh et al., 2009) were isolated and described, resulting in the current list of 86 *B. thuringiensis* serotypes (Table 13.1). The existence of strains that agglutinate spontaneously and cross-reactivity of *B. cereus* strains with *B. thuringiensis* H-serotypes were recognized as emerging limitations with H-serotyping classification (Lecadet et al., 1999). The subspecies are differentiated by various methods other than H-serotyping technique such as Cry morphology, biochemical reactions, bioassay for host specificity, SDS-PAGE, cellular fatty acid analysis, and several molecular techniques.

TABLE 13.1   Updated List of *B. thuringiensis* Serovars (=Subspecies).

H Antigen	Serovar	H Antigen	Serovar	H Antigen	Serovar
1	*thuringiensis*	19	*tochigiensis*	45	*roskildiensis*
2	*finitimus*	20a:20b	*yunnanensis*	46	*chanpaisis*
3a:3c	*alesti*	20a:20c	*pondicheriensis*	47	*wratislaviensis*
3a:3b:3c	*kurstaki*	21	*colmeri*	48	*balearica*
3a:3b:3d	*mogi*	22	*shandongiensis*	49	*muju*
3a:3d	*sumiyoshiensis*	23	*japonensis*	50	*navarrensisb*
3a:3d:3e	*fukuokaensis*	24a:24b	*neoleonensis*	51	*xiaguangiensis*
4a:4b	*sotto*	24a:24c	*novosibirsk*	52	*kim*
4a:4c	*kenyae*	25	*coreanensis*	53	*asturiensis*
5a:5b	*galleriae*	26	*silo*	54	*poloniensis*
5a:5c	*canadensis*	27	*mexicanensis*	55	*palmanyolensis*
6	*entomocidus*	28a:28b	*monterrey*	56	*rongseni*
7	*aizawai*	28a:28c	*jegathesan*	57	*pirenaica*
8a:8b	*morrisoni*	29	*amagiensis*	58	*argentinensis*

TABLE 13.1   *(Continued)*

H Antigen	Serovar	H Antigen	Serovar	H Antigen	Serovar
8a:8c	*ostriniae*	30	*medellin*	59	*iberica*
8b:8d	*nigeriensis*	31	*toguchini*	60	*pingluonsis*
9	*tolworthi*	32	*cameroun*	61	*sylvestriensis*
10a:10b	*darmstadiensis*	33	*leesis*	62	*zhaodongensis*
10a:10c	*londrina*	34	*konkukian*	63	*boliviab*
11a:11b	*toumanoffi*	35	*seoulensis*	64	*azorensis*
11a:11c	*kyushuensis*	36	*malaysiensis*	65	*pulsiensis*
12	*thompsoni*	37	*andaluciensis*	66	*graciosensis*
13	*pakistani*	38	*oswaldocruzi*	67	*vazensisb*
14	*israelensis*	39	*brasiliensis*	68	*thailandensis*
15	*dakota*	40	*huazhongensis*	69	*pahangi*
16	*indiana*	41	*sooncheon*	70	*sinensis*
17	*tohokuensis*	42	*jinghongiensis*	71	*jordanica*
18a:18b	*kumamotoensis*	43	*guiyangiensis*		
18a:18c	*yosoo*	44	*higo*		

Source: Modified from Lecadet et al. (1999).

The insecticidal activity of different *B. thuringiensis* serotypes is however, not dependent on the serology but rather on the insecticidal proteins or toxins produced by them. Such insecticidal protein or toxins are encoded in Crys produced by *B. thuringiensis* which have different morphology, that is, bipyramidal, spherical, cuboidal, triangular, pyramidal, rhomboidal, rectangular, irregular, or amorphous in shape. Generally, bipyramidal inclusions are active against Lepidoptera, ovoidal inclusions to Diptera, rhomboidal inclusions to Coleoptera and cuboidal inclusions to both Lepidopetra and Diptera. Because of the crystalline nature of the proteins, the term Cry is used in gene (*cry)* and protein (Cry) nomenclature. Initially, the classification was based on the host specificity of the toxins and their structural similarities (Höfte & Whiteley, 1989) and four major classes of ICPs *viz.*, CryI (Lepidoptera specific toxin), CryII (Lepidoptera and Diptera specific toxin), CryIII (Coleoptera specific toxin) and CryIV (Diptera specific toxin) were proposed besides CytA (cytolytic and haemolytic toxin). The classification was further extended to include CryV proteins, effective against lepidopteran and coleopteran larvae (Tailor et al., 1992) and many more were also added.

Since most of the *B. thuringiensis* strains can synthesize more than one toxin, resulting in complex and overlapping host profile and the discovery of many new toxins, this classification system became onerous. Later on, Crickmore et al. (1998) revised the nomenclature for these toxins which is solely based on amino acid sequences and the hierarchical clustering of these sequences, and consists of four degrees of ranks by which toxins are ranked accordingly. The primary, secondary, tertiary, and quaternary ranks are represented by Arabic numerals, a upper case alphabetical letter, a lower case alphabetical letter and an Arabic numeral, respectively, for example, Cry1Aa1 or Cyt1Aa1. The boundaries for these ranks have been set at 45, 78, and 95% based on amino acid sequence homology. This nomenclature is still prevailing and adopted by the researchers all over the world. To date, Cry toxins are classified into 74 primary ranks (Cry1 through Cry74) and Cyt toxins into three primary ranks (Cyt1, Cyt2, and Cyt3), based on the homology of their amino acid sequences (Crickmore et al., 2014). In addition to Cry and Cyt toxins, *B. thuringiensis* also produces insecticidal toxins during vegetative phase which are referred as Vip. Similarly, the same system of nomenclature (Vip1Aa1) is adopted and these are now classified into four primary ranks that is, Vip1, Vip2, Vip3, and Vip4 (Crickmore et al., 2014). Additionally, Sip toxins possessing insecticidal properties are also secreted into the culture medium and Sip1Aa1 is only the known member of this new group (Table 13.2).

13.6   TOXIN STRUCTURE, ACTION, AND SPECIFICITY

The more logical way to classify *B. thuringiensis* strains seems to be functionally on the basis of toxin protein secreted by respective strain. The toxins can be described in terms of their amino acid sequence, protein structures, and modes of activity (Crickmore et al., 1998). As described earlier, *B. thuringiensis* synthesizes various toxins like Cry, Cyt, Vip, and Sip which own insecticidal activities. Unlike chemical pesticides that are often contact poisons, the *B. thuringiensis* toxins are stomach poisons.

13.6.1   *Cry TOXINS*

Cry proteins are the largest group of crystalline proteins coded by *cry* genes and most of them possess insecticidal and other related properties. To

TABLE 13.2   List of Different *B. thuringiensis* Toxins Possessing Insecticidal Properties.

Cry toxins						Cyt toxins
Cry1Aa1 - 24	Cry2Aa1 - 17	Cry7Kb1	Cry14Ab1	Cry32Da1	Cry45Aa1	Cyt1Aa1 - 7
Cry1Ab1 - 35	Cry2Ab1 - 33	Cry7La1	Cry15Aa1	Cry32Ea1 - 2	Cry45Ba1	Cyt1Ab1
Cry1Ac1 - 38	Cry2Ac1 - 12	Cry8Aa1	Cry16Aa1	Cry32Eb1	Cry46Aa1 - 2	Cyt1Ba1
Cry1Ad1 - 2	Cry2Ad1 - 5	Cry8Ab1	Cry17Aa1	Cry32Fa1	Cry46Ab1	Cyt1Ca1
Cry1Ae1	Cry2Ae1	Cry8Ac1	Cry18Aa1	Cry32Ga1	Cry47Aa1	Cyt1Da1 - 2
Cry1Af1	Cry2Af1 - 2	Cry8Ad1	Cry18Ba1	Cry32Ha1	Cry48Aa1 - 3	Cyt2Aa1 - 4
Cry1Ag1	Cry2Ag1	Cry8Ba1	Cry18Ca1	Cry32Hb1	Cry48Ab1 - 2	Cyt2Ba1 - 15
Cry1Ah1 - 3	Cry2Ah1 - 5	Cry8Bb1	Cry19Aa1	Cry32Ia1	Cry49Aa1 - 4	Cyt2Bb1
Cry1Ai1 - 2	Cry2Ai1	Cry8Bc1	Cry19Ba1	Cry32Ja1	Cry49Ab1	Cyt2Bc1
Cry1Aj1	Cry2Aj1	Cry8Ca1 - 4	Cry19Ca1	Cry32Ka1	Cry50Aa1	Cyt2Ca1
Cry1Ba1 - 8	Cry2Ak1	Cry8Da1 - 3	Cry20Aa1	Cry32La1	Cry50Ba1 - 2	Cyt3Aa1
Cry1Bb1 - 3	Cry2Al1	Cry8Db1	Cry20Ba1 - 2	Cry32Ma1	Cry51Aa1 - 2	
Cry1Bc1	Cry2Ba1 - 2	Cry8Ea1 - 6	Cry21Aa1 - 2	Cry32Mb1	Cry52Aa1	**Vip toxins**
Cry1Bd1 - 2	Cry3Aa1 - 12	Cry8Fa1 - 4	Cry21Ba1	Cry32Na1	Cry52Ba1	Vip1Aa1 - 3
Cry1Be1 - 4	Cry3Ba - 3	Cry8Ga1 - 3	Cry21Ca1 - 2	Cry32Oa1	Cry53Aa1	Vip1Ab1
Cry1Bf1 - 2	Cry3Bb1 - 3	Cry8Ha1	Cry21Da1	Cry32Pa1	Cry53Ab1	Vip1Ac1
Cry1Bg1	Cry3Ca1	Cry8Hb1	Cry21Ea1	Cry32Qa1	Cry54Aa1 - 2	Vip1Ad1
Cry1Bh1	Cry4Aa1 - 4	Cry8Ia1 - 4	Cry21Fa1	Cry32Ra1	Cry54Ab1	Vip1Ba1 - 2
Cry1Bi1	Cry4Ba1 - 5	Cry8Ib1 - 3	Cry21Ga1	Cry32Sa1	Cry54Ba1 - 2	Vip1Bb1 - 3
Cry1Ca1 - 14	Cry4Ca1 - 2	Cry8Ja1	Cry21Ha1	Cry32Ta1	Cry55Aa1 - 3	Vip1Bc1
Cry1Cb1 - 3	Cry4Cb1 - 3	Cry8Ka1 - 3	Cry22Aa1 - 3	Cry32Ua1	Cry56Aa1 - 4	Vip1Ca1 - 2

TABLE 13.2   (*Continued*)

Cry toxins						Vip toxins
Cry1Da1 - 4	Cry4Cc1	Cry8Kb1 - 3	Cry22Ab1 - 2	Cry32Va1	Cry57Aa1	Vip1Da1
Cry1Db1 - 2	Cry5Aa1	Cry8La1	Cry22Ba1	Cry32Wa1 - 2	Cry57Ab1	Vip2Aa1 - 3
Cry1Dc1	Cry5Ab1	Cry8Ma1 - 3	Cry22Bb1	Cry33Aa1	Cry58Aa1	Vip2Ab1
Cry1Dd1	Cry5Ac1	Cry8Na1	Cry23Aa1	Cry34Aa1 - 4	Cry59Aa1	Vip2Ac1 - 2
Cry1Ea1 - 12	Cry5Ad1	Cry8Pa1 - 3	Cry24Aa1	Cry34Ab1	Cry59Ba1	Vip2Ad1
Cry1Eb1	Cry5Ba1 - 3	Cry8Qa1 - 2	Cry24Ba1	Cry34Ac1 - 3	Cry60Aa1 - 3	Vip2Ae1 - 3
Cry1Fa1 - 4	Cry5Ca1 - 2	Cry8Ra1	Cry24Ca1	Cry34Ba1 - 3	Cry60Ba1 - 3	Vip2Af1 - 2
Cry1Fb1 - 7	Cry5Da1 - 2	Cry8Sa1	Cry25Aa1	Cry35Aa1 - 4	Cry61Aa1 - 3	Vip2Ag1 - 2
Cry1Ga1 - 2	Cry5Ea1 - 2	Cry8Ta1	Cry26Aa1	Cry35Ab1 - 3	Cry62Aa1	Vip2Ba1 - 2
Cry1Gb1 - 2	Cry6Aa1 - 3	Cry9Aa1 - 5	Cry27Aa1	Cry35Ac1	Cry63Aa1	Vip2Bb1 - 4
Cry1Gc1	Cry6Ba1	Cry9Ba1 - 2	Cry28Aa1 - 2	Cry35Ba1 - 3	Cry64Aa1	Vip3Aa1 - 60
Cry1Ha1	Cry7Aa1	Cry9Bb1	Cry29Aa1	Cry36Aa1	Cry64Ba1	Vip3Ab1 - 2
Cry1Hb1 - 2	Cry7Ab1 - 9	Cry9Ca1 - 2	Cry29Ba1	Cry37Aa1	Cry64Ca1	Vip3Ac1
Cry1Hc1	Cry7Ac1	Cry9Da1 - 4	Cry30Aa1	Cry38Aa1	Cry65Aa1 - 2	Vip3Ad1 - 5
Cry1Ia1 - 35	Cry7Ba1	Cry9Db1	Cry30Ba1	Cry39Aa1	Cry66Aa1 - 2	Vip3Ae1
Cry1Ib1 - 11	Cry7Bb1	Cry9Dc1	Cry30Ca1 - 2	Cry40Aa1	Cry67Aa1 - 2	Vip3Af1 - 4
Cry1Ic1 - 2	Cry7Ca1	Cry9Ea1 - 10	Cry30Da1	Cry40Ba1	Cry68Aa1	Vip3Ag1 - 15
Cry1Id1 - 3	Cry7Cb1	Cry9Eb1 - 3	Cry30Db1	Cry40Ca1	Cry69Aa1 - 2	Vip3Ah1
Cry1Ie1 - 5	Cry7Da1 - 3	Cry9Ec1	Cry30Ea1 - 4	Cry40Da1	Cry69Ab1	Vip3Ai1
Cry1If1	Cry7Ea1 - 3	Cry9Ed1	Cry30Fa1	Cry41Aa1	Cry70Aa1	Vip3Aj1 - 2
Cry1Ig1	Cry7Fa1 - 2	Cry9Ee1 - 2	Cry30Ga1 - 2	Cry41Ab1	Cry70Ba1	Vip3Ba1 - 2

TABLE 13.2 (Continued)

Cry toxins						Vip toxins
Cry1Ja1 - 3	Cry7Fb1 - 3	Cry9Fa1	Cry31Aa1 - 6	Cry41Ba1 - 2	Cry70Bb1	Vip3Bb1 - 3
Cry1Jb1	Cry7Ga1 - 2	Cry9Ga1	Cry31Ab1 - 2	Cry41Ca1	Cry71Aa1	Vip3Ca1 - 4
Cry1Jc1 - 2	Cry7Gb1	Cry10Aa1 - 4	Cry31Ac1 - 2	Cry42Aa1	Cry72Aa1	Vip4Aa1
Cry1Jd1	Cry7Gc1	Cry11Aa1 - 4	Cry31Ad1 - 2	Cry43Aa1 - 2	Cry73Aa1	
Cry1Ka1 - 2	Cry7Gd1	Cry11Ba1	Cry32Aa1 - 2	Cry43Ba1	Cry74Aa1	**Sip toxin**
Cry1La1 - 3	Cry7Ha1	Cry11Bb1 - 2	Cry32Ab1	Cry43Ca1		Sip1Aa1
Cry1Ma1 - 2	Cry7Ia1	Cry12Aa1	Cry32Ba1	Cry43Cb1		
Cry1Na1 - 3	Cry7Ja1	Cry13Aa1	Cry32Ca1	Cry43Cc1		
Cry1Nb1	Cry7Ka1	Cry14Aa1	Cry32Cb1	Cry44Aa1		

Source: http://www.btnomenclature.info (accessed on December 15, 2015); Donovan et al. (2006). Hyphen indicates the range of quaternary rank in toxins (like Cry1Aa1-24 indicates Cry1Aa1 through Cry1Aa24); Cry toxins in red color do not possess any known invertebrate activity and also designation as parasporin.

date, the Cry toxins are classified into 74 primary ranking that is, Cry1–74 (Table 13.2). These proteins can be further sub-divided into a well-known three-domain Cry proteins and another distinct proteins families like bin- and ETX_MTX-like toxins (non-three-domain Cry proteins) produced by *Lysinibacillus sphaericus* (=*B. sphaericus*) (de Maagd et al., 2003; Berry, 2012). Figure 13.3 shows the three-dimensional structure of Cry8Ea, as a representative of a three-domain toxin produced by *Bt*. Domain I (perforating domain) located toward the *N* terminus, a bundle of seven α-helix, six of which are amhipathic encircling the seventh hydrophobic helix is subjected to proteolytic cleavage in all three-domain Cry proteins during toxin activation and this may be responsible for toxin membrane insertion and pore formation. Domain II (central or middle domain) consists of three anti-parallel β-sheets with exposed loop regions whereas domain III (galactose-binding domain) is a two anti-parallel β-sheet sandwich. Later two domains confer receptor binding specificity by which helping to define host the range (Boonserm et al., 2006). Table 13.3 shows the

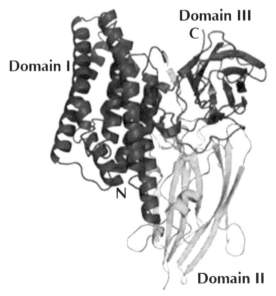

FIGURE 13.3 Three-dimensional structure of Cry8Ea determined at 2.20 Å (PBD code: 3EB7) (Guo, S.; Ye, S.; Liu, Y.; Wei, L.; Xue, J.; Wu, H.; Song, F.; Zhang, J.; Wu, X.; Huang, D.; Rao. Z. Crystal Structure of *Bacillus thuringiensis* Cry8Ea1: An Insecticidal Toxin Toxic to Underground Pests, the Larvae of *Holotrichia parallela*. *J. Struct. Biol.* **2009**, *168*(2), 259–266. Reprinted with permission via Creative Commons Attribution license, http://creativecommons.org/licenses/by/4.0/)

insecticidal activities of Cry toxins against a wide range of insect orders including lepidoptera, diptera, coleoptera, hemiptera, and hymenoptera besides their toxic properties against Rhabditids and Gastropods.

TABLE 13.3   Known Host Spectrum of Different Toxin Proteins Secreted by *B. thuringiensis*.

Type of the toxin	Toxin proteins	Target group(s)
Crystal protein	Cry1A-K; Cry2A; Cry7B; Cry8D; Cry9A-C,E; Cry15A; Cry22A; Cry32A; Cry51A	Lepidoptera
	Cry1A-C; Cry2A; Cry4A-B; Cry10; Cry11A-B; Cry16A; Cry19A-B; Cry16A; Cry19A-B; Cry20A; Cry24C; Cry27A; Cry32B-D; Cry39A; Cry44A; Cry47A; Cry48A; Cry49A	Diptera
	Cry1B,I; Cry3A-C; Cry7A; Cry8A-G; Cry9D; Cry14A; Cry18A; Cry22A-B; Cry23A; Cry34A-B; Cry35A-B; Cry36A; Cry37A; Cry43A-B; Cry55A	Coleoptera
	Cry2A; Cry3A; Cry11A	Hemiptera
	Cry3A; Cry5A; Cry22A	Hymenoptera
	Cry5A-B; Cry6A-B; Cry12A; Cry13A; Cry14A; Cry21A; Cry55A	Rhabditida
	Cry1Ab	Gastropoda
Cytolytic protein	Cyt1A-B; Cyt2A-B	Diptera
	Cyt1A; Cyt2C	Coleoptera
Vegetative insecticidal protein	Vip1; Vip2	Coleoptera &Hemiptera
	Vip3	Lepidoptera
Secreted insecticidal protein	Sip1	Coleoptera

Source: Adapted from Palma et al. (2014). Proteins separated by hyphen (-) indicates a group of Cry toxins; comma (,) indicates the different Cry toxins; and semicolon (;) separate groups or individual toxins.

The mode of action of three-domain Cry proteins has been studied mainly in Lepidopteran insects wherein three different models *viz.*, the classical model, the sequential binding model and the signaling pathway model have been proposed (Vachon et al., 2012). The classical model proposes lyses of midgut epithelial cells through various steps including

Cry inclusion ingestion and dissolution in the alkaline midgut lumen, pro-toxin proteolytic activation resulting protease-resistant toxic polypeptides, binding of toxin fragments to specific receptors on the surface of midgut epithelial cells, and formation of non-selective pores permeable to inorganic ions, amino acids and sugars which cause the lysis of epithelial cells leading to midgut disarrangements and further insect death. In addition, spores may colonize, germinate, and replicate in the hemolymph thereby resulting death of the larvae by septicemia. However, the sequential binding model suggests that Cry toxins, once activated by intestinal proteases, bind to cadherin-like proteins and undergo a conformational change that favors proteolytic removal of the α-1 helix from domain I and formation of an oligomeric pre-pore structure. Later, binding to a secondary receptor, such as an aminopeptidase, facilitates the insertion of the pre-pore structure into the membrane, leading to cell and insect death (Bravo et al., 2007; Pardo-López et al., 2013). In contrast, the signaling-pathway model suggests that the toxic activity is mediated by the specific binding to cadherin receptors, leading to undescribed Mg^{2+}-dependent and adenylyl cyclase/protein kinase A signaling pathway that produces necrotic cell death (Kirouac et al., 2002). The present available information supports the classical model postulating that Cry toxins act by forming pores, although most events leading to their formation and receptor binding remain still poorly understood (Zhang et al., 2013).

As discussed earlier, several other types of unrelated toxins which do not possess three-domain Cry protein structure are also covered in the Cry nomenclature. On the basis of their primary sequence identity, these toxins are classified under distinct primary ranking. Cry15, Cry23, Cry33, Cry38, Cry45, Cry51, Cry60, and Cry64 all show features of the ETX_MTX2 family and these toxins are likely to form beta-barrel pores in target cell (Popoff, 2011). In addition to Cry35 and Cry36 belonging to Toxin_10 family, Cry45, Cry46, and Cry23 toxins are found to have aerolysin-like fold (de Maagd et al., 2003; Kelker et al., 2014; Akiba et al., 2006, 2009). These toxins may also represent beta pore forming toxins, sharing aerolysin as a structural homolog with the ETX_MTX2 family. In contrast to above toxins, few toxins require a second protein to act as binary toxins including Cry23/Cry37, Cry34/Cry35, and Cry48/Cry49 (de Maagd et al., 2003; Jones et al., 2007). The precise interactions by which these binary toxins elicit their activity against coleopteran targets remains to be elucidated. Other non-three-domain proteins are Cry22, Cry6, and Cry55.

13.6.2 Cyt TOXINS

Cyt proteins constitute another relevant insecticidal protein family and coded by *cyt* genes. These proteins exhibit a general Cyt (haemolytic) activity *in vitro* and predominantly dipetran specificity *in vivo* (Butko, 2003; de Maagd et al., 2003). Table 13.2 demonstrates that the Cyt toxins are classified up to three primary ranking (Cyt1–3). The three-dimensional structure of Cyt2Ba (Fig. 13.4) shows that these proteins are to be single domain, three-layer alpha-beta protein. In contrast to Cry, Cyt toxins directly interact with membrane lipids and insert into the membrane. The Cyt1Ca synthesized by *B. thuringiensis* subsp. *isrealensis* indicated no larvicidal or haemolytic activity whereas *B. thuringiensis* subsp. *morrisoni* producing Cyt toxins exhibited toxicity against a wider range of insect-orders including diptera, lepidoptera and coleoptera (Guerchicoff et al., 2001) (Table 13.3).

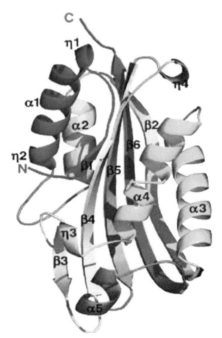

FIGURE 13.4 Crystal structure of a monomer of Cyt2Ba determined at 1.80 Å (PDB code: 2RCI) replace with: (Reprinted from Cohen, S.; Dym, O.; Albeck, S.; Ben-Dov, E.; Cahan, R.; Firer, M.; Zaritsky, A. High-Resolution Crystal Structure of Activated Cyt2Ba Monomer from Bacillus thuringiensis Subsp. Israelensis. J. Mol. Biol. 2008, 380(5), 820–827., © 2008 with permission from Elsevier.)

13.6.3   Vip TOXINS

Vip are secreted in the medium during vegetative stage and coded by *vip* genes. To date, these toxins are identified and classified into four different families namely Vip1, Vip2, Vip3, and a novel family of Vip proteins, recently identified and classified as Vip4 (Table 13.2). Amongst these, Vip1 and Vip2 constitute a binary toxin with high insecticidal activity against some coleopteran pests (Warren et al., 1998) and the sap-sucking insect pest *Aphis gossypii* (Hemiptera) (Sattar & Maiti, 2011) (Table 13.3). The Vip1 and Vip2 form typical A+B type binary toxins wherein Vip2 is the cytotoxic A-domain and Vip1 is the receptor-binding domain which is responsible for the translocation of the cytotoxic Vip2 into the host cell (de Maagd et al., 2003; Barth et al., 2004). The proposed mechanism of action involves the proteolytic activation of the cell-binding B precursor (Vip1) and its monomeric interaction with cell surface receptor(s) followed by formation of homoheptamers that subsequently translocate the A (Vip2) toxic component into the cytoplasm through acid endosomes (Barth et al., 2004). Once inside the cytoplasm, the A component destroys filamentous actin, likely by mono-ADP-ribosylation of the Arg177 residue of G-actin, blocking its polymerization and leading to cell death by cytoskeletal disarrangement (Shi et al., 2004, 2006). Vip3A proteins are single-chain (not binary) toxins and exhibit insecticidal activity against a wide variety of lepidopterans and, interestingly, certain species less susceptible to some Cry1A proteins (e.g., *Agrotis ipsilon*, *Spodoptera exigua* and *S. frugiperda*) (MacIntosh et al., 1990; Estruch et al., 1996). In 2010, a new protein under Vip4 was identified by Sun et al., and designated as Vip4Aa1 (Crickmore et al., 2014). The insecticidal properties (activity and host range) of this protein are unknown.

13.6.4   Sip TOXINS

Sip is the first and only member belonging to *B. thuringiensis* insecticidal family of secreted proteins and demonstrated toxicity against coleopteran larvae (Table 13.3). This protein was initially obtained from culture supernatants of the *B. thuringiensis* strain EG2158 and designated as Sip1Aa1 (Donovan et al., 2006). This toxin is lethal for *L. decemlineata* (Coleoptera: Chrysomelidae) and inhibits growth of *Diabrotica undecimpunctata howardi* (Coleoptera: Chrysomelidae) and *D. virgifera virgifera*.

13.7 *B. thuringiensis* TOPICAL PRODUCTS

In 1938, the first *B. thuringiensis* based pesticide "Sporeine" was used in France but such pesticides were registered after long time in the USA during 1961. The first commercial name was Thurincide, prepared from *B. thuringiensis* subsp. *kurstaki* (Beegle & Yamamoto, 1992). The poor performance of the products compared to chemical insecticides led to concentrate work on process development and strain improvement in order to increase the efficacy of *B. thuringiensis* products and intrinsic toxicity of the bacterium, respectively. In addition to improving fermentation and harvest procedures, the work on integrating suspension agents and preservatives for liquid formulations, and wetting agents for powder formulations was initiated. In order to minimize the problem of rapid photolysis after spraying, UV screening agents were also added. Further standardization of potency testing and more rigorous quality control resulted in manifold increase in efficacy under field conditions. (Burges & Jones 1998a, 1998b). On the other side, the strain improvement efforts during 1960s led to development of new *B. thuringiensis* strains with higher potency. Dulmage (1970) discovered a more active *B. thuringiensis* subsp. *kurstaki* (HD1), which was commercialized in the USA as Dipel. By the peroid, most of the *B. thuringiensis* products were based on subspecies *kustaki* strain HD1 including Thuricide, Biobit, and Dipel for managing Lepidopteran pests (Baum et al., 1999). Other strains used to manage Lepidopteran pests include *kurstaki* SA-11 and *kurstaki* SA-12 while the subspecies *israelensis* and *tenebrionis* were used to tackle dipteran and colepteran pest, respectively. Few of the *B. thuringiensis* topical products derived from natural strains are listed in Table 13.4.

Initially, the strain search and assessment programs were lying on bioassay and biochmical typing which was stimulated by the progress of biotechnology using PCR testing for specific gene. In addition to discovering the strains carrying novel toxins, new bacterial strains were developed to carry unkown combination of existing toxins through conjugation or direct transformation (González et al., 1982; Kronstad et al., 1983; Carlton & Gonzalez, 1985). Examples of such products are listed in Table 13.4. The strain EG2424 was developed through conjugation for synthesizing Cry1Ac and Cry3A against European corn borer, *O. nubilalis* and Colorado potato beetle, *L. decemlineta*, respectively, and it was marketed under the name 'Foil' (Carlton & Gawron-Bruke, 1993). The limitations for strain improvement through conjugation were overcome when Schnepf and Whiteley (1981) cloned a Cry toxin gene from *B. thuringiensis* subsp. *kurstaki* into *Escherichia coli* and since then it was convenient to introduce different genes of interest. Table 13.5 shows the currently available pesticides based on *B. thuringiensis* worldwide. Today, the *B. thuringiensis* biopesticide market is dominated by Abbott Laboratories, Chicago (USA) and Novartis, Basle (Switzerland), together accounting for 70% of the global *B. thuringiensis* production and other 30% is divided among ~30 companies with over 100 different formulations (Sanahuja et al., 2011). Most of the *B. thuringiensis* products are containing a single *B. thuringiensis* toxin but some can synthesize up to five toxins. There were ~180 registered *B. thuringiensis* products with U.S. EPA in 1998, 120 microbial products in European Union and ~276 registered *B. thuringiensis* formulation in China (EPA, 1998; Federici & Siegel, 2008; Huang et al., 2007; Hammond & Koch, 2012).

TABLE 13.4   *B. thuringiensis* Topical Products Based on Natural, Transconjugant, and Recombinant Strains.

Trade name/product	Subspecies/strain/genes	Target insect order
Natural strains		
Biobit	*kurstaki* HD-1	Lepidoptera
Dipel	*kurstaki* HD-1	Lepidoptera
Thuricide	*kurstaki* HD-1	Lepidoptera
Bactospeine	*kurstaki* HD-1	Lepidoptera
Javelin	*kurstaki* SA-11*	Lepidoptera
Delfin	*kurstaki* SA-11*	Lepidoptera
Costar	*kurstaki* SA-12	Lepidoptera
Florbac	*aizawai*	Lepidoptera
Tekar	*israelensis*	Diptera
Bactimos	*israelensis*	Diptera
Acrobe	*israelensis*	Diptera
Novodor	*tenebrionis*	Coleoptera
Trident	*tenebrionis*	Coleoptera
Transconjugant strains		
Agree	*aizawai*	Lepidoptera

TABLE 13.4 *(Continued)*

Trade name/product	Subspecies/strain/genes	Target insect order
Condor	*kurstaki*	Lepidoptera
Cutlass	*kurstaki*	Lepidoptera
Design	*aizawai*	Lepidoptera
Foil	*kurstaki*	Lepidoptera, Coleoptera
Recombinant strains		
Raven	Cry1Ac (x2), Cry3A Cry3Bb (imported)	Lepidoptera
CRYMAX	Cry1Ac (x3), Cry2A Cry1C (imported)	Lepidoptera
Lepinox	Cry1Aa, Cry1Ac (x2), Cry2A Cry1F-1Ac (imported)	Lepidoptera

Source: Baum et al. (1999); Kaur et al. (2000); *kurstaki* HD-12 has been renamed *kurstaki* SA-11.

TABLE 13.5   Currently Available *B. thuringiensis* Based Biopesticide Worldwide.

Trade names	*B. thuringiensis* subspecies and strain	Host range
Able, Bactospeina, Condor, Costar, CRYMAX, Cutlas, Futura, Lepinox, Thuricide, Steward	*kurstaki*	Lepidoptera
Florbac, Agree, Design, Xentari	*aizawai*	Lepidoptera
Costar	*kurstaki* SA-12	Lepidoptera
Foil, Raven	*kurstaki*	Lepidoptera, Coleoptera
Thuricide, Biobit, Dipel, Foray, Javelin, Vault	*kurstaki* HD-1	Lepidoptera
M-Trak	*tenebrionis* in *Pseudomonas fluorescens*	Coleoptera
Mattch, MVP	*kurstaki* HD-1 in *Pseudomonas fluorescens*	Lepidoptera
Novodor, Trident	*tenebrionis*	Coleoptera

Source: Kaur (2000), Sanahuja et al. (2011).

In India, Majumdar (1968) was the first to deveop an indigenous *B. thuringiensis* product under the name "Lepidopterocide" on a piolet scale at CFTRI, Mysore. Subsequently, various indigenous *B. thuringiensis* isolates were obtained from various crop pests at many places but no serious efforts were made to exploit these bacteria anywhere in India. Further, one locally available *B. thuringiensis* subspecies *kenyae* was isolated from *Ephestia cautella* (Walker) at BARC, Mumbai which has shown effectiveness against several economically importnant insect pest species. The piolet-scale production of *B. thuringiensis* subspecies *isrealensis* was also successfully accomplished in South India. Local strains of *B. thuringiensis* are being isolated and identified for the development of bacterial insecticide at many research institutes in India (Battu & Arora, 1997). Several *B. thuringiensis* based products are being marketed in India by various agrochemical industries, mostly effective against lepidopteran pests (Table 13.6).

TABLE 13.6 *B. thuringiensis* Based Products Available in India.

B. thuringiensis subspecies	Trade name	Target insect pest	Crop	Manufacturer
kurstaki	Halt	Diamondback moth, Lepidopteran caterpillars	Cabbage	Wockhardt Ltd., Mumbai
kurstaki	Biolep	Cotton bollworms	Cotton	Biotech International Ltd., New Delhi
kurstaki	Bioasap	Cotton bollworms	Cotton	Biotech International Ltd., New Delhi
kurstaki	Delfin WG	Cotton bollworms, tobacco caterpillar	Cotton, Castor	Sandoz India Ltd., Mumbai
kurstaki	Dipel 8L	Cotton bollworms, fruit borers	Cotton, Tomato, Brinjal, Okra	Lupin Agrochemicals Pvt. Ltd., Mumbai
galleriae	Spicturin	American boll-worm, leaf-folder, tobacco catrepillar, diamondback moth	Cotton, rice, cabbage, cauliflower, chillies	Chemical and Fertilizers Ltd., Chennai
kurstaki	Biobit	Lepidopteran caterpillars	Several	Rallis India Ltd., Bangalore

Source: Dhaliwal and Arora (2006).

B. thuringiensis products contain either Cry or spore-crystal mixture and they are only effective when retain on the plant organs on which insects do feed. The main advantages of *B. thuringiensis* products include their safety toward non-target organisms, specificity for a group of insects and potency compared to chemical sprays. The *B. thuringiensis* products are also considered as biodegradable and persist for only a few days on the leaf surface. Various factors like UV light, weather, the chemical environment of leaf surface, and the presence of proteases contribute to the degradation of Cry proteins whereas many *B. thuringiensis* spores are generally washed off the leaf surface into soil and thus can persist for long time. On the other hand, the rapid inactivation of Cry by UV light, heat and extreme pH are disadvantageous for *B. thuringiensis* sprays. For this reason, the Cry proteins have been encapsulated in *Pseudomonas fluorescens* (MVP and M-Trak) to protect the Cry proteins from UV light and chemical degradation (Table 13.5).

13.8   MASS PRODUCTION

B. thuringiensis is considered as a powerful biopesticides against various insect pest of economic importance and therefore, the special attention was paid to optimize growth conditions in order to harvest maximum toxins. In addition to strain efficacy, the commercial mass production depends on the cost of raw material, fermentation cycle, maintenance of process parameters, bioprocessing of fermentation fluid and formulation of the final product (Sachdeva et al., 1999). The cost of raw material is one of the principal costs involved in overall *B. thuringiensis* production and this varied between 30 and 40% of total cost depending on the capacity of production plant (Lisansky et al., 1993). Therefore, it is imperative to look for cheap raw materials which are locally available. Foda et al. (2002, 2003) used various agricultural and industrial by-products in *B. thuringiensis* production like citrus peels, wheat bran, corn meal, seeds of dates, beef blood, silkworm pupal skin, groundnut cake, cane molasses, fishmeal, cottonseed meal, soybean meal, residues from chicken slaughterhouse, fodder yeast, cheese whey, and corn steep liquor. The mass production of microbial products can be achieved through two-fermentation method that is submerged fermentation and solid-state fermentation.

Several workers including Megna (1963), Dulmage and de Barjac (1973), Salama et al. (1983), Pearson and Ward (1988), Widjaya et al.

(1992), Farrera et al. (1998), Montiel et al. (2001), and Zouari et al. (2002) have studied the submerged fermentation method for *B. thuringiensis* production utilizing different raw material. In general, the mass production through submerged fermentation method is quite costly as it requires costly instruments like well-equipped deep-tank fermentor, high-speed cooling centrifuge as well as drying facilities, for example spray dryer. On the other side, the solid-state fermentation method is advantageous in respect of its low cost methodology, low wastewater output, low capital investment, and some spore-forming microorganisms only sporulate when grown on a solid substrate (Capalbo, 1995). Mechalas (1963) was the first to report on possible application of solid-state fermentation in production of *B. thuringiensis* followed by Dulmag and Rhodes (1971). Wang (1998) and Foda et al. (2002) produced *B. thuringiensis* through this method using agricultural products as raw material.

13.9 *B. thuringiensis* TRANSGENIC CROPS

The first attempt to introduce *B. thuringiensis cry* genes into tobacco and tomato plants was initiated during mid-1980s for expression of Cry proteins directly in plant tissues. The earliest transgenic tomato plants were carrying gene from *B. thuringiensis* var. *kurstaki* HD-1 for conferring resistance against tomato fruit worm, *Heliothis zea* and tomato pinworm, *Keiferia lycopersicella*, and truncated forms of the gene from *B. thuringiensis* var. *kurstaki* HD-1 against lepidopteran larvae. Similarly, tobacco plants were developed to confer resistance against tobacco hornworm, *Manduca sexta,* and other lepidopteran insects through transfer of DNA encoding a toxin from *B. thuringiensis* var. *kurstaki* HD-1 (Krattiger, 1996). However, no Cry proteins could be detected in the first generation of experimental transgenic plants (Fischhoff et al., 1987; Vaeck et al., 1987; Perlak et al., 1991). The expression of recombinant Cry proteins at high levels in heterologous bacteria were identified as one reason for low expression mainly due to large difference in average GC content between *B. thuringiensis* and plant DNA, differences in codon preference and presence of frequent ATTTA sequence in bacterial genes. The modification in the sequence of *cry1Ab* and *cry1Ac* genes resulted in 100-fold increase in protein expression, that is 0.02% total soluble protein (TSP) but this was still insufficient for adequate pest control (Perlak et al., 1991). Further, the development of synthetic *cry* genes could achieve Cry protein expression at levels of 0.2–1% TSP (Koziel et al., 1993). The expression of such proteins was increased to over 5% when *cry* genes were introduced into the chloroplast genome (McBride et al., 1995).

In 1986, the first field trials with *B. thuringiensis* transgenic tobacco were conducted in the USA and France and the plants expressed a truncated gene encoding the N-terminal portion of Cry1A from *B. thuringiensis* var. *kurstaki* HD-73 against *Helicoverpa zea* (Hoffmann et al., 1992). Later on, transgenic potato plants expressing Cry3A from *B. thuringiensis* var. *tenebrionis* to confer resistance against Colorado potato beetle were field-tested and found more efficient than topical spray (Perlak et al., 1993). Subsequently, other crops *viz.*, cotton, maize, and rice expressing Cry1A were also tested in field trials against pink bollworm, European corn borer, and stem borers, respectively. *B. thuringiensis* potato, corn, and cotton crops were first registered by US Environmental Protection Agency (EPA) in 1995 expressing Cry3A, Cry1Ab, and modified Cry1Ac, respectively. In 1998, the EPA approved *B. thuringiensis* tomato expressing Cry1Ac and *B. thuringiensis* corn expressing Cry1F for protecting plants against black cutworm (*Agrotis ipsilon*), fall armyworm (*Spodoptera frugiferda*) and European corn borer (*O. nubilalis*). In 2002, the approval of *B. thuringiensis* cotton expressing Cry1Ac and Cry2Ab created a landmark and it was also commercialized in India. In 2009, China made a landmark decision that approved the safety of two *B. thuringiensis* transgenic rice cultivars (James, 2009). Recently, *B. thuringiensis* brinjal was commercialized in Bangladesh during 2014. Many other *B. thuringiensis* crops are under development for future agricultural use. Table 13.7 provides information on *B. thuringiensis* Cry proteins in transgenic crops which are authorized for cultivation. *B. thuringiensis* corn and cotton confer respective resistance against lepidopteran pests such as corn borers (mainly *Ostrinia nubilalis*) and the budworm-bollworm complex (*Heliothis virescens, Helicoverpa* spp., *Pectinophora gossypiella*) (Shelton et al., 2002; James, 2014).

Since the first approval of genetically modified (GM) crops in 1994 and subsequently large scale commercialization in the USA during 1996, the area under GM crops is increased remarkably from a mere 1.7 million hectares to over 180 million hectares in 2014 worldwide. Of which ~80 million hectares with crops containing the insecticidal *B. thuringiensis* traits, often in combination with the herbicide tolerance trait are grown in different countries (James, 2014). The success of *B. thuringiensis* crops including corn and cotton has resulted in economic benefits to growers and reduced the use of conventional insecticides (Brookes & Barfoot, 2006). The Table 13.8 indicates the crop-wise list of countries where *B. thuringiensis* transgenic crops were commercialized since 1996.

TABLE 13.7 *B. thuringiensis* Transgenic Crops Authorized for Cultivation in One or More Countries.

Protein	Target insect order	Approved crop	Product examples	Registrant company
Cry1Ab	Lepidoptera	Maize	YieldGard	Monsanto
			Agrisure CB/LL	Syngenta
Cry1Ac	Lepidoptera	Cotton	Bollgard	Monsanto
		Maize*	*B. thuringiensis* Xtra	Monsanto
		Soybean	Intacta Roundup Ready 2 Pro	Monsanto
		Brinjal	Bari *B. thuringiensis* Begun- 1, -2, -3, -4	Mahyco
Cry1A.105 + Cry2Ab2	Lepidoptera	Maize	Genuity VT Double Pro	Monsanto
Cry1Ac + Cry2Ab2	Lepidoptera	Cotton	Bollgard II	Monsanto
Cry1Ac + Cry1F	Lepidoptera	Cotton	WideStrike	Dow
		Soybean	DAS-81419-2	Dow
Cry1Fa2	Lepidoptera	Maize	Hercilex I	Dow
Cry1Ab + Cry2Ae	Lepidoptera	Cotton	TwinLink	Bayer
mCry3A**	Coleoptera	Maize	Agrisure RW	Syngenta
Cry3Bb1	Coleoptera	Maize	YieldGard Rootworm RW	Monsanto
eCry3.1Ab***	Multiple	Maize	Agrisure Duracade	Syngenta
Cry34Ab1 + Cry35Ab1	Coleoptera	Maize	Herculex RW	Dow and Dupont

trmobr, *with pinII (protease inhibitor) from *Solanum tuberosum*; **modified Cry3A protein; ***Chimeric Cry3A-Cry1Ab protein. (Reprinted from Koch, M. S.; Ward. J. M.; Levine, S. L.; Baum, J. A.; Vicini, J. L.; Hammond, B. G. The Food and Environmental Safety of Bt Crops. *Front. Plant Sci.* **2015,** *6,* 283. https://creativecommons.org/licenses/by/4.0/)

TABLE 13.8   *B. thuringiensis* Transgenic Crops and Their Commercializing Countries.

B. thuringiensis crop	Country
Cotton	Argentina, Australia, Brazil, Burkina Faso, Canada, China, Colombia, Costa Rica, European Union (EU), India, Japan, Mexico, Myanmar, New Zealand, Pakistan, Paraguay, Philippines, Singapore, South Africa, South Korea, United States of America (USA)
Eggplant	Bangladesh
Maize	Argentina, Australia, Brazil, Canada, Chile, China, Colombia, Egypt, El Salvador, EU, Honduras, Indonesia, Japan, Malaysia, Mexico, New Zealand, Panama, Paraguay, Philippines, Russian Federation, Singapore, South Africa, South Korea, Switzerland, Taiwan, Thailand, Turkey, USA, Uruguay
Poplar	China
Potato	Australia, Canada, Japan, Mexico, New Zealand, Philippines, Russian Federation, South Korea, USA
Rice	China, Iran
Soybean	Argentina, Australia, Brazil, Canada, China, Colombia, EU, Japan, Mexico, New Zealand, Paraguay, South Korea, Taiwan, Thailand, USA, Uruguay
Tomato	Canada, Chile, USA

Source: http://www.isaaa.org/gmapprovaldatabase/. (accessed on December 15, 2015).

13.10 INSECT RESISTANCE MANAGEMENT

The selection pressure on target population always creates evolution of resistant in the target population and it is a common phenomenon. The first case evidencing insect resistance to *B. thuringiensis* was observed in 1985 when mealmoths, *Plodia interpunctella* found in grain stores survived spray of *B. thuringiensis* spores. The laboratory experiments showed evolution of resistance after 15 generation of sub-lethal selection (McGaughey, 1985). Another case was reported in the wild population of diamondback moth, *Plutella xylostella* feeding on watercress in Hawaii, sprayed with *B. thuringiensis* up to 400 times (Liu & Tabashnik, 1997). *B. thuringiensis* transgenic crops producing single Cry proteins were more susceptible to evolution of insect resistance and therefore, it was mandatory to grow refugia strategy. Refugia strategy adoption on large-scale was responsible for resistance management in *B. thuringiensis* crops adopted area for long time. Other strategies like high-dose and low-dose approaches, target expression or temporally restricted expression were more difficult to implement in field and therefore, resistance pyramiding that is staking of multiple genes in the same plant that target the same pest via different mechanisms was adopted. This approach was also adopted for the improvement of *B. thuringiensis* topical product. Akhurst et al. (2003) revealed that pink bollworm, *Pectiniphora gossypiella* can evolve resistance to *B. thuringiensis* formulation like MVP based on single Cry protein but not to Dipel or XenTari, which contain multiple Cry proteins. *P. gossypiella* evolved resistance to Bollgard I cotton in Gujarat, India but no resistance was observed in cotton producing Cry1Ac and Cry2Ab (Bollgard II) (Monsanto, 2010). Additionally, attempts have also been made to enhance toxin activity by co-expression or proteins fusion. For instance, the hybrid Cry1Ab/Cry3A toxin (eCry3.1Ab) showed toxicity against the western corn rootworm, a pest which was unaffected by either of the parent toxin (Walters et al., 2010). Such efforts are not only helping in the management of insect resistance problem but also contributing a lot to extend the host spectrum.

13.11   CONCLUSION AND FUTURE PROSPECTS

B. thuringiensis synthesizes an extraordinary diversity of insecticidal proteins which has demonstrated its potential and safety in insect pest management programs for the development of topical biopesticides and transgenic crops. New strains of *B. thuringiensis* are being discovered on a regular basis and especially with the aid of proteomics, the novel toxins are also being identified on large scale. Additionally, the strain improvement program through strain engineering is giving promising results. Such attempts are opening avenues for expanding the known host spectrum of *B. thuringiensis* and subsequently the development of more advantageous products. Furthermore, the toxicity of Cry protein can be further enhanced through amino acid substitution, the introduction of cleavage sites in specific regions of the proteins and the deletion of small fragment from the N-terminal region. The *B. thuringiensis* topical formulation can be developed to enhance their efficacy and survival under various abiotic factors. Pyramiding of crops with many toxin genes can further expand the host spectrum and thereby manage the pest of broad group. *B. thuringiensis* can bring more economic benefits keeping in view the various ecological aspects of the associated environment which can be studied for a longer period.

KEYWORDS

- **entomopathogenic bacteria**
- **insecticidal proteins**
- **biopesticides**
- **insect-resistant crops**

REFERENCES

Aizawa, K. In *Shigetane Ishiwata: His Discovery of Sotto-Kin (Bacillus thuringiensis) in 1901 and Subsequent Investigations in Japan*, *Proceedings of a Centennial Symposium Commemorating Ishiwata's Discovery of Bacillus thuringiensis*, Kurume, Japan, November 1–3, 2001.

Akhurst, R. J.; James, W.; Bird, L. J.; Beard, C. Resistance to the Cry1Ac δ-Endotoxin of *Bacillusthuringiensis* in the Cotton Bollworm, *Helicoverpa Armigera* (Lepidoptera: Noctuidae). *J. Econ. Entomol.* **2003**, *96*, 1290–1299.

Akiba, T.; Abe, Y.; Kitada, S.; Kusaka, Y.; Ito, A.; Ichimatsu, T.; Katayama, H.; Akao, T.; Higuchi, K.; Mizuki, E.; Ohba, M.; Kanai, R.; Harata, K. Crystal Structure of the Parasporin-2 *Bacillus thuringiensis* Toxin that Recognizes Cancer Cells. *J. Mol. Biol.* **2009**, *386*, 121–133.

Akiba, T.; Higuchi, K.; Mizuki, E.; Ekino, K.; Shin, T.; Ohba, M.; Kanai, R.; Harata, K. Nontoxic Crystal Protein from *Bacillus thuringiensis* Demonstrates a Remarkable Structural Similarity to Beta-Pore-Forming Toxins. *Proteins* **2006**, *63*, 243–248.

Ammons, D.; Rampersad, J. Usefulness of Staining Parasporal Bodies When Screening for *Bacillus thuringiensis. J. Inverteb. Pathol.* **2002**, *79*, 203–204.

Angus, T. A. A Bacterial Toxin Paralyzing Silkworm Larvae. *Nature.* **1954**, *173*, 545–546.

Anil. 2013. Evaluation of Indigenous Entomopathogenic Bacteria against Some Iepidopterous Pests. Ph.D. Thesis, Department of Entomology, CSK HPKV, Palampur, India. p. 133.

Anitha, D.; Kumar, N. S.; Vijayan, D.; Ajithkumar, K.; Gurusubramanian, G. Characterization of *Bacillus thuringiensis* Isolates and Their Differential Toxicity against *Helicoverpa armigera* Populations. *J. Basic Microbiol.* **2011**, *51*, 107–114.

Aoki, K.; Chigasaki, Y. Uber die Pathogenität der Sog. Sotto-Bacillen (Ishiwata) Bei Seidenraupen. *Mitteil. Der. Med. Fakult. Der. Kaisser. Uni. Zu. Tokyo.* **1915**, *13*, 419–440.

Aoki, K.; Chigasaki, Y. Uber Die Pathogenitat Der Sog. Sottobacillen (Ishiwata) Bei Seidenraupen. *Bull. Imperial. Sericult. Exp. Sta.* **1916**, *1*, 97–139.

Barth, H.; Aktories, K.; Popoff, M. R.; Stiles, B. G. Binary Bacterial Toxins: Biochemistry, Biology, and Applications of Common *Clostridium* and *Bacillus* Proteins. *Microbiol. Mol. Biol. Rev.* **2004**, *68*, 373–402.

Battu, G. S.; Arora, R. Insect Pest Management through Microorganisms. In: *Biotechnological Approaches in Soil Microorganisms for Sustainable Crop Production;* Dadarwal, K. R., Eds.; Scientific Publishers: Jodhpur, India, 1997; pp 223–246.

Baum, J. A.; Johnson, T. B.; Carlton, B. C. *Bacillus thuringiensis:* Natural and Recombinant Bioinsecticide Products. *Methods Biotechnol.* **1999**, *5*, 189–209.

Bechtel, D. B.; Bulla, L. A. Ultrastructural Analysis of Membrane Development during *Bacillus thuringiensis* Sporulation. *J. Ultrastruct. Res.* **1982**, *79*(2), 121–132.

Beegle, C. C.; Yamamoto, T. Invitation Paper (CP Alexander Fund): History of *Bacillus thuringiensis* Berliner Research and Development. *Can. Entomol.* **1992**, *124*, 587–616.

Bernhard, K.; Jarrett, P.; Meadows, M.; Butt, J.; Ellis, D. J.; Roberts, G. M.; Pauli, S.; Rodgers, P.; Burges, H. D. Natural Isolates of *Bacillus thuringiensis* Worldwide Distribution, Characterization and Activity against Insect Pests. *J. Invertebr. Pathol.* **1997**, *70*, 59–68.

Berry, C. The Bacterium, *Lysinibacillus sphaericus*, as an Insect Pathogen. *J. Invertebr. Pathol.* **2012**, *109*, 1–10.

Boonserm, P.; Mo, M.; Angsuthanasombat, C.; Lescar, J. Structure of the Functional form of the Mosquito Larvicidal Cry4Aa Toxin from *Bacillus thuringiensis* at a 2.8-Angstrom Resolution. *J. Bacteriol.* **2006**, *188*, 3391–3401.

Bravo, A.; Gill, S. S.; Soberón, M. Mode of Action of *Bacillus thuringiensis* Cry and Cyt toxins and their Potential for Insect Control. *Toxicon.* **2007**, *49*, 423–435.

Bravo, A.; Sarabia, S.; Lopez, L.; Ontiveros, H.; Abarca, C.; Ortiz, A.; Ortiz, M.; Lina, L.; Villalobos, F. J.; Pena, G.; Nunez, V.; Maria, E.; Soberon, M.; Quintero, R. Characterization of *Cry* Genes in Mexican *Bacillus thuringiensis* Strain Collection. *Appl. Environ. Microbiol.* **1998**, *64*, 4965–4972.

Brookes, G.; Barfoot, P. Global Impact of Biotech Cops: Socio-Economic and Environmental Effects in the First Ten Years of Commercial Use. *Ag. Bio. Forum.* **2006**, *9*, 139–151.

Burges, H. D.; Jones, K. A. Formulation of Bacteria, Viruses and Protozoa to Control Insects. In *Formulation of Microbial Biopesticides, Beneficial Microorganisms. Nematodes and Seed Treatments;* Burges, H. D., Eds.; Kluwer Academic Publishers: Dordrecht, The Netherlands, 1998a; pp 33–127.

Burges, H. D.; Jones, K. A. Trends in Formulation of Microorganisms and Future Research Requirements. In *Formulation of Microbial Biopesticides, Beneficial Microorganisms, Nematodes and Seed Treatments;* Burges, H. D., Eds.; Kluwer Academic Publishers: Dordrecht, The Netherlands, 1998b; pp 311–332.

Butko, P. Cytolytic toxin Cyt1A and its Mechanism of Membrane Damage: Data and Hypotheses. *Appl. Environ. Microbiol.* **2003**, *69*, 2415–2422.

Capalbo, D. M. F. Fermentation Process and Risk Assessment. *Mem. Ins. Oswaldo. Cruz.* **1995**, *90*, 135–138.

Carlton, B. C.; Gawron-Burke, C. Genetic Improvement of *Bacillus thuringiensis* for Bioinsecticide Development. In *Advanced Engineered Biopesticides;* Kim, L., Eds.; Marcel Dekker Inc.: NY, 1993, pp 43–61.

Carlton, B. C.; Gonzalez, J. M. Plasmids and Delta-Endotoxin Production in Different Subspecies of *Bacillus thuringiensis*. In *Molecular Biology of Microbial Differentiation;* Hoch, J., A., Setlow, Q., Eds.; American Society for Microbiology: Washington DC, 1985, pp 246–252.

Chatterjee, S. N.; Bhattacharya, T.; Dangar, T, K.; Chandra, G. Ecology and Diversity of *Bacillus thuringiensis* in Soil Environment. *Afr. J. Biotechnol.* **2007**, *6,* 1587–1591.

Chaufaux, J.; Müller-cohn, J.; Buisson, C.; Sanchis, V.; Lereclus, D.; Pasteur, N. Inheritance of Resistance to the *Bacillus thuringiensis* Cry1A(c) Toxin in *Spodoptera littoralos* (Lepidoptera: Noctuidae). *J. Econ. Entomol.* **1997,** *90,* 873–878.

Christou, P.; Capell, T.; Kohli, A.; Gatehouse, J. A.; Gatehouse, A. M. R. Recent Developments and Future Prospects in Insect Pest Control in Transgenic Crops. *Trends Plant Sci.* **2006,** *11,* 302–308.

Cohen, S.; Dym, O.; Albeck, S.; Ben-Dov, E.; Cahan, R.; Firer, M.; Zaritsky, A. High-Resolution Crystal Structure of Activated Cyt2Ba Monomer from *Bacillus thuringiensis* Subsp. *Israelensis. J. Mol. Biol.* **2008,** *380*(5)*,* 820–827.

Crickmore, N.; Baum, J.; Bravo, A.; Lereclus, D.; Narva, K.; Sampson, K.; Schnepf, E.; Sun, M.; Zeigler, D. R. 2014. *Bacillus thuringiensis* Toxin Momenclature. http://www.btnomenclature.info/

Crickmore, N.; Zeigler, D. R.; Feitelson, J.; Schnepf, E.; Van Rie, J.; Lereclus, D.; Baum, J.; Dean, D. H. Revision of the Nomenclature for the *Bacillus thuringiensis* Pesticidal Crystal Proteins. *Microbiol. Mol. Biol. Rev.* **1998,** *62,* 807–813.

Crickmore, N. Beyond the Spore – Past and Future Developments of *Bacillus thuringiensis* as a Biopesticide. *J. Appl. Microbiol.* **2006,** *101,* 616–619.

Damgaard, P. H.; Hansen, B. M.; Pederson. J. C.; Eilenberg, J. Natural Occurrence of *Bacillus thuringiensis* on Cabbage Foliage and in Insects Associated with Cabbage Crops. *J. Appl. Microbiol.* **1997,** *82,* 253–258.

Das, J.; Das. B.; Dangar. T. K. Microbial Populations and *Bacillus thuringiensis* Diversity in Saline Rice Field Soils of Coastal Orissa, India. *African J. Microbiol. Res.* **2008,** *2,* 326–331.

de Barjac, H.; Bonnefoi, A. A Classification of Strains of *Bacillus thuringiensis* Berliner with a Key to their Differentiation. *J. Invertebr. Pathol.* **1968,** *11,* 335–347.

de Barjac, H.; Véron, M.; Cosmaodumanoir, V. Biochemical and Serological Characterization of *Bacillus sphaericus* Strains, Pathogenic or Non-pathogenic for Mosquitos. *Ann. Microbiol.* **1980,** *131,* 191–201.

de Maagd, R. A.; Bravo, A.; Berry, C.; Crickmore, N.; Schnepf, H. E. Structure, Diversity, and Evolution of Protein Toxins from Spore-forming Entomopathogenic Bacteria. *Annu. Rev. Genet.* **2003,** *37,* 409–433.

Deluca, A. J.; Simonson, J. G.; Larson, A. D. *Bacillus thuringiensis* Distribution in Soils of the United States. *Can. J. Microbiol.* **1981,** *27,* 865–870.

Dhaliwal, G. S.; Arora, R. *Integrated Pest Management: Concepts and Approaches.* Kalyani Publishers: Ludhiana, India, 2006; p 369.

Donovan, W. P.; Engleman, J. T.; Donovan, J. C.; Baum, J. A.; Bunkers, G. J.; Chi, D. J.; Clinton, W. P.; English, L.; Heck, G. R.; Ilagan, O. M.; Krasomil-Osterfeld, K. C.; Pitkin, J. W.; Roberts, J. K.; Walters, M. R. Discovery and Characterization of Sip1A: A Novel Secreted Protein from *Bacillus thuringiensis* with Activity against Coleopteran Larvae. *Appl. Microbiol. Biotechnol.* **2006,** *72,* 713–719.

Dulmag, H. T.; Rhodes, R. A. Production of Pathogen in Artificial Media. In *Microbial Control of Insects and Mites;* Burges, H.; Hussey, N. W., Eds.; Academic Press: London, UK, 1971; p 507.

Dulmage, H. T.; de Barjac, H. HD-187 A New Isolate of *Bacillus thuringiensis* that Produces High Yield of Delta-Endotoxin. *J. Invertbr. Pathol.* **1973,** *22,* 273–277.

Dulmage, H. T. Insecticidal Activity of HD-1, a New Isolate of *Bacillus thuringiensis* var. *Alesti. J. Invertbr. Pathol.* **1970,** *15,* 232–239.

EPA. 1998. R.E.D FACTS: *Bacillus thuringiensis. EPA-738-F-98-001,* United States Environmental Protection Agency: Washington, DC, 1998.

Estruch, J. J.; Warren, G. W.; Mullins, M. A.; Nye, G. J.; Craig, J. A.; Koziel, M. G. Vip3A, a Novel *Bacillus thuringiensis* Vegetative Insecticidal Protein with a Wide Spectrum of Activities against Lepidopteran Insects. *Proc. Natl. Acad. Sci. USA.* **1996,** *93,* 5389–5394.

Eswarapriya, B.; Gopalsamy, B.; Kameswari, B.; Meera, R.; Devi, P. Insecticidal Activity of *Bacillus thuringiensis* IBT-15 Strain Against *Plutella xylostella. Int. J. Pharmtech. Res.* **2010,** *2,* 2048–2053.

Farrera, R. R.; Pérez-Guevara, F.; De la Torre, M. Carbon: Nitrogen Ratio Interacts with Initial Concentration of Total Solids on Insecticidal Crystal Protein and Spore Production in *Bacillus thuringiensis* HD-73. *Appl. Microbiol. Biotechnol.* **1998,** *49*(6), 758–765.

Federici, B. A.; Siegel, J. P. Safety Assessment of *Bacillus thuringiensis* and Bt Crops Used in Insect Control. In *Food Safety of Proteins in Agricultural Biotechnology;* Hammond, B. G., Eds.; CRC Press: Boca Raton, FL, 2008.

Feitelson, J. S.; Payne, J.; Kim, L. *Bacillus thuringiensis*: Insects and Beyond. *Nature Biotechnol.* **1992,** *10,* 271–275.

Feitelson, J. S. The *Bacillus thuringiensis* Family Tree. In *Advanced Engineered Pesticides;* Kim, L., Eds.; Marcel Dekker, Inc.: New York, 1993, pp 63–72.

Fernando, H.; Valicente, E. D. S.; Tuelher, M. I. S. L.; Fernanda, L. F.; Corina, M. V. Production of *Bacillus thuringiensis* Biopesticides Using Commercial Laboratory Medium and Agricultural Byproducts as Nutrient Sources. *Revista. Brasileira. De. Milho. e. Sorgo.* **2010,** *9,* 1–11.

Fischhoff, D. A.; Bowdish, K. S.; Perlak, F. J.; Marrone, P. G.; McCormick, S. M.; Niedermeyer, J. G.; Dean. D. A.; Kusano-Kretzmer. K.; Mayer. E. J.; Rochester, D. E.; Rogers. S. G.; Fraley. R. T. Insect Tolerant Transgenic Tomato Plants. *Nature Biotechnol.* **1987,** *5,* 807–813.

Foda, M. S.; Ismail, I. M. K.; Moharam M. E.; Sadek Kh, H. A. A Novel Approach for Production of *Bacillus thuringiensis* by Solid State Fermentation. *Egypt J. Microbiol.* **2002,** *37,* 135–155.

Foda, M. S.; EL-Bendary, M. A.; Moharam, M. E. Salient Parameters Involved in Mosquitocidal Toxins Production from *Bacillus sphaericus* by Semi-Solid Substrate Fermentation. *Egypt J. Microbiol.* **2003,** *38,* 229–246.

Forsty, G.; Logan, N. A. Isolation of *Bacillus thuringiensis* from Northern Victoria Land, Antartica. *Lett. Appl. Microbiol.* **2000,** *30,* 263– 266.

Furlaneto, L.; Saridakis, H. O.; Arantes, O. M. N. Survival and Conjugal Transfer between *Bacillus thuringiensis* Strains in Aquatic Environment. *Braz. J. Microbiol.* **2000,** *31,* 233–238.

Goldberg, L. J.; Margalit, J. A Bacterial Spore Demonstrating Rapid Iarvicidal Activity against *Anopheles Sergentii, Uranotaenia unguicalata, Culex univitattus, Aedes aegyptii* and *Culex pipiens. Mosq. News.* **1977,** *37,* 355–358.

González, J. M.; Brown, Jr. B. J.; Carlton, B. C. Transfer of *Bacillus thuringiensis* Plasmids Coding for Delta-Endotoxin among Strains of *B. thuringiensis* and *B. cereus. Proc. Natl. Acad. Sci. USA.* **1982,** *79,* 6951–6955.

González, Jr.; Carton, B. J. A Large Transmissible Plasmid is Required for Crystal Toxin Production in *Bacillus thuringiensis* var. *Israelensis. Plasmid.* **1984,** *11,* 28–38.

Gould, F. Evolutionary Biology and Genetically Engineered Crops: Consideration of Evolutionary Theory Can Aid in Crop Design. *Bioscience.***1988,** *38,* 26–33.

Guerchicoff, A.; Delecluse, A.; Rubinstein, C. P. The *Bacillus thuringiensis Cyt* Genes for Hemolytic Endotoxins Constitute a Gene Family. *Appl. Environ. Microbiol.* **2001,** *67,* 1090–1096.

Guo, S.; Ye, S.; Liu, Y.; Wei, L.; Xue, J.; Wu, H.; Song, F.; Zhang, J.; Wu, X.; Huang, D.; Rao. Z. Crystal Structure of *Bacillus thuringiensis* Cry8Ea1: An Insecticidal Toxin Toxic to Underground Pests, the Larvae of *Holotrichia parallela. J. Struct. Biol.* **2009,** *168*(2), 259–266.

Hammond, B. G.; Koch, M. S. A Review of the Food Safety of Bt Crops. In *Bacillus thuringiensis Biotechnology;* Sansinenea, E., Eds.; Springer Science+Business Media BV: New York, 2012; pp 305–325.

Hannay, C. L.; Fitz-James, P. The Protein Crystals of *Bacillus thuringiensis* Berliner. *Can. J. Microbiol.* **1955,** *1,* 694–710.

Hannay, C. L. Crystalline Inclusions in Aerobic Spore-Forming Bacteria. *Nature.* **1953,** *172,* 1004–1006.

Heimpel, A. M.; Angus, T. A. The Taxonomy of Insect Pathogens Related to *Bacillus cereus* Frankland and Frankland. *Can. J. Microbiol.* **1958,** *4,* 531–541.

Heimpel, A. M.; Angus, T. A. The Site of Action of Crystalliferous Bacteria in Lepidoptera Larvae. *J. Insect. Pathol.* **1959,** *1,* 152–170.

Hoffmann, M. P.; Zalom F. G.; Wilson, L. T.; Smilanick, J. M.; Malyj, L. D.; Kiser J.; Hilder. V. A.; Barnes, W. M. Field Evaluation of Transgenic Tobacco Containing Genes Encoding *Bacillus thuringiensis* Delta-Endotoxin or Cowpea Trypsin Inhibitor: Efficacy against *Helicoverpa zea* (Lepidoptera: Noctuidae). *J. Econ. Entomol.* **1992,** *85,* 2516–2522.

Höfte, H.; Whiteley, H. R. Insecticidal Crystal Proteins of *Bacillus thuringiensis. Microbiol. Rev.* **1989,** *53,* 242–255.

Huang, D. F.; Zhang. J.; Song. F. P.; Lang, Z. H. Microbial Control and Biotechnology Research on *Bacillus thuringiensis* in China. *J. Invertebr. Pathol.* **2007,** *95,* 175–180.

Hunag, Y. X.; Hunag, R. R. Study of Isolating *Bacillus thuringiensis* from Soil. *Nat. Enemies Insects.* **1988,** *10,* 39–43.

Husz, B. On the Use of *Bacillus thuringiensis* in the Fight against the Corn Borer. *Int. Corn. Borer. Invest. Sci. Rept.* **1929,** *2,* 99–110.

Iriarte, J.; Dumanoir, V. C.; Bel, Y.; Porcar, M.; Ferrandis, M. D.; Lecadet, M.; Ferré, J.; Caballero, P. Characterization of *Bacillus thuringiensis* Ser. *Balearica* (Serotype H48) and Ser. *Navarrensis* (Serotype H50): Two Novel Serovars Isolated in Spain. *Curr. Microbiol.* **2000,** *40,* 17–22.

Ishiwata, S. On a Kind of Flacherie (Sotto Disease). *Dainihon Sanshi. Keiho.* **1901,** *114,* 1–5.

Ishiwata, S. About Sottokin, a Bacillus of a Disease of the Silkworm. *Dainihon Sanshi. Keiho.* **1905,** *161,* 1–5.

James, C. Global Status of Commercialized Biotech/GM Crops: 2009. *ISAAA Brief No. 41,* Ithaca, New York, USA, **2009.**

James, C. Global Status of Commercialized Biotech/GM Crops: 2014. *ISAAA Brief No. 49,* Ithaca, New York, USA, **2014.**

Jarrett, P.; Stephenson, M. Plasmid Transfer between Strains of *Bacillus thuringiensis* Infecting *Galleria mellonella* and *Spodoptera litoralis. Appl. Environ. Microbiol.* **1990,** *56,* 1608–1614.

Jones, G. W.; Nielsen-Leroux, C.; Yang, Y.; Yuan, Z.; Dumas, V. F.; Monnerat, R. G.; Berry, C. A New Cry Toxin with a Unique Two-Component Dependency from *Bacillus sphaericus. FASEB. J.* **2007,** *21,* 4112–4120.

Kaur, S. Molecular Approaches Towards Development of Novel *Bacillus thuringiensis* Biopesticides. *World J. Microbiol. Biotechnol.* **2000,** *16,* 781–793.

Kelker, M. S.; Berry, C.; Evans, S. L.; Pai, R.; McCaskill, D. G.; Wang, N. X.; Russell, J. C.; Baker, M. D.; Yang, C.; Pflugrath, J. W.; Wade, M.; Wess, T. J.; Narva, K. E. Structural and Biophysical Characterization of *Bacillus thuringiensis* Insecticidal Proteins Cry34Ab1 and Cry35Ab1. *PLoS One.* **2014,** *9*(11)*,* e112555.

Khetan, S. H. Microbial Pest Control. In *Bacterial Insecticides;* Khetan, S. H., Eds.; Marcel Decker: USA, 2001, pp 3–71.

Khyami-Horani, H.; Hajaij, M.; Charles, J. F. Characterization of *Bacillus thuringiensis* Ser. *Jordanica* (Serotype H71), a Novel Serovariety Isolated in Jordan. *Curr. Microbiol,* **2003,** *47,* 26–31.

Kirouac, M.; Vachon, V.; Noel, J. F.; Girard, F.; Schwartz, J. L.; Laprade, R. Amino Acid and Divalent Ion Permeability of the Pores Formed by the *Bacillus thuringiensis* Toxins Cry1Aa and Cry1Ac in Insect Midgut Brush Border Membrane Vesicles. *BBA-Biomembranes.* **2002,** *1561,* 171–179.

Koch, M. S.; Ward. J. M.; Levine, S. L.; Baum, J. A.; Vicini, J. L.; Hammond, B. G. The Food and Environmental Safety of Bt Crops. *Front. Plant Sci.* **2015,** *6,* 283.

Koziel, M. G.; Beland, G. L.; Bowman, C.; Carozzi, N. B.; Crenshaw, R.; Crossland, L.; Dawson, J.; Desai, N.; Hill. M.; Kadwell. S.; Launis, K.; Lewis, K.; Maddox, D.; McPherson, K.; Meghji, M. R.; Merlin, E.; Rhodes, R.; Warren, G. W.; Wright, M.; Evola, S. V. Field Performance of Elite Transgenic Maize Plants Expressing an Insecticidal Protein Derived from *Bacillus thuringiensis. Nature Biotechnol.* **1993,** *11,* 194–200.

Krattiger, A. F. Insect Resistance in Crops: A Case Study of *Bacillus thuringiensis* (Bt) and its Transfer to Developing Countries. *ISAAA Briefs No.* 2, Ithaca, New York, USA, 1996.

Krieg, V. A.; Huger, A. M.; Langenbruch, G. A.; Schnetter, W. *Bacillus thuringiensis* var. *Tenebrionis*: A new Pathotype Effective against Larvae of Coleoptera. *Z. Ang. Entomol.* **1983,** 96, 500–508.

Kronstad, J. W.; Schnepf, H. E.; Whiteley, H. R. Diversity of Locations for *Bacillus thuringiensis* Crystal Protein Genes. *J. Bacteriol.* **1983,** *154,* 419–428.

Kurstak, E. Le processus de l'infection par *Bacillus thuringiensis* Berl. d' *Ephestia kuhinella* Zell. Declenche par le Parasitisme de *Nemeritis canescens* Grav (Ichneumonidae) C.R. Hebd. Seanc. *Acad. Sci. Paris.* **1964,** *259,* 211–212.

Lambert, B.; Peferoen, M. Insecticidal Promise of *Bacillus thuringiensis* Facts and Mysteries about a Successful Biopesticide. *Bioscience* **1992**, *42*, 112–122.

Lecadet, M. M.; Frachon, E.; Dumanoir, V. C.; Ripouteau. H.; Hamon. S.; Laurent, P.; Thiery, I. Updating the H-Antigen Classification of *Bacillus thuringiensis*. *J. Appl. Microbiol.* **1999**, *86*, 660–672.

Li, R.; Dai, S.; Gao, M. A new Serovar of *Bacillus thuringiensis*: *Bacillus thuringiensis* Subsp. *Sinensis. Virol. Sin.* **2000**, *15*, 224–225.

Lin, Y.; Xiong, G. Molecular Cloning and Sequence Analysis of Chitinase Gene from *Bacillus thuringiensis* Serovar. *Alesti. Biotechnol. Lett.* **2004**, *26*, 635–639.

Lingren, P. D.; Green, G. L. *Suppression and Management of Cabbage Looper Populations;* Technical Bulletin No. 1684; US Department of Agriculture: US, 1984; pp152.

Lisansky, S. G.; Quinlan, R.; Tassoni, G. *The Bacillus thuringiensis Production Handbook;* CPL Perss: Newbury, UK, 1993.

Liu, Y. B.; Tabashnik, B. E. Experimental Evidence that Refuges Delay Insect Adaptation to *Bacillus thuringiensis. Proc. R. Soc. Lond. B.* **1997**, *264*, 605–610.

MacIntosh, S. C.; Stone, T. B.; Sims, S. R.; Hunst, P. L.; Greenplate, J. T.; Marrone, P. G.; Perlak, F. J.; Fischhoff, D. A.; Fuchs, R. L. Specificity and Efficacy of Purified *Bacillus thuringiensis* Proteins against Agronomically Important Insects. *J. Invertebr. Pathol.* **1990**, *56*, 258–266.

Maeda, M.; Mizuki, E.; Nakamura, Y.; Hatano, T.; Ohba, M. Recovery of *Bacillus thuringiensis* from Marine Sediments of Japan. *Curr. Microbiol.* **2000**, *40*, 418–422.

Majumdar, S. K. Production of Bacterial Lepidopterocide. In *Pesticides;* Madumdar, S. K, Eds.; Academy of Pest Control Sciences: Mysore, 1968, pp 167–174.

Manonmani, A.; Balaraman, K. A Highly Mosquitocidal *Bacillus thuringiensis* var. *Thompsoni. Curr. Sci.* **2001**, *80*, 770–781.

Martin, P. A. W.; Travers, R. S. Worldwide Abundance and Distribution of *Bacillus thuringiensis* Isolates. *Appl. Environ. Microbiol.* **1989**, *55*, 2437–2442.

Mattes, O. Parasitäre Krenkheiten der Mehlmottenlarven und Versuch über Ihre Verwendbarkeit als Biologisches Bekämpfungsmittel. *Ges. Beford. Gesamte, Naturwiss Marburg.* **1927**, *62*, 381–417.

McBride, K. E.; Svab, Z.; Schaaf, D. J.; Hogan, P. S.; Stalker, D. M; Maliga, P. Amplification of a Chimeric Bacillus Gene in Chloroplasts Leads to an Extraordinary Level of an Insecticidal Protein in Tobacco. *Nature Biotechnol.* **1995**, *13*, 362–365.

McGaughey, W. H. Insect Resistance to the Biological Insecticide *Bacillus thuringiensis. Science.* **1985**, *229*, 193–195.

Meadows, M. P. *Bacillus thuringiensis* in the Environment: Ecology and Risk Assessment. In *Bacillus thuringiensis, an Environmental Biopesticide: Theory and Practice;* Entwhistle, P. F., Corey, J. S., Bailey, M. J., Higgs, S., eds.; Wiley and Sons: Chichester, 1993;pp 193–220.

Mechalas, B. J. Method for the Production of Microbial Insecticides. U.S .Patent 3086922, 1963. Megna, J. C. Preparation of Microbial Insecticide. U. S. Patent 3073749, 1963.

Milner, R. J. History of *Bacillus thuringiensis. Agric. Ecosyst. Environ.* **1994**, *49*, 9–13.

Mitani, K.; Watarai, J. A New Method to Isolate the Toxin of *Bacillus sotto* Ishiwata by Passing through a Bacterial Filter and a Preliminary Report on the Toxic Action of This Toxin to the Silkworm Larva. *Aichi. Gensanshu. Seizojo. Hokoku.* **1916**, *3*, 33–42.

Monsanto, C. 2010. Press Release. Available online at: http://www.monsanto.com/news-views/Pages/india-pinkbollworm. aspx. (accessed Mar 5, 2010).

Montiel, M. D. L. T.; Tyagi, R. D.; Valero, J. R. Wastewater Treatment Sludge as a Raw Material for Production of *Bacillus thuringiensis* Based Bioinsecticides. *Wat. Res.* **2001,** *35,* 3807–3816.

Nethravathi, C. J.; Hugar, P. S.; Krishnaraj, P. U.; Vastrad, A. S. Bioefficacy of Native *Bacillus thuringiensis* Isolates against Cabbage Leaf Webber, *Crocidolomia binotalis* Z. *Karnataka J. Agril. Sci.* **2010a,** *23,* 51–55.

Nethravathi, C. J.; Hugar, P. S.; Krishnaraj, P. U.; Vastrad, A. S.; Awaknavar, J. S. Bioefficacy of Native Sikkim *Bacillus thuringiensis* (Berliner) Isolates against Lepidopteran Insects. *J. Biopesticides.* **2010b,** *3,* 448–451.

Norris, J. R. The Classification of *Bacillus thuringiensis. J. Appl. Bacteriol.* **1964,** *27,* 439–447.

Padua, L. E.; Gabriel, B. P.; Aizawa, K.; Obha, M. *Bacillus thuringiensis* Isolated from the Philippines. *Philippine Entomol.* **1982,** *5,* 199–208.

Palma, L.; Muñoz, D.; Berry, C.; Murillo, J.; Caballero, P. *Bacillus thuringiensis* Toxins: An Overview of their Biocidal Activity. *Toxins.* **2014,** *6,* 3296–3325.

Pardo-López, L.; Soberón, M.; Bravo, A. *Bacillus thuringiensis* Insecticidal Three-Domain Cry Toxins: Mode of Action, Insect Resistance and Consequences for Crop Protection. *FEMS. Microbiol. Rev.* **2013,** *37,* 3–22.

Patel, H. K.; Jani, J. J.; Vyas, H. G. Isolation and Characterization of Lepidopteran Specific *Bacillus thuringiensis. Int. J. Integr. Biol.* **2009,** *6,* 121–126.

Patel, K. D.; Bhanshali, F. C.; Ingle, S. S. Diversity and Characterization of *Bacillus thuringiensis* Isolates from Alluvial Soil of Mahi River Basin, India. *J. Adv. Dev. Res.* **2011,** *2,* 14–20.

Pearson, D.; Ward, O. P. Effect of Culture Conditions on Growth and Sporulation of *Bacillus thuringiensis* subsp. *Israelensis* and Development of Media for Production of the Protein Crystal Endotoxin. *Biotechnol. Lett.* **1988,** *10,* 451–465.

Perlak, F. J.; Fuchs, R. L.; Dean, D. A.; McPherson, S. L.; Fischoff, D. A. Modification of the Coding Sequence Enhances Plant Expression of Insect Control Genes. *Proc. Natl. Acad. Sci. USA.* **1991,** *88,* 3324–3328.

Perlak, F. J.; Stone, T. B.; Muskopf, Y. M.; Petersen, L. J.; Parker, G. B.; McPherson. S. A.; Wyman, J.; Love, S.; Reed, G.; Biever, D. Genetically Improved Potatoes: Protection from Damage by Colorado Potato Beetles. *Plant Mol. Biol.* **1993,** *22,* 313–321.

Popoff, M. R. Epsilon Toxin: A Fascinating Pore-Forming Toxin. *FEBS. J.* **2011,** *278,* 4602–4615.

Ramalakshmi, A.; Udayasuriyan, V. Diversity of *Bacillus thuringiensis* Isolated from Western Ghats of Tamil Nadu state, India. *Curr. Microbiol.* **2010,** *61,* 13–18.

Randhawa, H. S.; Khanna, V.; Kaur, R. Variation in Spore Population of *Bacillus thuringiensis* Berliner with Different Native Isolates. *Agric. Sci. Dig.* **2009,** *29,* 291–293.

Roh, J. Y.; Choi, J. Y.; Li, M. S.; Jin, B. R.; Je, Y. H. *Bacillus thuringiensis* as a Specific, Safe, and Effective Tool for Insect Pest Control. *J. Microbiol. Biotechnol.* **2007,** *17,* 547–559.

Roh, J. Y.; Liu, Q.; Lee, D. W.; Tao. X.; Wang. Y.; Shim. H. J.; Choi. J. Y.; Seo. J. B.; Ohba. M.; Mizuki, E.; Je, Y. H. *Bacillus thuringiensis* Serovar *Mogi* (Flagellar Serotype 3a3b3d), a Novel Serogroup with a Mosquitocidal Activity. *J. Invertebr. Pathol.* **2009,** *102,* 266–268.

Sachdeva, V.; Tyagi, R. D.; Valero, J. R. Factors Affecting the Production of *Bacillus thuringiensis* Bioinsecticides. *Rec. Res. Dev. Microbiol.* **1999,** *3,* 363–375.

Salama, H. S.; Foda, M. S.; Selim, M. H.; EL-Sharaby, A. Utilization of Fodder Yeast and Agro-Industrial By-Products in Production of Spores and Biologically Active Endotoxins from *Bacillus thuringiensis. Zbl. Microbiol.* **1983,** *138,* 553–563.

Sanahuja, G.; Banakar, R.; Twyman, R. M.; Capell, T. Christou, P. *Bacillus thuringiensis:* A Century of Research, Development and Commercial Applications. *Plant Biotechnol. J.* **2011,** *9*(3), 283–300.

Sanchis, V. From Microbial Sprays to Insect-Resistant Transgenic Plants: History of the Biopesticide *Bacillus thuringiensis.* A Review. *Agr. Sustain. Devel.* **2011,** *31*(1), 217–231.

Sattar, S.; Maiti, M. K. Molecular Characterization of a Novel Vegetative Insecticidal Protein from *Bacillus thuringiensis* Effective against Sap-Sucking Insect Pest. *J. Microbiol. Biotechnol.* **2011,** *21,* 937–946.

Schnepf, E.; Crickmore, N.; Van Rie, J.; Lereclus, D.; Baum, J.; Feitelson, J.; Zeigler, D. R; Dean, D. H. *Bacillus thuringiensis* and its Pesticidal Crystal Proteins. *Microbiol. Mol. Biol. Rev.* **1998,** *62,* 775–806.

Schnepf, H. E.; Whiteley, H. R. Cloning and Expression of the *Bacillus thuringiensis* Crystal Protein Gene in *Escherichia Coli. Proc. Natl. Acad. Sci. USA.* **1981,** *78,* 2893–2897.

Shelton, A. M.; Zhao, J. Z.; Roush, R. T. Economic, Ecological, Food Safety, and Social Consequences of the Deployment of Bt Transgenic Plants. *Annu. Rev. Entomol.* 2002, *47,* 845–881.

Shi, Y.; Ma, W.; Yuan. M.;, Sun, F.; Pang, Y. Cloning of *vip1/vip2* Genes and Expression of Vip1Ca/Vip2Ac Proteins in *Bacillus thuringiensis. World J. Microbiol. Biotechnol.* **2006,** *23,* 501–507.

Shi, Y.; Xu, W.; Yuan, M.; Tang, M.; Chen, J.; Pang, Y. Expression of *vip1/vip2* Genes in *Escherichia Coli* and *Bacillus thuringiensis* and the Analysis of Their Signal Peptides. *J. Appl. Microbiol.* **2004,** *97,* 757–765.

Smith, R. A.; Couche, G. A. The Phylloplane as a Source of *Bacillus thuringiensis. Appl. Environ. Microbiol.* **1991,** *57,* 311–315.

Steinhaus, E. A. Report on Diagnosis of Diseases of Insects, 1944–1950. *Hilgardia.* **1951,** *20,* 629–678.

Tailor, R.; Tippett, J.; Gibb, G., Pells, D.; Jordan, L.; Ely, S. Identification and Characterization of a Novel *Bacillus thuringiensis* Delta-Endotoxins Entomocidal to Coleopteran and Lepidopteran Larvae. *Mol. Microbiol.* **1992,** *6,* 1211–1217.

Tanda, Y.; Kaya, H, K. *Insect Pathology*; Academic Press: San Diego, 1993.

Thanabalu, T.; Hindley, J.; Brenner, S.; Oei, C.; Berri, C. Expression of the Mosquitocidal Toxins of *Bacillus sphaericus* and *Bacillus thuringiensis* Subsp. *Israelensis* by Recombinant *Caulobacter crecentus*, a Vehicle for Biological Control of Aquatic Insect Larvae. *Appl. Envir. Microbiol.* **1992,** *58,* 905–910.

Vachon, V.; Laprade, R.; Schwartz, J. L. Current Models of the Mode of Action of *Bacillus thuringiensis* Insecticidal Crystal Proteins: A Critical Review. *J. Invertebr. Pathol.* **2012,** *111,* 1–12.

Vaeck, M.; Reynaerts, A.; Höfte, H.; Jansens, S.; De Beuckeleer, M.; Dean, C.; Zabeau, M.; Van Montagu, M.; Leemans, J. Transgenic Plants Protected from Insect Attack. *Nature* **1987,** *328,* 33–37.

Vilas-Bôas, G. T.; Peruca, A. P. S.; Arantes, O. M. N. Biology and Taxonomy of *Bacillus cereus, Bacillus anthracis,* and *Bacillus thuringiensis. Can. J. Microbiol.* **2007,** *53,* 673–687.

Walters, F. S.; deFontes, C. M.; Hart, H.; Warren, G. W.; Chen, J. S. Lepidopteran-Active Variable-Region Sequence Imparts Coleopteran Activity in eCry3.1Ab, an Engineered *Bacillus thuringiensis* Hybrid Insecticidal Protein. *Appl. Environ. Microbiol.* **2010,** *76,* 3082–3088.

Wang, T. A Simple and Economic Semi-Solid Fermentation Process for *Bacillus thuringiensis.* In *International Symposium on Insecticide of Bacillus thuringiensis;* Hubei Academy of Agricultural Science: Wuhan,1988.

Warren, G. W.; Koziel, M. G.; Mullins, M. A.; Nye, G. J.; Carr, B.; Desai, N. M.; Kostichka, K.; Duck, N. B.; Estruch, J. J. Auxiliary Proteins for Enhancing the Insecticidal Activity of Pesticidal Proteins. U.S. Patent 5,770,696, June 23, 1998.

Widjaya, T. S.; Osborne, K. J.; Rogers, P. L. In The *Effect of Acetate on Growth and Sporulation of the Mosquito Pathogen Bacillus sphaericus 2362.* Proceedings of the 10th Australian Biotechnol Conference, 1992; pp 239–242.

Yara, K.; Kunimi, Y.; Iwahana, H. Comparative Studies of Growth Characteristic and Competitive Ability in *Bacillus thuringiensis* and *Bacillus cereus* in soil. *Appl. Entomol. Zool.* **1997,** *32,* 625–634.

Zhang, Q.; Hua, G.; Bayyareddy, K.; Adang, M. J. Analyses of Alpha-Amylase and Alpha-Alucosidase in the Aalaria Vector Mosquito, *Anopheles gambiae,* as Receptors of Cry11Ba toxin of *Bacillus thuringiensis* Subsp. *Jegathesan. Insect Biochem. Mol. Biol.* **2013,** *43,* 907–915.

Zouari, N.; Ali, S. B. S.; Jaoua S. Production of Delta-Endotoxins by *Bacillus thuringiensis* Strains Exhibiting Various Insecticidal Activities Toward Lepidoptera and Diptera in Gruel and Fish Meal Media. *Enz. Microbial. Technol.* **2002,** *31,* 411–418.

CHAPTER 14

ROLE OF PHYTOCHEMICALS IN INSECT PEST MANAGEMENT

TAMOGHNA SAHA[1*], NITHYA CHANDRAN[2], and SHYAMBABU SHA[1]

[1]*Department of Entomology, Bihar Agricultural University, Sabour 813210, Bhagalpur, India*

[2]*Division of Entomology, Indian Agricultural Research Institute (IARI), New Delhi 110012, India*

*Corresponding author. E-mail: tamoghnasaha1984@gmail.com

CONTENTS

ABSTRACT

Plant-based material has been proposed and potential alternatives to chemical pesticides for safe and environment-friendly pest management options. These plant-based products act as antifeedant, fumigant, growth regulator, repellent, and a direct mortality factor against insect pests. The chemical components present in plants have these activities and show promise in managing pests in agriculture. Recently, researchers' interest has focused on the detection of the chemical compounds present in plants that have insecticidal properties. Even though there is a lot of information available on this field of science, but the information is scattered in various journals and other publications. So, it requires the exhaustive compilation. This chapter will give an overview about some of the important phytochemicals, which play an important role to protect themselves against herbivory, pathogenic microbes, and various abiotic stresses as well as specific plant responses to pathogen attack.

14.1　INTRODUCTION

Plants produce massive and diverse kinds of chemical compounds within their system but not all these composites contribute directly in the growth and development of the plant. These phytochemicals, also known as secondary metabolites are allocated largely within the plant kingdom and the plant physiological form (Croteau et al., 2000). Plant chemicals are categorized as either primary or secondary plant metabolites. Approximately, 308,800-plant species only a few have been investigated and most remained unutilized for pesticidally active principles. Approximately, 2400-plant species have been reported till date to have pesticidal properties belonging to 189 families. Out of 189 families approximately 22 families hold more than 10 plant species in each family with anti-insect properties. Roughly, more than 350 insecticidal compounds, > 800 insect antifeedents and moderately a good amount of insect growth inhibitors and growth regulators have been cut off from various plant species but, it seems that only few have accomplished the commercial status (Misra, 2014). However, the exploit of plant-derived products to manage the insect pest is a well-known fact in the developing world and before the invention of synthetic pesticides; plant derived products were only pest-managing representative to farmer around the world (Owen, 2004). Yang and Tang (1988) reviewed the plants utilized for pest control and observed that there is a sturdy connection between medicinal and pesticidal plants. Particularly notable are tropical plants from which hundreds of secondary metabolites with pesticidal properties have been extorted. They are environmentally less harmful than synthetic pesticides and acting in many insects in diverse ways (Hiiesaar et al., 2001). Literature reports have indicated that many of these compounds have important significance in protection against herbivorous nature and the phytochemical diversity of insect defenses in tropical and temperate plant families has also been significantly proven (Croteau et al., 2000).

At presently marketed phytochemicals in the world major ones include pyrethrins, rotenone, nicotine, ryanodine, sabadilla, neem-based products, and toosendanin. Throughout last few years' plant essential oils consisting mono and sesquiterpenoids are being expanded as green pesticides. These oils are act as insect toxins, repellents, and deterrents (Misra, 2014).

Allelochemicals are also known as secondary metabolites or non-nutritional primary metabolites, which might be the parts of both extracts or essential oils that inhibit development, reproduction, or activities of individuals other than the ones producing them, or structure, and dynamics of populations of either plants or animals or microbes. The capacity of allelochemicals is extreme wider than that correlated with allelopathy and has a variety of interactions mediated by chemicals with the above properties (Ross et al., 1991). Even though allelochemicals mediate a wide variety of multifaceted interactions, allomonal chemical compounds fall into one of the two basic categories. Under first category include materials produced by the organisms and discharged into the environment, mostly unstable compounds, which perform their influence over some distance from the emitter. Such unstable compound includes a wide variety of short-chain alcohols and aldehydes, ketones, esters, aromatic phenols, mono- and sesquiterpenes, and a host of other secondary metabolites. Under second group of allomones comprises compounds produced or acquired for defense, which stay in the body of the producer. Such group comprises toxins sequestered by insects for protection and the huge range of phytochemicals. Actually in modern decades, creative writing has been flooded with umpteen studies where extracts, isolated compounds, or combination products have been assessed for their efficacy against a variety of pests (Copping et al., 2007; Koul, 2005; Koul & Dhaliwal, 2001; Parmar & Walia, 2001).

However, terpenes, phenolics, and N and S containing compounds are the tree major groups of secondary metabolites wherein terpenes composed of 5-C isopentanoid units are toxins and feeding deterrents to many herbivores. Phenolics produced mostly from products of the shikimic acid pathway, have numerous key defensive role in the plants. In N and S containing compounds are synthesized principally from common amino acids (Rosenthal et al., 1992; Van Etten et al., 2001).

In recent times, a variety of plant extracts and plant parts have been revealed to have insecticidal properties (Mahmood et al., 1984). A few of them have significant oviposition difference or antifeedant or toxic effects on selected tea pests (Hazarika et al., 2008). This article will help to bring together pertinent aspects of the basic and practical sciences of natural pesticides. During the last several years, it has been discovered that hundreds of compounds that plants make have significant ecological and chemical defensive roles, opening a new area of scientific endeavor, often called ecological biochemistry (Harborne, 1989).

14.2 INDIGENOUS PLANT PARTS AND PLANT EXTRACTS IN INSECT PEST MANAGEMENT

Botanical insecticides obtained from plants are currently distinguished as biodegradable, systemic, environmental friendly and non-toxic to mammals and are thus considered safe. The modes of action of botanical pesticides against insect pests are diverse. Some plant products (e.g., neem-based products, pyrethroids, and essential oils) are already used to control pest populations on a large scale. Botanical pesticides are eco-friendly, less hazardous, economically viable, and easily available (Mamun & Ahmed, 2011). Mamun (2011) reported that plants are rich source of bioactive organic chemicals and approximated that the plants may have as many as 4,000,000 secondary metabolites. The anti-pest plants recognized by Mugisha-Kamatenesi et al. (2008) and those are *Capsicum frutescens, Tagetes* spp., *Nicotiana tabacum, Cupressus* spp., *Tephrosia vogelii, Azadirachta indica (A. indica), Musa* spp., *Moringa oleifera, Tithonia diversifolia, Lantana camara, Phytolacca dodecandra, Vernonia amygdalina, Aloe* spp., *Eucalyptus* spp., *Cannabis sativa, Coffea* spp., and *Carica papaya*. Plant-based products are extracted from different parts (leaves, stems, seeds, roots, bulbs, rhizomes, unripe fruits, and flower heads, etc.) of diverse plant species. They hailed for having a broad spectrum of activity, being simple to process and utilize, possessing a short residual activity, and for not accumulating in the environment or in fatty tissues of warm-blooded animals (Philip & Robert, 1998). Some of them are neem, chrysanthemum, rotenone, annona, tobacco, etc.

14.2.1 NEEM (AZADIRACHTA INDICA)

Neem has come out as the single most vital source of botanical pesticides having a broad-spectrum control of several insects, mites, fungi, nematodes, and viral diseases. They are the natural source of environment friendly insecticides, pesticides, and agrochemicals (Brahmachari, 2004). Neem products have no harmful effects on humans and animals and no residual effect on agricultural produce. Azadirachtin, an oxygenated triterpenoid, obtained from the seed kernels of the neem tree, it is very effective against pink and purple mites and certain leaf-folding caterpillars. Approximately, 100s of secondary metabolites like limonoids, protolimonoids, tetranortriterpenoids, pentanoortripenoids, hexanor-tripenoids, and some non-terpenoids have been obtained from diverse parts of neem tree and still more are being acquired. These compounds and neem extracts show nearly each kind of biological activity conceivable against a large number of insects. More than 400 species of insects have been reported to be affected by neem compounds (Schmutterer & Singh, 1995). The leaves and seeds of neem comprise azadirachtin, miliantriol etc. have insecticidal properties. These properties are more in seeds rather than in other parts of the plant. Currently, neem is being available in diverse formulations, such as neem oil, neem seed kernels extracts (NSKE), neem cake, etc. for the management of insect pests. *A. indica* also shows considerable effect on feeding to tea mosquito bug (Sarmah & Bhola, 2008). Therefore, neem-based product can be utilized for the control of Red spider mites, Helopeltis, Thrips, Scales, Termites, Looper caterpillar, and Nematodes, etc. Mamun and Iyengar (2010) reported that application of neemcake @ 2kg/ bush was found superior for the plants suffering from the attack of root knot nematodes, *Meloidogyne brevicauda*.

14.2.2 KARANJA (PONGAMIA PINNATA)

Karanja seeds are also deadly to the insect pests. Their seed extracts will be very helpful to control Helopeltis, Red spider mites, and other minor

insects of tea. The oilcakes of karanja also play an important role in reduction the infestation of root knot nematodes (Mamun & Ahmed, 2011).

14.2.3   TOBACCO (NICOTIANA TABACUM)

Tobacco was known since long as the insecticidal properties due to presence of Nicotine. It is a quick acting insecticide. Nicotine is an alkaloid;
therefore, it is used as nicotine sulfate for insect control. Nicotine sulfate is
accessible in 40% liquid form. It can be mixed with soap and diluted with
water and spray on plants. Tobacco solutions can be prepared by 400 g
small pieces tobacco is to be cut and immersed in four liter of water for
one day. Then collect the fresh liquid and put in soft soap @ 1ml per liter.
This concentrated decoction may be diluted with water and utilized for
spraying against aphids, thrips, scales, looper caterpillars, etc. (Mamun &
Ahmed, 2011).

14.2.4   ROTENONE (DERRIS ELLIPTICA)

Rotenone obtained from the roots of *Derris elliptica* a leguminous plant
comprising 4–11% rotenone can be utilized by drying and dusting or
by mixing with water. The active ingredient concentration in the spray
liquid should be 0.002–0.004%. Rotenone is effective against sucking
pests, caterpillars, and some beetles. It performs as a contact and stomach
poison. In one hectare of Derris, the yields about 1.5 liters of roots with
5% rotenone content, enough active ingredients to take care of 500–700 ha
of tea plantation (Mamun & Ahmed, 2011).

14.2.5   ANNONA (ANNONA SQUAMOSA)

Annona leaves and seeds having insecticidal properties because this plant
comprises lanolin and anonaine. The seeds are to be dried, dusted, and
prepared into a solution by adding with water or alcohol for application.
They are effective against stem borer, sucking pests, scale insects, etc. The
aqueous leaf extract of annona has feeding deterrent activity (66–82%)
against Helopeltis (Gurusubramanian et al., 2008).

14.2.6 CHRYSANTHEMUM (CHRYSANTHEMUM CINERARIAEFOLIUM)

Chrysanthemum is obtained from the dried flowers of Chrysanthemum. Pyrethrum or natural pyrethroid is a combination of six different substances. Stachydrine, the alkaloid is the main active principle of pyrethrum. They have the capability to paralyse the pests on contact. They can be used to control sucking pests like thrips, aphids, scale insects, tea mosquito bugs, etc. (Mamun & Ahmed, 2011).

14.3 SECONDARY PLANT METABOLITES AS A SOURCE OF DEFENCE AGAINST MANY INSECT PESTS

There are two types of Plant metabolites—primary metabolites and secondary metabolites. The primary metabolites contain of all plant cells components that are directly engaged in development, expansion, or repro-duction processes, these consist of proteins, amino acids, sugars, and nucleic acids. Secondary metabolites are those plant components that are not directly engaged in the improvement process of the plant but in the physiological portion including the plant protection system. Secondary metabolites get their source from the metabolism of glucose via two inter-mediaries' main shikimic acid and ethyl acetate. They create huge molecules with efficient groups, such as fatty acids, hydrocarbons, esters, aldehydes, ketones, alcohols, acetylenic compounds, alkaloids, phenols, and couma-rins (Santos, 2010). Secondary metabolites are different from primary metabolites, because besides not being straight linked to the purpose of sustaining life, they have limited distribution within and between plants (Taiz & Zeiger, 2013; Gonzaga et al., 2008). Secondary metabolites have been distinguished several adaptive functions of these metabolites as protec-tion against herbivores and microorganisms, UV protection, attraction of pollinators, and attraction of seed dispersers animals, besides, functions allelopathic (Simoes et al., 2007). The production and storage of secondary metabolites are limited to certain stages of plant development, organ and tissue-specific, or specialized cells (Panda & Khush, 1995).

Plants have developed two types of protection against herbivores, that is, direct and indirect. In the direct protection, substances which are directly involved are silica, secondary metabolites, enzymes and proteins, and organs, such as trichomes and thorns that straight affect the performance of the insect. In the indirect protection, there are involved substances produced by the plant itself that attract parasites and predators of phytophagous insect. Volatile terpenes and phenylpropanoids produced by plant species can have, depending on the insect in case, attractive properties (supply, pollination) and/or feeding deterrents, and insecticides (Simas et al., 2004). It can affect herbivores in several ways, such as deterrence, indigestible, inhibitors of oviposition, and mortality of young adults (Gutierrez & Villegas, 2008).

The secondary metabolites that show biological activity are recognized as active ingredients and have produced attention for a promising market for the discovery of therapeutic activities. Secondary metabolites are separated into three major groups: terpenes, phenols, and nitrogenous compounds (Taiz & Zeiger, 2013).

Principal plant secondary metabolites can be divided into four chemically distinct groups, viz., Terpenoids, Phenolics, Alkaloids, and Nitrogenous compounds. A few detailed examples of well-documented secondary metabolites which perform as insect antifeedents obtained from terrestrial plants are listed in Table 14.1.

TABLE 14.1   Some Important Examples of Effective Insect Antifeedants Isolated from Plants.

Chemical type	Compound	Plant source
Monoterpene	Thymol	*Thymus vulgaris* (Lamiaceae)
Sesquiterpene	Lactone (germacranolide type) Glaucolide A	*Vernonia* species (Asteraceae)
Sesquiterpene (drimane type)	Polygodial	*Persicaria hydropiper* (Polygonaceae)
Diterpene (abietane type)	Abietic acid	*Pinus* species (Pinaceae)
Diterpene (clerodane type)	Ajugarin I	*Ajuga remota* (Lamiaceae)
Flavonoid	Quercetin	*Buddleja madagascariensis* (Caesalpiniaceae)
Triterpene (limonoid type)	Azadirachtin	*Azadirachta indica* (Meliaceae)
Triterpene (cardenolide type)	Digitoxin	*Digitalis purpurea* (Scrophulariaceae)
Triterpene (ergostane type)	Withanolide E	*Withania somnifera* (Solanaceae)

TABLE 14.1 *(Continued)*

Chemical type	Compound	Plant source
Triterpene (spirostane type)	Aginosid	*Allium porrum* (Liliaceae)
Alkaloid (indole type)	Strychnine	*Strychnos nux-vomica* (Loganiaceae)
Alkaloid (steroidal glycoside)	Tomatine	*Lycopersicon esculentum* (Solanaceae)
Phenolic (furnanocoumarin)	Xanthotoxin (= 8-methoxypsoralen)	*Pastinaca sativa* (Apiaceae)
Phenolic (lignan)	Podophyllotoxin	*Pelargonium peltatum* (Berberidaceae)
Phenolic (benzoate ester)	Methyl salicylate	*Gaultheria procumbens* (Ericaceae)

(Reprinted from Adeyemi, M. M. Hassan, The Potential of Secondary Metabolites in Plant Material as Deterents against Insect Pests: A Review. *Afr. J. Pure Appl. Chem.* **2010,** 4(11), 243–246. https://creativecommons.org/licenses/by/4.0/)

14.3.1 TERPENES

They cover a large variety of materials of plant origin and its ecological significance as protective plants are well established. Garcia and Carril (2009) reported that, terpenoids comprise the largest set of secondary metabolites, with more than 40,000 different molecules, which role in plant is to defend or to attract beneficial organisms. They are normally insoluble in water and attached to protection against herbivore owing to the fact of conferring bitter or having the same molecular configuration of the molting hormone of insects, to suspend that process function, or by creation complex sterol precursors of animal hormones (Taiz & Ziger, 2013).

A triterpenoid achieved from the neem tree, *A. indica* and several monoterpenes, such as citronella, pinene, linalool, geraniol, citronelol, limonene, myrecene, and citral from essential oils and pyrethrins from various *Chrysanthemum* spp. are common terpenoids known to have insecticidal properties. The potential effect of plant chemicals against insect pests ranges from repellency expressed as feeding and oviposition deterrence, to interfering with growth and development (Mann & Kaufan, 2012). The triterpenoids have parallel molecular structures to plant and animal sterols and steroid hormones, the phytoecdysones are imitates of insect molting hormones. The subclasses of terpenoides are discussed below.

14.3.1.1 MONOTERPENE (C_{10})

Several monoterpenes were obtained and assessed for toxicity to different insects. These studies included α-pinene, β-pinene, 3-carene, limonene, myrcene, α-terpinene, and camphene (Junior, 2003). On the lower level, the comparatively easy composition of monoterpenes, such as limonene, myrcene, and the epoxy-pulegone, they functions to the plants that generate them. Marangoni et al. (2012) reported that actually monoterpenes insecticidal activity was due to inhibition of acetylcholinesterase in insects, which is the case of epoxy-pulegone, which causes consequences, such as growth retardation, reduction of reproductive capacity, appetite suppressants, and may lead the predatory insects to death by starvation or direct toxicity.

14.3.1.2 DITERPENE (C_{20})

Diterpenes are a promising group of compounds that inhibit the feeding actions of insect pests. Both clerodane and neoclerodane group of diterpenoids are well recognized for their insecticidal and antifeedant activity (Belles et al., 1985). Numerous natural neoclerodane diterpenoids obtained from *Linaria saxatilis* and some semisynthetic derivatives were investigated against several insect species with diverse feeding adaptations. The antifeedant assessments illustrated that the oligophagous *Leptinotarsa decemlineata* was the most sensitive insect, followed by the aphid *Myzus persicae*. The polyphagous *Spodoptera littoralis* was not inhibited by these diterpenoids; still, following oral administration, some of these compounds did have post-ingestive antifeedant effects on this insect. In general terms, the antifeedant activities of these compounds were species-dependent and more selective than their toxic/post-ingestive effects (Lajide et al., 1995).

14.3.1.3 TRITERPENES (C_{30})

A lot of well-recognized insect antifeedants are triterpenoids. Particularly, well considered in this connection are the limonoids from the neem (*A. indica*) and chinaberry (*Melia azedarach*) trees, exemplified by azadirachtin, toosendanin and limonin from citrus species. Other anti-feedant triterpenoids comprise cardenolides, steroidal saponins, and with anolides (Hassan, 2010). Another group of triterpene, limnoid, a bitter substances in citrus fruits and execute as antiherbivore compounds in members of family Rutaceae and some other families also. For example, Azadirachtin, a complex limnoid from *A. indica*, performs as a feeding deterrent to some insects and exerts various toxic effects (Mordue & Blackwell, 1993).

14.3.1.4 SESQUITERPENE

Sesquiterpenes are one of the important sources of insect antifeedants. Numerous insecticidal and antifeedant sesquiterpenoids are recognized as major deterrents in insect–plant interactions (Ivie & Witzel, 1982). Some feeding deterrents, such as sesquiterpene lactone angelate argophyllin-A and 3-O-methyl niveusin-A have been isolated from inflorescences of cultivated sunflower. Alpha-cyperone, a sesquiterpene obtained from the *Cyperus rotundus* (nutgrass) tubers is insecticidal against diamondback moth *Plutella xylostella* (Thebtaranonth et al., 1995). The antifeedant and insecticidal activity of polyol esters against *Pieris rapae* and *Ostrinia furnacalis* have been characterized to the ester moieties connected to the decalin portion of the molecule. Some recent records also prove such terpenes obtained just from Rutales are effective antifeedants for stored grain pests, mostly the spirocaracolitones, which are absolute antifeedants (Omar et al., 2007).

14.3.2 ALKALOIDS

They are nitrogenous compounds that portray insecticidal properties at low concentration and are often toxic to vertebrates. Nicotine, anabasine, and ryanodine are common alkaloids utilized as pesticides. The mode of action of alkaloids vector transmitted diseases differs from formation of their molecules, but many are reported to affect acetylcholinesterase or sodium channels. New alkaloidal compounds (+)-11-methoxy-10-oxo-erysotramidine, and (+)-10,11-dioxoerysotramidine recognized in the seeds, seed pods, and flowers of *Erythrina latissima* proved strong anti-feedant activities against third-instar *Spodoptera littoralis* (Boisduval) larvae. Erythrinaline alkaloids confer a signal of its usefulness in post-harvest storage and crop protection due to its antifeedant effect. It was also examined that farms with maize growing under the *E. latissima* tree were meagerly attacked by the stem borer. As the tree is a prevalent flowering plant, its seeds and flowers can be harnessed and utilized as a potential bio-pesticide or antifeedant in agricultural post-harvest processes (Wanjala et al., 2009). Application of essential oils of *Matricaria recutita* known to have precocenes as an active ingredient also inhibit with insect glands that generate juvenile hormones resulting in the inhibition of insect growth while molting (Ndungu et al., 1995). Now a days essential oils are acquiring significant acceptance for apply in post-harvest storage, crop protection, and fumigation because of the virtual safety, biodegradable status, and their utilization for other multi-purpose uses.

14.3.3   PHENOLIC COMPOUND

These compounds are another group of secondary metabolites observed within plant tissues that act as defense mechanism against pathogens, produced through the shikimic acid, and malonic acid pathways in plant systems; they cover a broad range of defense-related secondary metabolites, such as tannins, lignin, flavonoids, anthocyanins, and furanocoumarins. Among the secondary metabolites, phenolic, or polyphenols are present themselves as one of the most diverse groups of substances, with more than 8000 phenolic structures known and widely allocated throughout the plant kingdom (Harborne, 1993). These phenolic compounds are very diversify natural products, having common characteristic the presence of at least one aromatic ring, wherein at least one hydrogen is altered by one free hydroxyl group or other derivative function as ester or heteroside (Carvalho et al., 2007).

The phenolic compounds are known to be chemically a heterogeneous group, while some phenolic compounds have attractive function to pollinators or to fruit dispersers, other have reject function to herbivores. They also perform as defense against UV radiation or to have allelochemistry function in competing adjacent plants (Taiz & Ziger, 2013).

Tannins are one of the important phenolic compound show toxicity to insects through their action of binding to the salivary proteins and digestive enzymes including trypsin and chymotrypsin, which consequences in protein inactivation in the organisms, this reasons them to ingest high amounts of tannins but they fail to gain weight and become emaciated and may eventually die. Lignin, another one more important phenolic compound is a well-branched heterogeneous polymer observed in the secondary cell walls of plants, although the primary walls could also be lignified. Lignin comprises of a large number of the phenolic monomers (in hundreds or thousands) and structures the primary component of wood. The insoluble, stiff and nearly indigestible structural frame certify lignin presents a strong hurdle against insect attack. Flavonoids show one of the most significant and diverse groups amongst the products of plant origin and are extensively allocated in the plant kingdom. The biochemical actions of flavonoids and their metabolites depend on their chemical composition, which can vary with substitutions including hydrogenation, hydroxylation, methylation, malonylation, sulfation, and glycosylation. Flavonoids and isoflavones commonly happen as esters, ethers or derivatives of glycosides, or yet mixture of them (Brito et al., 2001).

14.3.4 NITROGENOUS COMPOUNDS

These compounds are well recognized in plant protection against herbivory, mainly alkaloids and cyanogenic glycosides. Nitrogenous compounds have deadly or medicinal action for humans (Taiz & Ziger, 2013). Alkaloids comprise a large set of secondary metabolites with structural diversity and be defined as "cyclic organic compounds of natural origin comprising one nitrogen in a negative oxidation state with limited distribution among organisms" as methylxanthine, theophylline, theobromine, codeine, thebaine, papaverine, and narcotine, caffeine (Pelletier, 1983). Amongst the most imperative natural alkaloid used to control insect pests are nicotine and nor-nicotine. Those alkaloids were used in the sixteenth century and reached 2500 tons in the middle of the nineteenth century. Since then, the annual production has been deteriorating and at present covers about 1250 tons of nicotine sulfate and 150 tons of nicotine, because of their high cost of production, poor odor, acute toxicity to mammals, and limited insecticidal activity (Viegas et al., 2003).

14.4 ROLES OF PHYTOCHEMICALS IN INSECT PEST MANAGEMENT

The existence of certain chemicals in plants stops insects from feeding on them resulting starvation of the insects and, in some cases, eventual death.

Some examples like, *Ocimum viride* showed to be an efficient insect repellent that compare very healthy with the widely explored *A. indica*. It showed strong repellency toward *Tribolium castaneum* (Herbst) (Coleoptera: Tenebrionidae) and the rice weevil, *Sitophilus oryzae* (Linnaeus) (Coleoptera: Curculionidae), which are both hazardous pests of maize and rice (*Oryzae sativa* L.). Treatment of rice grain with Leaf extract of *O. viride* treated in rice grain showed in less than 25% survival of the two insects after 10 days (Owusu, 2000). Wood ash is another important plant product which has been observed effective in the control of insect pests. It is usually believed that inert substances result a loss of body moisture in insects and also one of the traditional ways of protecting grains from storage pests amongst low-resource farmers in some parts of the world, including Africa, China, Japan, and India (Dennis, 2002). However, a large number of plant extracts from different herbs have been evaluated for fumigant toxicity against insects. Essential oils from a few members of the family Lamiaceae made 90% mortality in adult populations of the maize weevil, rice weevil, cowpea weevil, and *Sitotroga cerealella* (Linnaeus) (Lepidoptera: Gelechiidae) after 24 h of exposure to a concentration of 1.4–4.5 µL l–1 (Shaaya et al., 1997). Similarly, 70–80% mortality was recorded in populations of *Callosobruchus maculatus* after 12 h exposure to essential oils from basil (*Ocimum basilicum* L.).

Some of the phytochemicals and their active component, mode of action and commercial application as plant protection agents are listed in Table 14.2.

Research has revealed that botanical insecticides can play an important role in integrated pest management (IPM) (Scott et al., 2002). Recent study has focused on the opportunity of incorporating botanical insecticides into IPM systems. This comprised the impact of botanicals on natural enemies, that is, beneficial organisms that attack or kill pest organisms. Even though joint research by the International Foundation of Science, Wageningen University, the Netherlands, and the ARC-Plant Protection Research Institute in South Africa, pointed out that neem and China-berry plant extracts effectively managed diamondback moth, with no harmful effect on two of its natural parasitoid species (Charleston et al., 2005). This further confirmed by Goudegnon et al. (2000), who reported the effect of deltamethrin and neem kernel solution on diamondback moth and *Cotesia plutellae* (Kurdjumov) (Hymenoptera: Braconidae) parasitoid

TABLE 14.2 Some Commercial Natural Plant Products, Their Active Compound and Mode of Action.

Name of the plant products	Plant source	Active compounds	Biological activity	Mode of action	References
Neem (neem oil, medium polarity extracts)	*Azadirachta indica* A. Juss	azadirachtin, dihydroaza-dirachtin, variety of triterpenoids (nimbin, salannin, and others)	insecticide, acaricide, fungicide	molting inhibitors (ecdysone antagonists), antifeedant/ repellent	Isman (2006); Copping and Duke (2007); Isman and Paluch (2011)
Pyrethrum	*Tanacetum cinerariifolium* Schultz-Bip.	esters of chrysanthemic acid and pyrethric acid (pyrethrins I and II, cinerins I and II, jasmolins I and II)	insecticide, acaricide	axonic poisons (sodium channels agonists)	Isman (2006); Copping and Duke (2007); Isman and Paluch (2011)
Nicotine	*Nicotiana* spp.	(S)-isomer (RS)-isomers, and (S)-isomer of nicotine sulfate	insecticide	neurotoxin (acetylcholine agonist)	Isman (2006); Copping and Duke (2007); Isman and Paluch (2011)
Rotenone	*Derris, Lonchocarpus,* and *Tephrosia* species	rotenone, deguelin, (isoflavonoids)	insecticide, acaricide	mitochondrial cytotoxin	Isman (2006); Copping and Duke (2007); Isman and Paluch (2011)
Ryania	*Ryania* spp. (*Ryania speciosa* Vahl)	ryanodine, ryania, 9,21- didehydroryanodine (alkaloids)	insecticide	neuromuscular poison (calcium channel agonist)	Isman (2006); Copping and Duke (2007); Isman and Paluch (2011)
Sabadilla	*Schoenocaulon* spp. (*Schoenocaulon officinale* Gray)	mixture of alkaloids (cevadine, veratridine)	insecticide	axonic poisons (sodium channels agonists, heart and skeletal muscle cell membranes)	Isman (2006); Copping and Duke (2007); Isman and Paluch (2011)
Quassia	Quassia, Aeschrion, Picrasma	quassin (triterpene lactone)	insecticide	unknown	Isman (2006); Isman and Paluch (2011)

TABLE 14.2 (Continued)

Name of the plant products	Plant source	Active compounds	Biological activity	Mode of action	References
Extract of giant knotweed	*Reynoutria sachalinensis* (Fr. Schm.) Nakai	physcion, emodin	fungicide, bactericide	induction of SAR (phenolic phytoalexins)	Dayan et al. (2009); Regnault (2012)
Karanjin	*Derris indica* (Lam.) Bennet	Karanjin	insecticide, acaricide	antifeedant/repellent, insect growth regulator	Copping and Duke (2007);
Clove essential oil	*Syzygium aromaticum*, *Eugenia caryophyllus* Spreng	eugenol (mixture of several predominantly terpenoid compounds)	insecticide, herbicide	Neurotoxic, interference with the neuromodulator octopamine	Isman (2006); Copping and Duke (2007); Regnault (2012)
Cinnamon essential oil	*Cinnamomum zeylanicum*	cinnamaldehyde	insecticide, herbicide	octopamine antagonists; membrane disruptors, others	Dayan et al. (2009)
Lemon grass essential oil	*Cymbopogon nardus*, *Cymbopogon citratus*, Stapf., *Cymbopogon flexuosus* D.C	citronellal, citral	insecticide, herbicide	octopamine antagonists; membrane disruptors, others	Dayan et al. (2009)
Mint essential oil	*Mentha* species (mint)	menthol	insecticide	octopamine antagonists; membrane disruptors, others	Dayan et al. (2009)
Citronella oil	*Cymbopogon* spp.	citronellal, geraniol, and other terpenes	repellent, herbicide		Dayan et al. (2009)

populations. Therefore, neem products can be appearing to have one of the potential to be incorporated in IPM for some crops.

14.5   CONCLUSION AND FUTURE PROSPECTS

Usually, farmers will purchase chemical pesticides still at higher price because of good advertisement of the product, rapid reply in term of pest mortality, and readily available in the market. So, there is need to educate/teach the farmers about the ecological/environment friendly benefits of plant products during awareness campaigns. The practice of using plant products from sources allows us to expand and use naturally occurring plant protection mechanisms, thereby reducing the use of conventional pesticides. However, most of these new strategies require to be developed with four basic facts: organize the natural sources, develop quality control, adopt standardization strategies, and modify regulatory constraints.

There is a wide scope for the use of plant-based pesticides in the integrated management of different insect pests. Production of botanical antifeedants would reduce the high cost of importation in developing countries. When neem products are mixed with kerosene, the combined action increases bio-efficacy of the combination and is safe to predators and parasitoids of mealybugs (Gokuldas et al., 1989). Synergism of synthetic pesticides and neem seed powder extract or vegetable oils has also been reported (Tripathi & Singh, 2003). Moreover, IPM is environment friendly and socially acceptable to consumers because it protects and preserves environment including biodiversity; it prevents unfavorable consequences to target and non-target organisms of ecosystems and thus checks imbalance in nature; it gives effective and cheaper control and thus makes plant protection feasible, safe, and economical even for small farmers. Presently, numeros botanicals have been formulated for large-scale application as bio-pesticides and antifeedants in environment friendly management of plant pests and are being utilizes as substitutes to synthetic pesticides in crop protection and post-harvest usage. These products have low-mammalian toxicity and are cost effective. Such products of higher plant origin may be exploited as eco-chemicals and integrated into plant protection programs.

KEYWORDS

- phytochemicals
- pest management
- metabolites
- antifeedant
- repellent

REFERENCES

Belles, X.; Camps, F.; Coll, J.; Dollars, P. M. Insect Antifeedant Activity of Clerodane Diterpenoids against Larvae of Spodoptera Littoralis (Lepidoptera). *J. Chem. Ecol.* **1985**, *11*, 1439–45.

Brahmachari, G. Neem-a Omnipotent Plant. A Retrospection. *Chem. Biochem.* **2004**, *5*, 408–421.

Brito, J. P.; Oliveira, J. E. M.; Bortoli, S. A. Toxicidade de óleos essenciais de Eucalyptus Spp. Sobre *Callosobruchus maculatus* (Fabr., 1775) (Coleoptera: Bruchidae). *Revista de Biologia e Ciência da Terra,* **2001**, *6*(1), 96–103.

Carvalho, J. C. T.; Gosmann, G.; Schenkel, E. P. Compostos Fenólicos Simples e Hetero-sídicos. In *Farmacognosia: da plantaao Medicamento;* Simões, 6th Ed.; C. M. O., Schenkel, E. P., Gosmann, G., de Mello, J. C. P., Mentz, L. A., Petrovick, P. R. (org.) Eds.; UFRGS: Porto Alegre, Editora da, 2007; pp 519–536.

Charleston, D. S.; Kfir, R.; Dicke, M.; Vet, L. E. M. Impact of Botanical Pesticides Derived from *Melia azedarach* and *Azadirachta indica* on the Biology of two Parasitoid Species of the Diamondback Moth. *Biol. Control.* **2005**, *33*, 131–142.

Copping, L. G.; Duke, S. O. Natural Products that have been Used Commercially as Crop Protection Agents. *Pest Manage. Sci.* **2007**, *63*, 524–54.

Croteau, R.; Kutchan, T. M.; Lewis, N. G. "Natural Products (Secondary Metabolites)". *Biochem. Mol. Biol. Plants.* **2000**, 1250–1318.

Dayan, F. E.; Cantrell, C. H. L.; Duke, S. O. Natural Products in Crop Protection. *Bioorg. Med. Chem.* **2009**, *17*, 4022–4034.

Dennis, S. H. *Pests of Stored Foodstuffs and their Control;* Kluwer Academic: Dordrecht, Netherlands, 2002.

García, A. A, Carril, E. P. U. Metabolismo Secundario de Plantas. Reduca (Biología*). Serie Fisiología Vegetal.* **2009**, *2*(3), 119–145.

Gokuldas, K. M.; Bhat, P. K.; Ramaiah, P. K. *J. Coffee Res.* **1989**, *19*, 17–29.

Gonzaga, A. D.; Garcia, M. V. B.; Souza, S. G. A.; PyDaniel, V.; Correa, R. S.; Ribeiro, J. D. Toxicidade de Manipueira de Mandioca (*Manihotes culenta* Crantz) e Erva-de-Rato (*Palicourea marcgravii* St. Hill) a Adultos de *Toxoptera citricida* Kirkaldy (Homoptera: Aphididae). *Acta Amazônica.* **2008**, *38*(1): 101–106.

Goudegnon, A. E.; Kirk, A. A.; Schiffers, B.; Bordat, D. Comparative Effects of Delta-methrin and Neem Kernel Solution Treatments on Diamondback Moth and *Cotesia plutellae* (Hym. Braconidae) Parasitoid Populations in the Cotonou Peri-Urban Area in Benin. *J. Appl. Entomol.* **2000,** *124,* 141–144.

Gurusubramanian, G.; Rahaman, A.; Samrah, M.; Roy, S.; Bora, S. Pesticide Usage Pattern in Tea Ecosystem, their Retrospect and Alternative Measures. *J. Environ. Biol.* **2008,** *29*(6), 813–826.

Gutierrez, G. P. A.; Villegas, M. C. Efecto tóxico de Verbena Officinalis (Familiaverbena-ceae) em Sitophilus Granarius (Coleoptera: Curculionidae). *Rev. Lasallista Investig.* **2008,** *5*(2), 1–7.

Harborne, J. B. Recent advances in chemical ecology. *Nat. Prod. Rep.* **1989,** 6, 85–109.

Harborne, J. B. *The Flavonoids: Advances in Research Since 1986;* Chapman & Hall: London, 1993.

Adeyemi, M. M. Hassan. The Potential of Secondary Metabolites in Plant Material as Deterents against Insect Pests: A Review. *Afr. J. Pure Appl. Chem.* **2010,** *4*(11), 243–246. https://creativecommons.org/licenses/by/4.0/)

Hazarika, L. K.; Gogoi, N.; Barua, N. C.; Kalita, S.; Gogoi, N. Search of Green Pesticides for Tea Pest management: *Phlogocanthus thyrsiflorus* Experience. In *Recent Trends in Insect Pest Management;* Ignacimuthu, S., Jayraj, S., Eds.; Elite Publication: New Delhi, 2008; pp 79–90.

Hiiesaar, K. L.; Metspalu, A.; Kuusik, S. An Estimation of Influences Evoked by Some Natural Insecticides on Greenhouse Pest Insects and Mites. In *Practice Oriented Results on the Use of Plant Extracts and Pheromones in Pest Control,* Proceedings of the international workshop, Estonia, Tartu, Jan 24–25, 2001; Metspalu, L., Mitt, S., Eds.; pp 17–27.

Isman, M. B.; Paluch, G. Needles in the Haystack: Exploring Chemical Diversity of Botanical Insecticides. In *Green trends in insect control;* López, O., Fernández-Bolaños, J. G., Eds.; RSC: London, 2011; pp 248–265.

Isman, M. B. Botanical Insecticides, Deterrents, and Repellents in Modern Agriculture and an Increasingly Regulated World. *Annu. Rev. Entomol.* **2006,** *51,* 45–66.

Ivie, G. W.; Witzel, D. A. Sesquiterpene Lactones. In *Structure, Biological Action and Toxicological Significance;* Keeler, R. F., Tu, A. T., Eds.; Handbook of Natural Toxins, Marcel Dekker Inc.: New York, NY, 1982; pp 543–584.

Júnior, V. Terpenos Com Atividade Inseticida: Uma Alternativa Para o Controle Químico de Insetos. *Química Nova.* **2003,** *26*(3), 390–400.

Koul, O.; Dhaliwal, G. S. *Phytochemical Biopesticides;* Harwood Academic Publishers: Amsterdam, The Netherlands, 2001.

Koul, O. *Insect Antifeedants;* CRC Press: Bota Racon, FL, 2005.

Lajide, L.; Escoubas, P.; Mizutani, J. Termite Antifeedant Activity in *Xylopia aethiopica.* *Phytochemistry.* **1995,** *40,* 1105–1112.

Mahmood, I.; Saxena, S. K.; Zakiuddin, M. 1984. Effect of Certain Plant Extracts on the Mortality of Rotylenchulus Reniformis and Meloidogyne incognita. *Bangladesh J. Bot.* *4*(2), 154–157.

Mamun, M. S. A. Development of Tea Science and Tea Industry in Bangladesh and Advances of Plant Extracts in Tea Pest Management. *Int. J. Sustain. Agril. Tech.* **2011,** *7*(5), 40–46.

Mamun, M. S. A.; Ahmed, M. Prospect of Indigenous Plant Extracts in Tea Pest Management. *Int. J. Agril. Res. Innov. Tech.* **2011,** *1*(1, 2), 16–23.

Mamun, M. S. A.; Iyengar, A. V. K. Integrated Approaches to Tea Pest Management in South India. *Int. J. Sustain. Agril. Tech.* **2010,** *6*(4), 27–33.

Mann, R. S.; Kaufman, P. E. Natural Product Pesticides: Their Development, Delivery and Use against Insect Vectors. *Min-Rev. Org. Chem.* **2012,** *9,* 185–202.

Marangoni, C.; Moura, N. F.; Garcia, F. R. M. Utilização de Óleos Essenciais e Extratos de Plantas no Controle de Insetos. *Revista de Ciências Ambientais.* **2012,** *6*(2), 95–112.

Misra, H. P. Role of Botanicals, Biopesticides and Bioagents in Integrated Pest Management. *Odisha Rev.* **2014,** 62–67.

Mordue, A. J.; Blackwell, A. Azadirachtin: An Update. *J. Insect Physiol.* **1993,** *39,* 903–924.

Mugisha-Kamatenesi, M.; Deng, A. L.; Ogendo, J. O.; Omolo, E. O.; Mihale, M. J.; Otim, J. P.; Buyungo, Bett, P. K. Indigenous Knowledge of Field Insect Pests and their Management Around Lake Victoria Basin in Uganda. *Afr. J. Environ. Sci. Tech.* **2008,** *2*(8), 342–348.

Ndungu, M.; Lwande, W.; Hassanali, A.; Moreka, L.; Chhabra, S. C. "Cleome Monophylla Essential Oil and its Constituents as Tick *Rhipicephalus appendiculatus* and Maize Weevil (*Sitophilus zeamais*) Repellents". *Entomologica Experimentalis et Applicata.* **1995,** *76,* 217–222.

Omar, S.; Marcotte, M.; Fields, P.; Sanchez, P. E.; Poveda, L.; Mata, R.; Jimenez, A.; Durst, T.; Zhang, J.; MacKinnon, S.; Leaman, D.; Arnason, J. T.; Philogène, B. J. R. Antifeedant Activities of Terpenoids Isolated from Tropical Rutales. *J. Stored Prod. Res.* **2007,** *43,* 92–96.

Owen, T. Geoponika: Agricultural Pursuits. Moore, S. J. & Langlet, A. 2004. In *An Overview of Plants as Insect Repellents;* Wilcox, M., Bodeker, G., Eds.; Traditional Medicine, Medicinal Plants and Malaria. Taylor& Francis: London, 2004; http://www. ancientlibrary.com/geoponica/index.html 1805.

Owusu, O. O. Effect Oo Some Ghanaian Plant Components on the Control of two Stored Product Insect Pests of Cereals. *J. Stored Prod. Res.* **2000,** *37,* 85–91.

Panda, N.; Khusch, G. S. *Host Plant Resistance to Insects;* CAB International: Wallingford, Oxon, UK, in Association with International Rice Research Institute, 1995; p 431.

Parmar, B. S.; Walia, S. Prospects and Problems of Phytochemical Biopesticides. In *Phytochemical Biopesticides;* Koul, O., Dhaliwal, G. S., Eds.; Hardwood Academic Publishers: Amsterdam, The Netherlands, 2001; pp 133–210.

Pelletier, S. W. The Nature and Definition of and Alkaloid. In *Chemical and Biological Perspectives;* Pelletier, S. W., Ed.; John Wiley & Sons Inc.: Alkaloids, New York, 1983; Vol. 1, pp 1–31.

Philip, G. K.; Robert, A. B. Florida Pest Control Association. 1998.

Regnault-Roger, C. Trends for Commercialisation of Biocontrol Agent (Biopesticide) Products. In *Plant Defence: Biological Control, Progress in Biological Control;* Mérillon, J M., Ramawat, K. G., Eds.; Springer: Dordrecht, Netherlands, 2012; Vol. 12, pp 139–160.

Rosenthal, G. A.; Berenbaum, M. R. *Herbivores: In Their Interaction with Secondary Plant Metabolites, Ecological and Evolutionary Processes;* 2nd edition, Academic press: San Diego, CA, Vol. II, 1992.

Ross, J. D.; Sombrero, D. Environmental Control of Essential Oil Production in Mediterranean Plants. In *Ecological Chemistry and Biochemistry of Plant Terpenoids;* Harborne, J. B., Tomes-Barberan, F. A., Eds.; Clarendon Press: New York, 1991; pp 64–94.

Santos, V. A. F. F. M.; Santos, D. P.; Castro-Gamboa, I.; Zanoni, M. V. B.; Furlan, M. Evaluation of Antioxidant Capacity and Synergistic Associations of Quinonemethide-triterpenos and Phenolic Substances from *Maytenus ilicifolia* (Celastraceae). *Molecules.* **2010,** *15,* 6956–6973.

Sarmah, M.; Bhola, R. K. Antifeedant and Repellent Effects of Aqueous Plant Extracts of *Azadirachta indica* and *Xanthium strumarium* on Tea Mosquito Bug, Helopeltis Theivora Waterhouse. *Two and a bud.* **2008,** *55,* 36–40.

Schmutterer, H.; Singh, R. P. Lists of Insects Susceptible to Neem Products. In *The Neem Trees Sources of Unique Natural Products for Integrated Pest Management, Medicine, Industry and Other Purposes;* Schmutterer, H., Ed.; VCH Publishers: New York, 1995.

Scott, I. M.; Puniani, E.; Durst, T.; Phelps, D.; Merali, S.; Assabgui, R. A.; Sanchez-Vindas, P. P. L.; Philogene, B. J. R.; Arnason, J. T. Insecticidal Activity of *Piper tuberculatum* Jacq. Extracts: Synergistic Interaction of Piperamides. *Agric. For. Entomol.* **2002,** *4,* 137–144.

Shaaya, E.; Kostjukovski, M.; Eilberg, J.; Suprakarn, C. Plant Oils as Fumigants and Contact Insecticides for the Control of Stored-Product Insects. *J. Stored Prod. Res.* **1997,** *33,* 7–15.

Simas, N. K.; Lima, E. C.; Conceição, S. R.; Kuster, R. M.; Filho, A. M. O.; Lage, C. L. S. Produtos Naturais Para o Controle da Transmissão da Dengue – Atividadelarvicida de Myroxylon Balsamum (Óleovermelho) e de Terpenoides e Fenilpropanoides. *Química Nova.* **2004,** *27*(1), 46–49.

Simões, C. M. O.; Schenkel, E. P.; Gosmann, G.; Mello, J. C. P.; Mentz, L. A.; Petrovick, P. R. Farmacognosia. In *da Planta ao Medicamento. 5ª Edição;* Porto Alegre, R S., Ed.; da UFSC: Brazil, 2004; p 1102.

Taiz, L.; Zeiger, E. *Fisiologia Vegetal.* 5ª Ed. Artmed: Porto Alegre, 2013; p 954.

Thebtaranonth, C.; Thebtaranonth, Y.; Wanauppathamkul, S.; Yuthavong, Y. Antimalarial Sesquiterpenes from Tubers of *Cyperus rotundus*: Structure of 10,12-Peroxycalamenene, a Sesquiterpene Endoperoxide. *Phytochemistry.* **1995,** *40,* 125–128.

Tripathi, M. K.; Singh, H. N.; *Indian J. Entomol.* **2003,** *65,* 373–378.

Van Etten, H.; Temporini, E.; Wasmann, C. Phytoalexin (and Phytoanticipin) Tolerance as a Virulence Trait: Why is it not Required by all Pathogens? *Physiol. Mol. Plant Pathol.* **2001,** *59,* 83–93.

Viegas, J. R. C.; Bolzani, V. S.; Barreiro, E. J. Os Produtos Naturais e a Química Medicinal Moderna. *Química Nova.* **2006,** *29*(2), 326–337.

Wanjala, W.; Cornelius, T. A.; Obiero, G. O.; Lutta, K. P. "Antifeedant Activities of the Erythrinaline Alkaloids From *Erythrina latissima* against *Spodoptera littoralis* (*Lepidoptera noctuidae*)" *Rec. Nat. Prod.* **2009,** *3*(2), 10–96.

Yang, R. Z.; Tangs, C. S. Plants Used for Pest Control in China: A Literature Review. *Eco. Bot.* **1988,** *42,* 376–406.

INDEX

T - #0815 - 101024 - C418 - 229/152/18 - PB - 9781774636763 - Gloss Lamination